D1256554

Methods in Enzymology

Volume 337
MICROBIAL GROWTH IN BIOFILMS
Part B
Special Environments and Physicochemical Aspects

METHODS IN ENZYMOLOGY

EDITORS-IN-CHIEF

John N. Abelson   Melvin I. Simon

DIVISION OF BIOLOGY
CALIFORNIA INSTITUTE OF TECHNOLOGY
PASADENA, CALIFORNIA

FOUNDING EDITORS

Sidney P. Colowick and Nathan O. Kaplan

Methods in Enzymology

Volume 337

Microbial Growth in Biofilms

Part B

Special Environments and Physicochemical Aspects

EDITED BY

Ron J. Doyle

UNIVERSITY OF LOUISVILLE
LOUISVILLE, KENTUCKY

ACADEMIC PRESS

San Diego   London   Boston   New York   Sydney   Tokyo   Toronto

Academic Press
A Harcourt Science and Technology Company
525 B Street, Suite 1900, San Diego, California 92101-4495, USA
http://www.academicpress.com

Academic Press
Harcourt Place, 32 Jamestown Road, London NW1 7BY, UK
http://www.academicpress.com

International Standard Book Number: 0-12-182238-9

PRINTED IN THE UNITED STATES OF AMERICA
01 02 03 04 05 06 07 SB 9 8 7 6 5 4 3 2 1

Table of Contents

Section I. Biofilms on Plant Tissues

Section II. Flow Systems and Biofilm Development and Characterization

Section III. Biofilm Growth in Special Environments

Section IV. Physical–Chemical Characterization of Biofilms

Section V. Susceptibility Testing of Biofilm Microbiota

Section VI. Oral Microbial Biofilms

Contributors to Volume 337

Article numbers are in parentheses following the names of contributors.
Affiliations listed are current.

YUEHUEI H. AN (6), *Department of Orthopaedic Surgery, Orthopaedic Research Laboratory, Medical University of South Carolina, Charleston, South Carolina 29425*

JEFFREY A. BANAS (30), *Center for Immunology and Microbial Disease, Albany Medical College, Albany, New York 12208*

BRIAN K. BEDNARSKI (6), *Department of Orthopaedic Surgery, Orthopaedic Research Laboratory, Medical University of South Carolina, Charleston, South Carolina 29425*

HALUK BEYENAL (23), *Department of Civil Engineering, Montana State University, Bozeman, Montana 59717*

CYNTHIA G. BLOOMQUIST (27), *School of Dentistry, University of Minnesota, Minneapolis, Minnesota 55455*

T. REG. BOTT (7), *School of Chemical Engineering, University of Birmingham, Birmingham B15 2TT, United Kingdom*

ROBERT D. BOYD (16), *Microbiology Department, Manchester Metropolitan University, Manchester, United Kingdom*

DAVID J. BRADSHAW (29), *Centre for Applied Microbiology and Research (CAMR), Porton Down, Salisbury SP4 0JG, United Kingdom*

JAMES D. BRYERS (17), *The Center for Biomaterials, University of Connecticut Health Center, Farmington, Connecticut 06030*

ANDRE BURET (25), *Biofilm Research Group, University of Calgary, Calgary, Alberta T2N 1N4, Canada*

BERND BURGER (29), *ESPE Dental Medizin GmbH and Co., D-82229 Seefeld, Germany*

ROBERT A. BURNE (28), *Department of Microbiology and Immunology, and Center for Oral Biology, University of Rochester School of Medicine and Dentistry, Rochester, New York 14642*

HENK J. BUSSCHER (18), *Department of Biomedical Engineering, University of Groningen, 9713 AV Groningen, The Netherlands*

CLIVE M. BUSWELL (5), *Research Centre for Applied Microbiology and Research, Porton Down, Salisbury SP4 0JG, United Kingdom*

HOWARD CERI (25), *Biofilm Research Group, University of Calgary, Calgary, Alberta T2N 1N4, Canada*

GUSTAVO CURUTCHET (11), *Centro de Investigación y Desarrollo de Fermentaciones Industriales (Cindefi-Conicet), Facultad de Ciencias Exactas (UNLP), La Plata, Argentina*

JOOP DE VRIES (18), *Department of Biomedical Engineering, University of Groningen, 9713 AV Groningen, The Netherlands*

EDGARDO DONATI (11), *Centro de Investigación y Desarrollo de Fermentaciones Industriales (Cindefi-Conicet), Facultad de Ciencias Exactas (UNLP), La Plata, Argentina*

RON J. DOYLE (31), *Department of Microbiology and Immunology, University of Louisville, Louisville, Kentucky 40292*

DAVID DRAKE (26), *Dows Institute for Dental Research, University of Iowa, College of Dentistry, Iowa City, Iowa 52242*

F. GRANT FERRIS (15), *Department of Geology, University of Toronto, Toronto, Ontario M5S 3B1, Canada*

RICHARD J. FRIEDMAN (6), *Department of Orthopaedic Surgery, Orthopaedic Research Laboratory, Medical University of South Carolina, Charleston, South Carolina 29425*

CLAY FUQUA (1), *Department of Biology, Indiana University, Bloomington, Indiana 47405*

BERND GANGNUS (29), *ESPE Dental Medizin GmbH and Co., D-82229 Seefeld, Germany*

DARLA GOERES (24), *Center for Biofilm Engineering, Montana State University, Bozeman, Montana 59717*

SHARON GORDON (27), *National Institute of Dental and Craniofacial Research, National Institutes of Health, Bethesda, Maryland 20892*

DEON M. GRANT (7), *Nalco Europe BV, 2300 AP Leiden, The Netherlands*

LAURANCE D. HALL (20), *Herschel Smith Laboratory for Medicinal Chemistry, University Forvie Site, Cambridge CB2 2PZ, United Kingdom*

LUANNE HALL-STOODLEY (21), *Center for Biofilm Engineering, Montana State University, Bozeman, Montana 59715*

MARTIN HAMILTON (24), *Center for Biofilm Engineering, Montana State University, Bozeman, Montana 59717*

KARSTEN R. O. HAZLETT (30), *Center for Microbial Pathogenesis, University of Connecticut Health Center, Farmington, Connecticut 06030*

JOANNA HEERSINK (24), *Center for Biofilm Engineering, Montana State University, Bozeman, Montana 59717*

PATRICIA A. HOLDEN (9), *Donald Bren School of Environmental Science and Management, University of California, Santa Barbara, California 93106*

JANA JASS (4), *Department of Microbiology, Umeå University, S-901 87 Umeå, Sweden*

C. WILLIAM KEEVIL (8), *School of Biological Sciences, University of Southampton, Southampton SO16 7PX, United Kingdom*

PASCAL KIERS (18), *Department of Biomedical Engineering, University of Groningen, 9713 AV Groningen, The Netherlands*

MOGENS KILIAN (27), *Department of Medical Microbiology and Immunology, Aarhus University, DK-8000 Aarhus, Denmark*

PAUL E. KOLENBRANDER (27), *National Institute of Dental and Craniofacial Research, National Institutes of Health, Bethesda, Maryland 20892*

KAREN A. KROGFELT (2), *Department of Gastrointestinal Infections, Statens Serum Institut, DK-2300 Copenhagen S, Denmark*

HILARY M. LAPPIN-SCOTT (21), *School of Biological Sciences, Hatherly Laboratories, Exeter University, Exeter EX4 4PS, United Kingdom*

YUAN KUN LEE (13), *Department of Microbiology, Faculty of Medicine, National University of Singapore, Singapore 117597, Republic of Singapore*

ZBIGNIEW LEWANDOWSKI (23), *Center for Biofilm Engineering, Montana State University, Bozeman, Montana 59717*

TINE R. LICHT (2), *Institute of Food Safety and Toxicology, Division of Microbial Safety, DK-2860 Søborg, Denmark*

WILLIAM F. LILJEMARK (27), *School of Dentistry, University of Minnesota, Minneapolis, Minnesota 55455*

CHRISTOPHER J. LINTON (3), *Department of Pathology and Microbiology, School of Medical Sciences, University of Bristol, Bristol BS8 3LN, United Kingdom*

JIRAPON LUENGPAILIN (31), *Department of Microbiology and Immunology, University of Louisville, Louisville, Kentucky 40292*

SOMKIAT LUENGPAILIN (31) *Department of Microbiology and Immunology, University of Louisville, Louisville, Kentucky 40292*

LYNNE E. MACASKIE (20), *School of Biosciences, University of Birmingham, Birmingham B15 2TT, United Kingdom*

ROBERT E. MARQUIS (28), *Department of Microbiology and Immunology, and Center for Oral Biology, University of Rochester School of Medicine and Dentistry, Rochester, New York 14642*

PHIL D. MARSH (29), *Centre for Applied Microbiology and Research (CAMR), Porton Down, Salisbury SP4 0JG, United Kingdom*

KYLIE L. MARTIN (6), *Department of Orthopaedic Surgery, Orthopaedic Research Laboratory, Medical University of South Carolina, Charleston, South Carolina 29425*

ANN G. MATTHYSSE (1), *Department of Biology, University of North Carolina, Chapel Hill, North Carolina 27599*

JOSEPH E. MAZURKIEWICZ (30), *Center for Neuropharmacology and Neuroscience, Albany Medical College, Albany, New York 12208*

JONATHAN B. MCGLOHORN (6), *Department of Orthopaedic Surgery, Orthopaedic Research Laboratory, Medical University of South Carolina, Charleston, South Carolina 29425*

MICHAEL R. MILLAR (3), *Department of Microbiology, Barts and The London NHS Trust, London E1 1BB, United Kingdom*

SØREN MOLIN (2), *Molecular Microbial Ecology Group, Department of Microbiology, Technical University of Denmark, DK-2800 Lyngby, Denmark*

DOUGLAS MORCK (25), *Biofilm Research Group, University of Calgary, Calgary, Alberta T2N 1N4, Canada*

HELEN S. NICHOLL (5), *The Queen's University, Belfast BT9 5PX, Ireland*

KEVIN P. NOTT (20), *Herschel Smith Laboratory for Medicinal Chemistry, University Forvie Site, Cambridge CB2 2PZ, United Kingdom*

SATOSHI OKABE (14), *Department of Urban and Environmental Engineering, Graduate School of Engineering, Hokkaido University, Sapporo 060-8626, Japan*

CRISTIAN OLIVER (11), *Centro de Investigación y Desarrollo de Fermentaciones Industriales (Cindefi-Conicet), Facultad de Ciencias Exactas (UNLP), La Plata, Argentina*

BARBARA OLSON (25), *Biofilm Research Group, University of Calgary, Calgary, Alberta T2N 1N4, Canada*

MERLE OLSON (25), *Biofilm Research Group, University of Calgary, Calgary, Alberta T2N 1N4, Canada*

J. GARY O'NEILL (4), *Public Health Scientist, Yorkshire Water, Bradford BD1 5PZ, United Kingdom*

ARTHUR C. OUWEHAND (13), *Department of Biochemistry and Food Chemistry, University of Turku, 20014 Turku, Finland*

ROBERT J. PALMER, JR. (27), *National Institute of Dental and Craniofacial Research, National Institutes of Health, Bethesda, Maryland 20892*

MARION PATERSON-BEEDLE (20), *School of Biosciences, University of Birmingham, Birmingham B15 2TT, United Kingdom*

STEVEN PERCIVAL (12, 16), *Pathogen and Biofilm Microbiology Group, The University of Central Lancashire, Preston PR1 2HE, United Kingdom*

CRISTINA POGLIANI (11), *Centro de Investigación y Desarrollo de Fermentaciones Industriales (Cindefi-Conicet), Facultad de Ciencias Exactas (UNLP), La Plata, Argentina*

MICHEL QUINTARD (22), *Institut de Mécanique des Fluides de Toulouse, Allee du Professeur Camille Soula, 31400 Toulouse, France*

CAYO RAMOS (2), *Molecular Microbial Ecology Group, Department of Microbiology, Technical University of Denmark, DK-2800 Lyngby, Denmark*

ANNETA RAZATOS (19), *Department of Chemical and Materials Engineering, Arizona State University, Tempe, Arizona 85287*

RONALD READ (25), *Biofilm Research Group, University of Calgary, Calgary, Alberta T2N 1N4, Canada*

SEPPO SALMINEN (13), *Department of Biochemistry and Food Chemistry, University of Turku, 20014 Turku, Finland*

HISASHI SATOH (14), *Department of Civil Engineering, Faculty of Engineering, Hachinohe Institute of Technology, Hachinohe 031-8501, Japan*

ANDREA SHERRIFF (3), *Department of Paediatric Epidemiology, Royal Hospital for Sick Children, University of Bristol, Bristol BS2 8BJ, United Kingdom*

D. SCOTT SMITH (15), *Department of Geology, University of Toronto, Toronto, Ontario M5S 3B1, Canada*

CLAUS STERNBERG (2), *Molecular Microbial Ecology Group, Department of Microbiology, Technical University of Denmark, DK-2800 Lyngby, Denmark*

PAUL STOODLEY (21), *Center for Biofilm Engineering, Montana State University, Bozeman, Montana 59715*

DOUGLAS STOREY (25), *Biofilm Research Group, University of Calgary, Calgary, Alberta T2N 1N4, Canada*

ELINA M. TUOMOLA (13), *Novatreat, Inc., 20700 Turku, Finland*

HENNY C. VAN DER MEI (18), *Department of Biomedical Engineering, University of Groningen, 9713 AV Groningen, The Netherlands*

JOANNE VERRAN (16), *Microbiology Department, Manchester Metropolitan University, Manchester, M1 5GD, United Kingdom*

MARISA R. VIERA (11), *Centro de Investigación y Desarrollo de Fermentaciones Industriales (Cindefi-Conicet), Facultad de Ciencias Exactas (UNLP), La Plata, Argentina*

CHRISTIAN J. VOLK (10), *Indiana-American Water Company, Inc., Muncie, Indiana 47302*

JAMES T. WALKER (4, 5, 12, 16, 29), *Centre for Applied Microbiology and Research (CAMR), Porton Down, Salisbury SP4 0JG, United Kingdom*

YOSHIMASA WATANABE (14), *Department of Urban and Environmental Engineering, Graduate School of Engineering, Hokkaido University, Sapporo 060-8626, Japan*

STEPHEN WHITAKER (22), *Department of Chemical Engineering and Material Science, University of California at Davis, Davis, California 95616*

BRIAN D. WOOD (22), *Pacific Northwest National Laboratory, Richland, Washington 99352*

ROSEMARY WU (27), *National Institute of Dental and Craniofacial Research, National Institutes of Health, Bethesda, Maryland 20892*

NICK ZELVER (24), *MSU TechLink, Montana State University, Bozeman, Montana 59718*

Preface

Biofilms are usually characterized as a consortia of microorganisms surrounded by a protecting matrix of secreted polymers that are in most cases acidic polysaccharides, but may possess various functional groups other than carboxylate. In biofilms, the microorganisms possess regulatory molecules distinct from those produced by planktonic microorganisms. It is now possible not only to detect specific species and strains in a biofilm matrix, but also to identify which of their genes are up- or down-regulated in the planktonic cell to biofilm cell transition.

The advances in molecular biology methods have been paralleled by advances in instrumental and chemical probes used to define biofilm properties. Literature has burgeoned on all aspects of biofilms in the past few years. This growing literature, coupled with the recognition that biofilms are important in disease, industry, agriculture, and biotechnology, has prompted the development of a series of *Methods in Enzymology* volumes. The first, Volume 310, was concerned with the general approaches to biofilm molecular biology and the physical methods to probe biofilm structures. Volumes 336 and 337 focus on microbial growth in biofilms. In this volume focus is on special environments and specific microorganisms contributing to biofilms. Its companion Volume 336 emphasizes the genetics and molecular biology of biofilm genesis. Collectively, the three volumes comprise methods from the leading researchers in the world. The following decades of research on biofilms will borrow heavily from these volumes.

I thank Shirley Light of Academic Press for her competent handling of numerous questions related to the development of the volumes on biofilms.

RON J. DOYLE

METHODS IN ENZYMOLOGY

VOLUME LV. Biomembranes (Part F: Bioenergetics)
Edited by SIDNEY FLEISCHER AND LESTER PACKER

VOLUME LVI. Biomembranes (Part G: Bioenergetics)
Edited by SIDNEY FLEISCHER AND LESTER PACKER

VOLUME LVII. Bioluminescence and Chemiluminescence
Edited by MARLENE A. DELUCA

VOLUME LVIII. Cell Culture
Edited by WILLIAM B. JAKOBY AND IRA PASTAN

VOLUME LIX. Nucleic Acids and Protein Synthesis (Part G)
Edited by KIVIE MOLDAVE AND LAWRENCE GROSSMAN

VOLUME LX. Nucleic Acids and Protein Synthesis (Part H)
Edited by KIVIE MOLDAVE AND LAWRENCE GROSSMAN

VOLUME 61. Enzyme Structure (Part H)
Edited by C. H. W. HIRS AND SERGE N. TIMASHEFF

VOLUME 62. Vitamins and Coenzymes (Part D)
Edited by DONALD B. MCCORMICK AND LEMUEL D. WRIGHT

VOLUME 63. Enzyme Kinetics and Mechanism (Part A: Initial Rate and Inhibitor Methods)
Edited by DANIEL L. PURICH

VOLUME 64. Enzyme Kinetics and Mechanism (Part B: Isotopic Probes and Complex Enzyme Systems)
Edited by DANIEL L. PURICH

VOLUME 65. Nucleic Acids (Part I)
Edited by LAWRENCE GROSSMAN AND KIVIE MOLDAVE

VOLUME 66. Vitamins and Coenzymes (Part E)
Edited by DONALD B. MCCORMICK AND LEMUEL D. WRIGHT

VOLUME 67. Vitamins and Coenzymes (Part F)
Edited by DONALD B. MCCORMICK AND LEMUEL D. WRIGHT

VOLUME 68. Recombinant DNA
Edited by RAY WU

VOLUME 69. Photosynthesis and Nitrogen Fixation (Part C)
Edited by ANTHONY SAN PIETRO

VOLUME 70. Immunochemical Techniques (Part A)
Edited by HELEN VAN VUNAKIS AND JOHN J. LANGONE

VOLUME 71. Lipids (Part C)
Edited by JOHN M. LOWENSTEIN

VOLUME 72. Lipids (Part D)
Edited by JOHN M. LOWENSTEIN

VOLUME 91. Enzyme Structure (Part I)
Edited by C. H. W. HIRS AND SERGE N. TIMASHEFF

VOLUME 92. Immunochemical Techniques (Part E: Monoclonal Antibodies and General Immunoassay Methods)
Edited by JOHN J. LANGONE AND HELEN VAN VUNAKIS

VOLUME 93. Immunochemical Techniques (Part F: Conventional Antibodies, Fc Receptors, and Cytotoxicity)
Edited by JOHN J. LANGONE AND HELEN VAN VUNAKIS

VOLUME 94. Polyamines
Edited by HERBERT TABOR AND CELIA WHITE TABOR

VOLUME 95. Cumulative Subject Index Volumes 61–74, 76–80
Edited by EDWARD A. DENNIS AND MARTHA G. DENNIS

VOLUME 96. Biomembranes [Part J: Membrane Biogenesis: Assembly and Targeting (General Methods; Eukaryotes)]
Edited by SIDNEY FLEISCHER AND BECCA FLEISCHER

VOLUME 97. Biomembranes [Part K: Membrane Biogenesis: Assembly and Targeting (Prokaryotes, Mitochondria, and Chloroplasts)]
Edited by SIDNEY FLEISCHER AND BECCA FLEISCHER

VOLUME 98. Biomembranes (Part L: Membrane Biogenesis: Processing and Recycling)
Edited by SIDNEY FLEISCHER AND BECCA FLEISCHER

VOLUME 99. Hormone Action (Part F: Protein Kinases)
Edited by JACKIE D. CORBIN AND JOEL G. HARDMAN

VOLUME 100. Recombinant DNA (Part B)
Edited by RAY WU, LAWRENCE GROSSMAN, AND KIVIE MOLDAVE

VOLUME 101. Recombinant DNA (Part C)
Edited by RAY WU, LAWRENCE GROSSMAN, AND KIVIE MOLDAVE

VOLUME 102. Hormone Action (Part G: Calmodulin and Calcium-Binding Proteins)
Edited by ANTHONY R. MEANS AND BERT W. O'MALLEY

VOLUME 103. Hormone Action (Part H: Neuroendocrine Peptides)
Edited by P. MICHAEL CONN

VOLUME 104. Enzyme Purification and Related Techniques (Part C)
Edited by WILLIAM B. JAKOBY

VOLUME 105. Oxygen Radicals in Biological Systems
Edited by LESTER PACKER

VOLUME 106. Posttranslational Modifications (Part A)
Edited by FINN WOLD AND KIVIE MOLDAVE

VOLUME 107. Posttranslational Modifications (Part B)
Edited by FINN WOLD AND KIVIE MOLDAVE

VOLUME 266. Computer Methods for Macromolecular Sequence Analysis
Edited by RUSSELL F. DOOLITTLE

VOLUME 267. Combinatorial Chemistry
Edited by JOHN N. ABELSON

VOLUME 268. Nitric Oxide (Part A: Sources and Detection of NO; NO Synthase)
Edited by LESTER PACKER

VOLUME 269. Nitric Oxide (Part B: Physiological and Pathological Processes)
Edited by LESTER PACKER

VOLUME 270. High Resolution Separation and Analysis of Biological Macromolecules (Part A: Fundamentals)
Edited by BARRY L. KARGER AND WILLIAM S. HANCOCK

VOLUME 271. High Resolution Separation and Analysis of Biological Macromolecules (Part B: Applications)
Edited by BARRY L. KARGER AND WILLIAM S. HANCOCK

VOLUME 272. Cytochrome P450 (Part B)
Edited by ERIC F. JOHNSON AND MICHAEL R. WATERMAN

VOLUME 273. RNA Polymerase and Associated Factors (Part A)
Edited by SANKAR ADHYA

VOLUME 274. RNA Polymerase and Associated Factors (Part B)
Edited by SANKAR ADHYA

VOLUME 275. Viral Polymerases and Related Proteins
Edited by LAWRENCE C. KUO, DAVID B. OLSEN, AND STEVEN S. CARROLL

VOLUME 276. Macromolecular Crystallography (Part A)
Edited by CHARLES W. CARTER, JR., AND ROBERT M. SWEET

VOLUME 277. Macromolecular Crystallography (Part B)
Edited by CHARLES W. CARTER, JR., AND ROBERT M. SWEET

VOLUME 278. Fluorescence Spectroscopy
Edited by LUDWIG BRAND AND MICHAEL L. JOHNSON

VOLUME 279. Vitamins and Coenzymes (Part I)
Edited by DONALD B. MCCORMICK, JOHN W. SUTTIE, AND CONRAD WAGNER

VOLUME 280. Vitamins and Coenzymes (Part J)
Edited by DONALD B. MCCORMICK, JOHN W. SUTTIE, AND CONRAD WAGNER

VOLUME 281. Vitamins and Coenzymes (Part K)
Edited by DONALD B. MCCORMICK, JOHN W. SUTTIE, AND CONRAD WAGNER

VOLUME 282. Vitamins and Coenzymes (Part L)
Edited by DONALD B. MCCORMICK, JOHN W. SUTTIE, AND CONRAD WAGNER

VOLUME 283. Cell Cycle Control
Edited by WILLIAM G. DUNPHY

Section I

Biofilms on Plant Tissues

[1] Methods for Studying Bacterial Biofilms Associated with Plants

By CLAY FUQUA and ANN G. MATTHYSSE

Introduction

Although many researchers have examined what are in effect biofilms on the surface of plants, the word "biofilm" is rarely used in the reporting of these studies. Bacteria may be present on the surface of plants as individual organisms, as isolated microcolonies, or as complex layers of bacteria surrounded by an extracellular matrix (a biofilm). The leaf surface often represents a dry, nutrient-poor environment. For this reason many bacteria on leaves are found as isolated microcolonies or individual organisms rather than as a large complex biofilm. The gradation between a colony and a biofilm is not always distinct, and on leaves some researchers consider all multicellular assemblies as biofilms. However, large biofilms have been observed on leaves. They are particularly prevalent on leaves growing in high humidity. On the surfaces of roots, bacteria and fungi often form large biofilms. In searching the literature the reader needs to be aware that studies of layers of microorganisms associated with roots are generally referred to as studies of root colonization rather than root biofilms.

General Considerations

The surfaces of a plant present very different environments for the establishment of a biofilm. On the surface of leaves the absence of water may limit the development of a biofilm. The surfaces of the upper parts of the plant (leaves, stems, flowers, and fruits) are generally covered with a waxy cuticle and are hydrophobic. The surface of roots generally lacks such a layer; the hydrophilic cell wall is often directly available for bacterial colonization. Roots grown in solution or agar usually have a limited number of relatively straight root hairs. Roots grown in soil or soil substitute (quartz sand, vermiculite, and similar substances) have many more root hairs that are longer and much more convoluted than those of water or agar-grown roots. Thus, the exposed surface area of roots is generally very large and is composed mainly of the surface of root hairs.

Plant surfaces are not the same as the inert substrates often used in laboratory studies of biofilm formation. The surface may be subject to alteration or digestion by the microorganisms. In addition, the extracellular matrix that surrounds the bacteria and fungi in a biofilm may be provided in part by the plant. The root tip produces a layer of mucigel that may be incorporated into the biofilm. In addition, the root secretes substances including sugars, amino acids, and dicarboxylic acids that support microbial growth. Other parts of the plant also supply chemical substrates that support bacterial growth, although these are usually secreted in smaller amounts than the concentrations typically found surrounding roots. The plant may respond actively to the presence of microorganisms growing in its vicinity or on its surface. Changes in chemicals secreted by the plant and in cell wall composition in response to the presence of microorganisms have been observed. The interactions between plants and microorganisms can be very specific and depend on the species and cultivar of the plant and species and strain of the microorganism. A well-studied example of this specificity is the interaction between members of the rhizobia and legumes.

Biofilms on plant surfaces as on other surfaces are fragile, and disturbing them to make observations or measurements may give a misleading picture of the biofilm in the absence of disturbance. Therefore, it is important to attempt to confirm any observations made by techniques that disturb the biofilm with additional observations of undisturbed material.

Establishing a Biofilm

Methods for Establishing a Biofilm on a Plant Surface

Plants taken from nature or from a greenhouse are likely to have a well-established biofilm on their leaf and root surfaces. Thus, if one wishes to establish a biofilm on an uncolonized plant surface it is necessary to start with aseptically grown plants. Alternatively, one can attempt to remove the microorganisms present on the plant surface using sterilizing chemicals or mechanical treatments. This is not generally recommended as these treatments damage the plant surface and/or fail to remove all of the resident organisms (note that nonviable organisms may still remain attached to the surface and alter its properties for biofilm formation).

To grow plants aseptically the seeds are surface sterilized; generally bleach is used for this purpose. A typical protocol is as follows: soak seeds for 30 sec in 95% ethanol, then soak the seeds in a 20 to 35% dilution of household bleach in water for 20 to 30 min, wash the seeds 3 to 5 times with sterile distilled water, and germinate them in sterile water or on sterile water agar.[1] For small seeds that supply few minerals to the developing plant, a salt solution such as those used for the growth of plant tissue cultures may need to be added (for example, MS salts[2]). If it is desired to establish a biofilm that originates from the seed, the bacteria can be inoculated directly onto the seed before it is planted or germinated. Seeds may be planted in sterilized quartz sand,[3] microwaved soil,[4] or natural soil. Quartz

---

[1] A. G. Matthysse and S. McMahan, *Appl. Environ. Microbiol.* **64**, 2341 (1998).

[2] T. Murashige and F. Skoog, *Physiol. Plant* **15**, 473 (1962).

[3] M. Simons, A. J. van der Bij, I. Brand, L. A. de Wegner, C. A. Wijffelman, and B. J. J. Lugtenberg, *Mol. Plant–Microbe Interact.* **9**, 600 (1996).

[4] S. Ferriss, *Phytopathology* **74**, 121 (1984).

sand has the advantage that there will be no competing organisms. In natural soil there may be so many competing organisms that the inoculated organism has difficulty in becoming established on the plant surface. Isolation and identification of the inoculated organism from natural soil systems can be difficult. However, soil provides physical conditions and chemicals not available in quartz sand. As a compromise between these two extremes microwaved soil is often used. Soil is microwaved for 10 min in 2-kg lots in sealable plastic bags. If the soil contains *Trichoderma* species or other fungi whose spores are likely to be activated by the heat, it should be stored for more than 2 weeks before use (during this time the *Trichoderma* spores germinate, grow, fail to find a plant host, and die).[4] The physics and chemistry of the soil are relatively undisturbed by this treatment, but the number of viable microorganisms is greatly reduced. Autoclaved soil is not recommended because of the large chemical alterations in the soil caused by autoclaving.

Bacteria are generally inoculated onto seeds, leaves, or roots by dipping the plant into a suspension of bacteria in buffer. In some cases divalent cations, for example, 10 mM MgSO$_4$, are added to the buffer.[5] Alternatively, the bacterial suspension can be sprayed onto the surface of leaves or flowers. If the surface of the plant is difficult to wet or if the microorganisms appear to have difficulty adhering to the surface, 0.1% Triton X-100 may be added as a wetting agent or 300 mg/ml of carboxymethylcellulose[6] or 20 mg/ml methylcellulose[7] added to the suspension of microorganisms as an adhesive. Adhesives are often used when inoculating bacteria onto seeds. In general when plants are inoculated using these techniques, some of the bacteria do not survive the inoculation. The number of viable bacteria recovered an hour after inoculation may be only one-tenth of those inoculated.[1] Methods to study the initial attachment of bacteria to the plant surface have been described in a previous volume in this series.[8] There has been relatively little change in these methods since the earlier article.

Adding to an Existing Biofilm

The problem of adding to an existing biofilm is different depending on the part or organ of the plant involved. If the organ is growing, then the existing biofilm will also be expanding to cover the new surface. In this case introducing a new organism into the growing region of the biofilm will probably present little difficulty. For example, bacteria can be sprayed onto newly opened leaves or flowers.[9] However,

[5] R. Tombolini, D. J. van der Gaag, B. Gerhardson, and J. K. Jansson, *Appl. Environ. Microbiol.* **65,** 3674 (1999).
[6] E. A. Rattray, J. I. Prosser, L. A. Glover, and K. Killham, *Appl. Environ. Microbiol.* **61,** 2950 (1995).
[7] W. F. Mahaffee, E. M. Bauske, J. W. van Vuurde, J. M. van der Wolf, M. van der Brink, and J. W. Kloepper, *Appl. Environ. Microbiol.* **63,** 1617 (1997).
[8] A. G. Matthysse, *Methods Enzymol.* **253,** 189 (1995).
[9] J. Mercier and S. E. Lindow, *Appl. Environ. Microbiol.* **66,** 369 (2000).

if the biofilm is established on a mature part of the plant, then adding to an existing biofilm may be difficult. Additional nutrient can be supplied with the introduced organism to attempt to aid it in becoming established, or a portion of the biofilm can be removed chemically or mechanically to give the new organism space to become established. If the organism to be introduced has some selective advantage, then its establishment is likely to be easier to accomplish.

Description of the Biofilm

Microscopic Methods

It is important to examine the structure of a biofilm microscopically. Unless this is done the determination of the numbers and types of microorganisms present is not very informative. Even if the desired data require the examination of fixed material, it is best to observe the living biofilm as it changes over time as well as the fixed material. In this way fixation artifacts can be minimized and the investigator can optimize the sampling times and locations for material to be fixed or extracted.

Observation of Unstained Live Materials. Thin biofilms can be examined directly using living material and the light microscope. DIC (Nomarski) optics are helpful if no staining is employed. For roots or other organs that are of moderate thickness, live material can be observed using cover slips with a gasket such as probe-clip press-seal chambers (Sigma-Aldrich Co.). As the biofilm develops and becomes thicker, satisfactory resolution of organisms in unstained material may be difficult to achieve. Fluorescence microscopy can be used to overcome this problem.

Observation of Fluorescent Material. Fluorescent labels can be observed using a fluorescence microscope. However, fluorescent labeling of bacteria is most useful with scanning confocal laser microscopy (SCLM). This technique allows the investigation of deep biofilms and gives better resolution than can be obtained with ordinary epifluorescence microscopy. It also allows the construction of detailed three-dimensional images of the biofilm.

In order to observe the general structure of the biofilm, it can be flooded with a fluorescent dye such as 0.1% fluorescein and the areas where the access of the dye is unrestricted can be observed.[10] In a variation of this method, dextrans of various sizes linked to FITC (fluorescein isothiocyanate) can be used to estimate the size limitations for the penetration of molecules into the biofilm.[10] Bacteria can also be stained with fluorescent dyes such as DAPI or acridine orange. A major consideration in the use of fluorescence microscopy to examine biofilms attached to plants is the autofluorescence of the plant tissue. Leaves and roots have

---

[10] R. Lawrence, G. M. Wolfaardt, and D. R. Korber, *Appl. Environ. Microbiol.* **60**, 1166 (1994).

different autofluorescence. This is due to the presence of chloroplasts in leaves and to many different fluorescent compounds that are differentially localized in leaves or roots.[11] DAPI fluorescence is often visible on roots, whereas acridine orange may be easier to detect on leaves.[12-14] The autofluorescence of plant tissues can sometimes be used to advantage as it allows the easy visualization of the plant surface.

When working with leaves that have an easily removable cuticle (for example, endive or *Bryophyllum*), cuticle peels can be used to eliminate the problem of autofluorescence. The cuticle with the attached biofilm can be peeled off the leaf, soaked in 0.01% acridine orange in pH 4 acetate buffer, rinsed in water, and examined microscopically using ultraviolet light and a blue filter.[14]

The bacteria themselves can be made fluorescent by the introduction of the gene for green fluorescent protein (*gfp*). Gfp is available in a variety of colors and with a variety of half-lives ranging from the native protein, which is stable indefinitely in many species of bacteria, to proteins with half-lives as short as a few minutes (the actual half-life depends on the species and strain of bacterium in which *gfp* is expressed). The *gfp* gene can be introduced into the bacterium on a plasmid or on a transposon or placed in a defined chromosomal location by marker exchange of a *gfp* inserted into a cloned bacterial gene. The gene can be expressed from a constitutive promoter and used to track and visualize bacteria during biofilm formation. Alternatively, *gfp* can be placed behind a regulated promoter and used to monitor the activity of that promoter. If the promoter chosen is responsive to some environmental signal such as the availability of iron or sugar or acid pH, then Gfp fluorescence can be used to study the internal conditions inside the biofilm. If the promoter activity reflects the growth and metabolic activity of the bacteria (for example, the ribosomal RNA promoter), then measurement of a short half-life Gfp can be used to examine the metabolic state of the bacteria in the biofilm.[15] In any *gfp* construction it must be remembered that excessive expression of *gfp* may interfere with the normal functioning of the bacteria. High constitutive expression of the stable form of *gfp* has been found in some cases to reduce bacterial growth and binding to the plant surface.[16] Thus, the choice of the particular *gfp* gene with respect to color and protein stability and the promoter used to make a construction is important. There is a trade-off between obtaining enough fluorescence for easy

[11] F. W. D. Rost, "Fluorescence Microscopy," Vol. II. Cambridge University Press, Cambridge, U.K., 1995.

[12] B. Nomander, N. B. Hendriksen, and O. Nybroe, *Appl. Environ. Microbiol.* **65,** 4646 (1999).

[13] K. Kusel, H. C. Pinkart, H. L. Drake, and R. Devereux, *Appl. Environ. Microbiol.* **65,** 5117 (1999).

[14] C. E. Morris, J.-M. Monier, and M.-A. Jacques, *Appl. Environ. Microbiol.* **63,** 1570 (1997).

[15] C. Ramos, L. Molbak, and S. Molin, *Appl. Environ. Microbiol.* **66,** 801 (2000).

[16] R. Chabot, H. Antoun, J. W. Kloepper, and C. J. Beauchamp, *Appl. Environ. Microbiol.* **62,** 2767 (1996).

visualization and interfering with normal bacterial functions. The particular construction to be used will need to be optimized experimentally for the system to be studied.

In addition to *gfp*, *lux* genes can also be introduced into bacteria to render them luminescent. The proteins encoded by the *lux* genes are not themselves luminescent and require a substrate and cellular energy to produce light. In one method, roots that were inoculated with bacteria carrying *lux* genes were shaken to remove soil particles, washed in water, air dried for 5 to 10 min, and placed on tryptic soy agar plates. The bacteria were allowed to grow for 24 hr and then the location of bacteria carrying *lux* genes was determined using a camera with a charge coupled device cooled to −115° with liquid nitrogen and exposures of 1 to 5 min. Alternatively, the plants can be grown in Plexiglas boxes with soil, clay, sand, or vermiculite. The cover is removed to add a few drops of *n*-decyl aldehyde to the soil about 1 cm from the roots and the roots photographed as above using 10 to 20 min exposures.[6] Inoculated seeds can also be grown on Whatman number 42 filter paper or soil between plastic sheets.[6,17] The use of *gfp* and *lux* can be combined using a Gfp whose fluorescence does not interfere with that of *lux*. This allows the determination of bacterial cell number and location (*gfp*) and metabolic activity (*lux*) on the same sample.[18]

Fluorescent antibodies can also be used to examine biofilms. This is most conveniently done with epifluorescence for early stage biofilms or SCLM for later deeper biofilms. FITC-coupled antibodies can be examined with SCLM using an excitation wavelength of 488 nm and detection at 515 nm. This can be combined with root autofluorescence to detect the root surface, which can be seen using excitation at 543 nm and detection at 590 nm. In addition, DAPI staining can be used to determine the position of all of the bacteria (not just those that react with the antibody) using excitation at 340 nm and detection at 390 nm.[19] This triple combination of fluorescent observations should give a good picture of the structure of the biofilm. Rhodamine-coupled antibodies can be used, although some authors have experienced difficulty examining rhodamine fluorescence because of interference from root autofluorescence.[13]

Observation of Fixed Materials. The use of fixed materials to study biofilms presents unusual difficulties due to the frailty of the matrix in which the biofilm is embedded. Considerable care must be taken that parts of the biofilm are not loosened and washed away during the fixation, staining, and dehydration necessary for the preparation of fixed material. It is important to compare fixed preparations

[17] A. Unge, R. Tombolini, L. Molbak, and J. K. Jansson, *Appl. Environ. Microbiol.* **65,** 813 (1999).

[18] M. Schloter, W. Wiehe, B. Assmus, H. Steidl, H. Becke, G. Höflich, and A. Hartmann, *Appl. Environ. Microbiol.* **63,** 2038 (1997).

[19] M. K. Chelius and E. W. Triplett, *Appl. Environ. Microbiol.* **66,** 787 (2000).

with live specimens to be sure that they look the same. Fixation artifacts are difficult to avoid with such tenuous material.

Except for the early stages of biofilm formation, scanning electron microscopy (SEM) does not give a useful picture of a biofilm. The procedures for SEM of the early stages of biofilm formation are the same as those described previously for SEM studies of bacterial attachment.[8] For more mature, deeper biofilms the metal coating required for SEM tends to produce a large amorphous blob rather than the intricate set of interlinked microorganisms seen in SCLM.

Fixed material containing deep biofilms can be examined using light microscopy, fluorescence microscopy, SCLM, and transmission electron microscopy (TEM).

Roots can be fixed overnight in 3 to 4% paraformaldehyde in 10 mM phosphate buffer, pH 7.2, with 0.1% glutaraldehyde at room temperature. The fixative is washed out with phosphate buffer and the sample dehydrated in an ethanol series to 80% ethanol. The sample is embedded in LR White resin that is polymerized at 60° for 24 hr, and thin sections are cut. This material is suitable for reaction with fluorescent or gold-coupled antibodies.[18,19] It can also be used for hybridization with oligonucleotide probes that have been labeled with ^{32}P or coupled to rhodamine.[13]

Enumeration and Physiological Typing of Biofilm Bacteria

General Considerations. Direct microscopic observation of biofilms has been an important approach toward understanding the microbial biofilm communities associated with plant tissues. However, other methodologies involving disruption of the plant-associated microbes have complemented the microscopic data as well as supplying information that could not otherwise be obtained.

A simple approach often employed in studies of biofilms is to physically separate free, dispersed cells from biofilm cells, followed by disruption of the biofilm into individual cells. The constituents in the dispersed and biofilm phases can then be compared using a number of methods such as viable cell counts, direct cell counts, and total protein measurements. Although these approaches have been successful in examining aquatic and marine biofilms on inert surfaces, biofilms associated with living tissues such as plants represent a unique problem. Total measurements of biological macromolecules such as protein are not reliable because of interference from the plant tissue. Therefore, direct and viable cell counting methods are typically used. In addition, because of irregularities in the surfaces of plant tissues, it is often difficult to be certain that the cells removed by a given treatment are representative of the *in situ* populations. Furthermore, as with all techniques that rely on cultivation, only those organisms that grow under the conditions and media employed will be enumerated. In spite of these caveats, there are several cases where meaningful information regarding the composition and organization of biofilm bacteria in association with host plants has been obtained from this approach.[14,20]

Separating Single Cells from Larger Order Aggregates (Biofilms). Biofilms associated with plant tissues are usually moderately sized with lateral dimensions of 20–500 μm. This size range allows single, dispersed cells to be separated from the larger order aggregates by passage through filters with highly uniform pore sizes greater than 5 μm. The best examples of this approach have used leaf tissue, although root and shoot tissues can be processed in the same manner (see Morris *et al.* for an excellent report on implementing this general approach[20]).

Removal of Bacteria from Plant Tissue. The first step is to remove the bacterial growth from the plant tissue. The plant tissue can be directly suspended in a buffer solution, followed by extended incubation on a rotary shaker (a gentle prewash step to remove loosely bound cells may be beneficial if the biofilms are not removed by this treatment). The release of more tenaciously adhering biofilm material may require gentle sonication with a sonicator probe set to low output or a bath sonicator. The removal of bacteria using these wash treatments should be assessed, if possible, by harvesting the washed plant material, homogenizing the tissue, and enumerating the remaining bacteria by viable cell counts of the homogenate on the appropriate growth medium. The number and diversity of bacteria recovered from the homogenate can be compared to those released by washing. Direct examination of plant tissue using DIC (Nomarski optics) can also provide information on the number of remaining bacteria (see above).

Filtration of the Bacterial Suspension. The bacterial suspension obtained by washing (see above) is filtered with the aim of retaining biofilm aggregates on the filter and allowing solitary bacteria to pass through the filter. The choice of filter material and pore size is crucial to the success of this approach and should be optimized for each specific application. Filters used for this approach must have large pore sizes (>5 μm) that allow unhindered passage of solitary cells, but effectively retain larger order aggregates. Consequently, an important consideration is the propensity of individual cells to associate with the filter matrix. C. Morris and colleagues compared several different filter matrices of approximately the same pore size for retention of dispersed cells from several leaf-associated genera and found that very thin (10 μm thickness) polycarbonate (TMTP) filters bound the lowest number of individual cells.[20] However, the appropriate filter should be determined for each specific application. The number of individual cells retained by the filtration can be assessed by passing the initial filtrate through a second, identical filtration system and comparing the cell numbers in the single-passed filtrate with those in the double-passed filtrate. The material retained on the filter can be suspended in a small volume of buffer for further processing (this biofilm-enriched

---

[20] C. E. Morris, J.-M. Monier, and M.-A. Jacques, *Appl. Environ. Microbiol.* **64**, 4789 (1998).

material can also be directly examined using a variety of techniques including microscopy, see above, and molecular probing, see below).

Disruption of Biofilm Material and Analysis. For accurate enumeration of retained bacteria by viable cell counts, the aggregates must be fully dissociated. The most common technique is to use ultrasonication under carefully controlled conditions with a sonicator that has a microtip. Care must be taken in choosing the conditions for sonication so that aggregates are effectively disrupted but the overall viability of individual cells is not affected. These conditions will vary substantially for different samples and microbial populations and need to be empirically determined. The effect on cell viability can be assessed using the preparation of dispersed cells that passed through the 5 μm pore size filter. A range of sonication regimens can be tested using these cells to define the maximum boundaries. The efficacy of sonication on biofilm disruption can be assessed microscopically by direct counting of multicellular aggregates. Perfect disruption should convert 100% of the aggregates to solitary cells. Therefore, sonication of biofilm suspensions should release cells that pass through the filter. Sonication should be continued until no further increase in filterable cell numbers is noted or until the maximum treatment at which cell viability is unaffected is reached. Once an effective sonication regimen is determined, biofilms retained on filters can be disrupted and subjected to standard plate counting procedures. If different bacteria in the biofilm population form diagnostic colony types, the average biofilm constituency can be determined directly from the plate.

Variations. Separation of biofilm (i.e., aggregated) cells from solitary cells is clearly an important step in characterizing biofilm material enriched from any environment, including plants. Although filtration satisfies some of the requirements of this separation, the interaction of individual cells with the filtration matrix, aggravated by clogging of the filter with the biofilm material itself, forces users to run a series of extensive controls. An alternative approach that has been successfully employed for separation of microbial cell populations with different physical densities is centrifugation through colloidal silica gradients (e.g., Ludox, Dupont Chemicals).[21,22] Solitary cells and other small particulates will traverse the gradient rapidly, while the biofilm material of lower overall density should migrate very slowly. This procedure can be optimized to separate single bacterial cells rapidly and may aid in characterizing the size range of multicellular aggregates.

Another interesting variation developed by Morris *et al.* involves the harvesting and characterization of individual biofilm aggregates.[14] Rather than immediate disruption, the material retained on the large-pore-size filters is serially diluted and applied to agar plates. The plate surfaces are inspected using phase contrast microscopy. Dilution plates with well separated "biofilm-like" particles released

---

[21] M. Evinger and N. Agabian, *J. Bacteriol.* **132,** 294 (1977).
[22] R. W. Shulman, L. H. Hartwell, and J. R. Warner, *J. Mol. Biol.* **73,** 513 (1973).

during processing are identified. Individual biofilms are collected using an angled Pasteur pipette or similar device to core the biofilm and the underlying agar. This process can be aided by adjusting the microscope light beam to a fine point and using this point to guide harvesting of the aggregate. The core carrying the cells is transferred to sterile buffer and homogenized. The homogenate is plated in dilutions on the appropriate growth medium. Following a fixed growth period, colonies are enumerated on the plates and are typed using a battery of physiological tests. This method provides information on the number and diversity of bacteria within specific biofilms, but does not enumerate the number of solitary cells relative to those in biofilms.

Plant-associated bacteria also can be removed from plant tissue by directly imprinting it on to an agar plate (a tissue print). Loosely associated bacteria are washed from the tissue surface with buffer. Those tightly adherent bacteria that remain on the surface are then used to inoculate cultivation media, typically an agar plate. The spatial organization of bacterial growth on the tissue is somewhat preserved after transfer. In order to apply this technique to biofilms, the large aggregates (e.g., biofilms) must be sufficiently separated on imprinting to allow visual identification and harvesting as described above. The imprinting method, although much simpler than the laborious separation described above, does not physically separate solitary bacteria adherent to the leaf from those in biofilms. Therefore, the enumeration of bacteria from these experiments is suspect, although in cases where the processing described above causes loss of viability, tissue prints may be useful.

Measuring Metabolic Activity in Plant-Associated Microbial Communities

General Considerations. As with the physiology of planktonic bacteria, metabolic activity within a biofilm can vary temporally and as a function of physical and chemical conditions. In addition, there is significant metabolic variation at different positions within the biofilm. Although there have been very few studies characterizing the metabolic status of plant-associated biofilms, there is every reason to assume that this will be addressable using available techniques. The primary technical problem for plant-associated biofilms is that the metabolic capacity of the plant tissue may itself interfere with measurements. This problem may be circumvented by (i) separating the bacterial biofilms from plant material (see above), (ii) monitoring metabolism specific to the bacteria, (iii) probing with oligonucleotides and antibodies specific for the bacteria, and (iv) utilizing bacterial strains harboring metabolic reporter genes.

Viability Staining. An important issue in the study of any microbial population is the distribution of living compared to nonviable cells. In many cases the overall viability of the population is of greater importance than any specific aspect of metabolism or microbial diversity. There are a number of vital staining approaches that have been used to determine the level of viability in microbial cultures or samples, among them laboratory-grown biofilms. Many of these approaches rely on commercially available, fluorescent dyes that label nucleic acids. The LIVE-DEAD viability stain (Molecular Probes, Eugene, OR) combines two stains, one that stains all cells green (a membrane permeant dye, SYTO9) and a dye that only stains dead cells red and quenches the green dye (propidium iodide, a dye that is impermeant to membranes of living cells). Optimally, it would be desirable to apply these stains to biofilms *in planta*. However, the vital stains currently available also effectively stain plant cells (as well as animal cells), increasing the background fluorescence to unacceptable levels. In order to use this vital staining approach, biofilm material must be dislodged from the plant tissue (see above). The vital stain solution can be added to suspensions containing dislodged microbial aggregates and the number of viable cells compared to dead cells within the biofilms using three-dimensional reconstructions of the biofilm by SCLM (see above). One important consideration is the penetration of the stain into the biofilm. Biofilms from different sources will exhibit different propensities to exclude the stain. A simple control for accessibility to stain is to disrupt the aggregates using techniques such as gentle ultrasonication (see above) followed by vital staining of the released cells. Although the information regarding spatial distribution is lost, the relative ratio of living to nonliving cells should correlate with that determined for the intact biofilms.

Adenylate Energy Charge. A general measure of metabolic activity in a variety of systems is the relative amounts of adenosine phosphates, the AEC or adenylate energy charge ($([ATP] + [ADP])/([ATP] + [ADP] + [AMP])$). The AEC values are reflective of the level of bacterial metabolic activity. Several studies on model, steady-state biofilms have estimated the AEC under different conditions.[23] Adenylate concentrations are measured from horizontal or vertical sections of the biofilm, using standard assay procedures in a reaction coupled to firefly luciferase (an ATP-requiring enzyme), and detected as a change in luminescence.[24] In general, a gradient of increasing AEC has been observed, from the base of a biofilm closest to the substratum toward the surface in full contact with the environment. This classic physiological assay has not been employed to date to study plant-associated biofilms. It would require either separation of the biofilm from the plant material or some other way to identify the microbial contribution to the AEC. In some cases, it might be worthwhile to measure the AEC for the plant and microbial community together.

Enzymatic Assays. Traditional assays of microbial enzymatic activity are generally performed on whole samples and thus always represent the average activity of a given population. The heterogeneity of biofilms makes these measurements

[23] S. L. Kinniment and J. W. T. Wimpenny, *Appl. Environ. Microbiol.* **58,** 1629 (1992).
[24] A. Lundin and A. Thore, *Appl. Microbiol.* **30,** 713 (1975).

particularly problematic, as average enzymatic activities misrepresent the range of activity within the biofilm. However, with the advent of fluorescent substrates coupled with SCLM, specific enzymatic activities can be monitored for individual cells. Fluorescent substrates for a variety of enzymatic activities are available. Only recently have these *in situ,* high-resolution approaches been applied to biofilms. An example of such an approach employed a fluorescent alkaline phosphatase substrate (ELF-97, Molecular Probes) to spatially map the activity of alkaline phosphatase within flow-chamber grown biofilms of *Pseudomonas aeruginosa* and *Klebsiella pneumoniae.*[25] These approaches should be amenable to plant-associated microbial communities, but the enzymatic activity of the plant tissue must be considered. Although there are some unique bacterial enzyme systems not found in plant tissue, many important microbial enzymes are also active in plant tissue. Therefore, separation of biofilms may be necessary prior to analysis.

Probing Biofilm Physiology with Fluorescently Labeled Nucleic Acids and Antibodies. The ability to probe for the expression of specific genes and/or proteins under a given set of conditions is an essential tool in molecular biology. Biofilms that have been harvested intact and chemically fixed by a variety of methods can be directly examined microscopically using nucleic acid or antibody probes conjugated to fluorescent labels (see above). These fixed films retain much of their original structure, and therefore direct examination provides spatial information. For biofilms associated with plant tissue, autofluorescence in a variety of plant tissues is a major consideration.[11] Therefore, the choice of fluorescent labels with an emission spectrum that minimizes interference from the plant autofluoresence is essential. An alternative approach is to remove the biofilm material from the plant to avoid interfering fluorescence.

Fluorescently labeled oligonucleotide probes conjugated to a variety of dyes are commercially available. Methods for conjugating fluorescent dyes to the oligonucleotides are relatively straightforward and have been described in detail elsewhere.[26] Using standard hybridization techniques it is extremely difficult to detect mRNA *in situ* that is expressed at moderate to low levels. However, rRNA is easily detected *in situ.* Determination of ribosome content in a biofilm community using rRNA-specific, fluorescently tagged oligonucleotide probes allows measurements at the level of single cells and surmounts some of the problems of biofilm heterogeneity.[27] The level of rRNA (and hence ribosomes) present in a single cell is reflective of the growth rate of the cell, and measurements of rRNA in biofilm constituents can provide information on the spatial distribution of metabolic activity within the film. In addition, rRNA probes that allow

[25] C. Huang, K. Xu, G. McFeters, and P. Stewart, *Appl. Environ. Microbiol.* **64,** 1526 (1998).
[26] W. Manz, *Methods Enzymol.* **310,** 79 (1999).
[27] L. K. Poulsen, G. Ballard, and D. A. Stahl, *Appl. Environ. Microbiol.* **59,** 1354 (1993).

phylogenetic differentiation can be similarly employed to gauge microbial diversity in biofilms. Levels of specific mRNA transcripts have been assessed by using *in situ* RT-PCR with fluorescently labeled oligonucleotides.[28] However, the efficiency of PCR amplification in complex samples, such as biofilms, may vary and needs to be carefully controlled.

Proteins can be detected *in situ* using antibodies labeled with a range of fluorescent dyes. The antibody specific for the protein of interest can be labeled directly, or a secondary antibody conjugated to a fluorescent dye can be used. In general, the latter allows a variety of primary antibodies to be used, but may aggravate problems of nonspecific background due to multiple antibodies. Fixed biofilm samples can be readily probed with immunofluorescent antibodies using standard protocols.[29] Likewise, immunofluorescent antibodies have been extensively used in plant biology. Several studies have examined plant–microbe interactions using FITC or rhodamine labeled antibodies, to visualize and quantify the number of specific bacteria and bacterial exoproducts (e.g., exopolysaccharide) associated with plant tissue.[18,30] Although immunofluorescence has not been extensively used to analyze levels of specific proteins within microbial assemblages *in planta,* there is every reason to think that, with the appropriately chosen fluorescent labels, this approach should be readily adaptable to plant-associated biofilms.

Reporter Systems. Recent studies have begun to examine artificial biofilms composed of bacteria carrying an easily assayed reporter gene. Reporter genes including *lacZ* (β-galactosidase), *lux* (luciferase), *ice* (ice nucleation protein), and *gfp* have been successfully employed to measure gene expression within biofilms. Noninvasive detection of reporter gene activity with intact biofilms facilitates *in situ* observation and has been particularly revealing.[31–34] Although many applications have used constitutive, strongly expressed *gfp* fusions to visualize bacteria colonizing surfaces including plant surfaces, monitoring the expression of specific gene products using *gfp* fusions is also a highly tractable approach[32,35] (see above).

[28] R. E. Hodson, W. A. Dustman, R. P. Garg, and M. A. Moran, *Appl. Environ. Microbiol.* **61,** 4074 (1995).

[29] D. R. Korber, G. M. Wolfaardt, V. Brozel, R. MacDonald, and T. Neipel, *Methods Enzymol.* **310,** 3 (1999).

[30] J. A. McGarvey, T. P. Denny, and M. A. Schell, *Phytopathol.* **89,** 1233 (1999).

[31] C. Sternberg, B. B. Christensen, T. Johansen, A. T. Nielsen, J. B. Andersen, M. Givskov, and S. Molin, *Appl. Environ. Microbiol.* **65,** 4108 (1999).

[32] G. V. Bloemberg, G. A. O'Toole, B. J. J. Lugtenberg, and R. Kolter, *Appl. Environ. Microbiol.* **63,** 4543 (1997).

[33] D. G. Davies, A. M. Chakrabarty, and G. G. Geesey, *Appl. Environ. Microbiol.* **59,** 1181 (1993).

[34] D. G. Davies, M. R. Parsek, J. P. Pearson, B. H. Iglewski, J. W. Costerton, and E. P. Greenberg, *Science* **280,** 295 (1998).

[35] D. J. Gage, T. Bobo, and S. R. Long, *J. Bacteriol.* **178,** 7159 (1996).

Genetic Analysis of Biofilms Composed of Plant-Associated Bacteria

General Considerations

Mutational analysis allows identification of bacterial genes required for biofilm formation. Simple mutational screens for biofilm formation deficiencies have been employed to isolate such genes.[36–40] Likewise, a number of bacterial genes required for adhesion to plants have been identified by direct, microscopic screening of transposon mutants for reduced interactions with plant tissue culture cells (for examples see refs. 41–43). However, these studies were primarily focused on initial adhesion events and not the development of biofilms. Initial adhesion is an obvious first step for physical interaction of the plant and microbe and therefore essential for biofilm formation. Although some overlap is expected, many of those functions that allow the bacteria to proceed from colonization by single cells to formation of multicellular assemblages are likely to be distinct from adhesion functions.

Although genetic factors in the host plant that promote biofilm formation have not yet been examined, identification of these genes through mutant screening should be possible. Plants amenable to genetic manipulation such as *Arabidopsis thaliana* are clearly the model systems of choice in this respect. In addition, the genome sequence for *A. thaliana* should provide likely candidates for host genes that might be important for long term plant–microbe interactions.

Screening for Biofilm Deficient Mutants in Plant-Associated Microbes

There are two basic approaches that can be utilized to identify mutants deficient for biofilm formation. The first is simply to use the approach developed extensively by O'Toole and Kolter and applied to *P. fluorescens*.[44] The second is to adapt currently existing adhesion assays to screen for incomplete biofilm formation.

In vitro growth conditions and model surfaces can be optimized for the plant-associated microbe of interest. For example, *A. tumefaciens* forms extensive biofilms on a number of different inert surfaces such as polyvinyl chloride (PVC)

---

[36] C. Y. Loo, D. A. Corliss, and N. Ganeshkumar, *J. Bacteriol.* **182,** 1374 (2000).
[37] G. A. O'Toole and R. Kolter, *Mol. Microbiol.* **30,** 295 (1998).
[38] G. A. O'Toole and R. Kolter, *Mol. Microbiol.* **28,** 449 (1998).
[39] L. A. Pratt and R. Kolter, *Mol. Microbiol.* **30,** 285 (1998).
[40] P. I. Watnick and R. Kolter, *Mol. Microbiol.* **34,** 586 (1999).
[41] C. J. Douglas, W. Halperin, and E. W. Nester, *J. Bacteriol.* **152,** 1265 (1982).
[42] A. G. Matthysse, *J. Bacteriol.* **169,** 313 (1987).
[43] J. L. Robertson Crews, S. Colby, and A. G. Matthysse, *J. Bacteriol.* **172,** 6182 (1990).
[44] G. A. O'Toole, L. A. Pratt, P. I. Watnick, D. K. Newman, V. B. Weaver, and R. Kolter, *Methods Enzymol.* **310,** 91 (1999).

and cellulose acetate (Goforth and Fuqua, unpublished results). These surfaces lend themselves to large scale screening formats to test for biofilm formation mutants using simple staining methodologies (see O'Toole *et al.* for a general discussion of this approach[44]). It is incorrect to assume that all bacterial mutants unable to form biofilms on inert surfaces will also be defective in plant interactions. Likewise, not all functions required for biofilm formation on plant surfaces will also be required on inert surfaces. However, it is almost certain that there will be overlapping functions, particularly those downstream of initial adhesion events, that are shared irrespective of the surface composition. Mutants identified by these screens can subsequently be tested for plant-associated biofilm formation using the microscopic, molecular, and bacteriological methods described above.

A more direct, yet substantially more difficult approach is to screen directly for biofilm deficiencies on excised plant tissues or plant cells. In several of the genetic screens employed to identify plant association mutants, differences in cell aggregation were noted in phase contrast micrographs as well as scanning electron micrographs.[45] This same screening method can be applied to identification of biofilm mutants. Although a strictly visual screen is subject to a great deal of variability and user subjectivity, there is a well developed set of statistical tools for analyzing aggregation state (originally developed for geostatistical applications) that can be applied to image analysis.[46] Recently, a suite of algorithms for assessing a range of biofilm structural attributes has been developed that, in automated form, could dramatically enhance the efficiency and mathematical robustness of direct microscopic screens.[47]

Analysis of Preexisting Mutants. Mutant identification such as that discussed above, as well as more directed approaches, has begun to reveal those gene products involved in biofilm formation. It is clear there are many general features that are often involved in biofilm development for different bacteria (see Pratt and Kolter for a review[48]). In the age of molecular genetics and ever-proliferating genome sequences, null mutations in specific genes are relatively straightforward to generate in a variety of bacteria. Therefore, it will become increasingly facile to generate a suite of mutations in genes suspected to be involved in biofilm formation, and subsequently test these mutants *in situ*. The techniques used to analyze biofilms formed by wild-type bacteria can be directly applied to specific mutants. Of course, this analysis will not readily identify new biofilm genes. However, the development of genome-wide, nonessential null mutation sets will make it possible to screen every nonessential gene from a given microbe with a sequenced genome for a role

[45] A. G. Matthysse, *Crit. Rev. Microbiol.* **13**, 281 (1986).
[46] L. M. Dandurand, D. J. Schotzko, and G. R. Knudsen, *Appl. Environ. Microbiol.* **63**, 3211 (1997).
[47] X. Yang, H. Beyenal, G. Harkin, and Z. Lewandowski, *J. Microbiol. Methods* **39**, 109 (2000).
[48] L. A. Pratt and R. Kolter, *Curr. Opin. Microbiol.* **2**, 598 (1999).

in biofilm development. Among the countless subdisciplines of microbiology that will benefit from these genomic reagent sets will be the area of plant-associated biofilms.

Expression Analysis of Biofilm-Specific Genes

Mutational analysis of biofilms provides information on those genes essential for biofilm formation. However, there is broader group of genes that are differentially regulated by conditions within the biofilm. Included among these are some of the genes essential for biofilm formation, but also genes that may have important, yet nonessential functions (at least under the chosen screening conditions). Expression analysis encompasses a range of techniques allowing the identification of genes that are differentially expressed under a specific set of conditions. Strategies for gene identification based on reporter genes include the use of differential reporter genes such as *lacZ* and *gfp*, selectable reporter genes that impart growth advantages to the microbe, and genetic reporters resulting in heritable changes that provide a historical record of expression.[49–51] In all these approaches the genes are tagged and can be readily isolated and sequenced for identification. Using random *lacZ* promoter fusions Prigent-Combaret *et al.* have examined differential expression in biofilms, revealing a large number of genes that significantly differ in expression levels in a model biofilm compared to the planktonic state.[49] Additionally, genomic and proteomic techniques that allow direct physical detection of mRNA transcripts and differentially-expressed proteins, are gaining popularity for analysis of microbial gene expression.[52–54] Although application of this wave of new gene expression technology to microbial biofilms has only just begun, these types of expression analyses should be readily adapted to identifying biofilm-regulated genes *in planta*.

[49] C. Prigent-Combaret, O. Vidal, C. Dorel, and P. Lejeune, *J. Bacteriol.* **181,** 5993 (1999).
[50] S. L. Chiang, J. J. Mekalanos, and D. W. Holden, *Annu. Rev. Microbiol.* **53,** 129 (1999).
[51] A. Camilli, D. T. Beattie, and J. J. Mekalanos, *Proc. Natl. Acad. Sci. U.S.A.* **91,** 2634 (1994).
[52] C. S. Richmond, J. D. Glasner, R. Mau, H. Jin, and F. R. Blattner, *Nucl. Acids Res.* **27,** 3821 (1999).
[53] R. A. VanBogelen, E. E. Schiller, J. D. Thomas, and F. C. Neidhardt, *Electrophoresis* **20,** 2149 (1999).
[54] J. R. Yates III, *Trends Genet.* **16,** 5 (2000).

Section II

Flow Systems and Biofilm Development and Characterization

[2] Monitoring Bacterial Growth Activity in Biofilms from Laboratory Flow Chambers, Plant Rhizosphere, and Animal Intestine

By Cayo Ramos, Tine R. Licht, Claus Sternberg, Karen A. Krogfelt, and Søren Molin

Introduction

It has become evident from investigations carried out in recent years that bacteria growing on surfaces tend to build up heterogeneous communities in which different species may reside, interact, and perhaps exchange information (genes, signals). In natural as well as in artificial settings, community organization results in structured surface-bound populations, and the apparent nonrandomness of these developments is a challenge to our understanding of how the most primitive of all organisms live and sustain in complex, stressful and sometimes even hostile environments. The distribution of biomass is one aspect of this heterogeneity; activity distribution is yet another. On one hand, it is clear that the distribution of biomass is caused by one or more of several biological activities: growth of some individuals but not of others, transfer of cells by the flow of liquid surrounding the community, bacterial motility coupled to chemotaxis, detachment and attachment processes, and probably others. Cellular activities, on the other hand, reflect interactions between cells and the environment, be it nutrients in the outer environment or metabolic interactions between individual community members. The integrated understanding of such structure/function relationships is the target of a significant part of modern microbial ecology, and investigations of bacterial biofilms to a large extent are focused on this topic.

We have decided to put a strong emphasis on the analysis of cellular growth activity in our biofilm studies. The reason is that growth of the individual cell is a fairly reliable indicator of positive interactions between the cell and its local environment from which further investigations may reveal the underlying resource distribution. Distribution of growth activity in a microbial community is also an important cause of heterogeneity. In studies of community dynamics, monitoring of growth activity throughout a community may provide a good explanation for differentiation and development of the community. In complex surface communities it is therefore of great importance to be able to record growth activity at all levels, including the single-cell level, before initiating more detailed and specific investigations of structure/function relationships in the community.

Many cellular activities reflect the growth status of the cell, but one process more than most others has become the choice of target for quantification when estimating the growth activity of the cell: the translation process. This part of the cellular macromolecular synthesis is the most complicated and the most resource-demanding, and is obviously a highly regulated process. The translation apparatus comprises a large number of factors, which all seem to be produced in accord with the energy supply sensed by the cell. If optimal resources are available, protein synthesis goes on at its maximum rate, which requires a high cellular pool of the factors involved in translation. In contrast, poor environmental conditions do not allow a high rate of protein synthesis, and consequently the translation factors are only synthesized at low rates. The ribosome—and in particular its content of RNA—is by far the easiest translation factor to monitor in single cells, and since this component is also a phylogenetic marker for species identification, rRNA has become a very popular marker for cellular growth activity. In the following sections we describe a number of ribosome-related measurements of cellular growth in relation to bacterial biofilm formation in laboratory flow chambers, plant root systems, and mammal gastrointestinal tracts.

Biofilm Model Systems

The Flow Chamber

A frequently used laboratory biofilm system involves a flow-chamber setup. Several variations exist, with two prominent major types, the modified Robbins device (MRD) and the mini flow chamber. In a previous volume of this series, a number of detailed descriptions of variants of these flow chambers are presented.[1] We have mainly worked with the mini flow cell, which is easy to handle in terms of setup, operation, and reuse. It has also been shown, using a newly developed statistical program,[2] that these flow chambers are producing reproducible biofilms with respect to the structural properties of the microbial community, when run conditions are kept identical.

The MRD consists of flow chambers with immersed sample holders, each holding a piece of substratum (a "coupon") for biofilm growth. The disadvantage of the MRD is that it is not possible to study the biofilm on the coupons unless they are removed from the device. The mini flow chambers are established as small milled channels in Plexiglas covered with microscope cover-slip glasses serving as substrata. The biofilms growing in the mini flow chambers can easily be observed microscopically. The disadvantages of the mini-flow-chamber system are that it may be difficult to get physical access to the biofilm without an embedding step,

---

[1] B. B. Christensen, C. Sternberg, J. B. Andersen, R. J. Palmer Jr., A. Toftgaard Nielsen, M. Givskov, and S. Molin, *Methods Enzymol.* **310,** 20 (1999).

[2] A. Heydorn, A. T. Nielsen, M. Hentzer, B. Ersbøll, C. Sternberg, M. Givskov, and S. Molin, unpublished.

and flow rates are somewhat limited because of the thin cover-slip substratum. Further details of these flow-chamber devices have been described previously.[1]

The Plant Rhizosphere

Before investigations of root colonization begin; a model test system for the assays must be selected. Choices must be made regarding the substrate supporting plant growth, the use of sterile versus nonsterile systems, the addition of water and nutrients, the incubation conditions (i.e., growth chambers, greenhouses, or field tests), and the age of the plant to be analyzed. Ideally, determination of bacterial activity in rhizosphere samples should consider competition of the introduced strain with the indigenous soil microbiota, and thus should be performed in nonsterile field soils. However, the complexity of the plant–microbe interactions in a soil system is immense,[3] and very little is known about the sum of factors contributing to the overall activity of microbes in the rhizosphere. In many instances, it is therefore necessary to design less complex and presumably more reproducible systems allowing studies of the effects of variations of single biotic or abiotic factors on the activities of specific bacterial strains. It should be noted, however, that results obtained with one system normally cannot be extrapolated to any other plant/bacterial system, since it is known that rhizosphere colonization is influenced by the nature of the microorganisms, the type of substrate, and the seeds used.[4,5] Various substrates have been used to develop sterile systems for studies of microbial behavior in the surfaces of plant roots, such as filter paper,[6] quartz and sand,[7] agar tubes,[8] vermiculite,[9] and sterile soil.[10] Under sterile conditions, we have mainly worked with up to 3-day-old barley seedlings growing on 1% agar plates consisting of M9-minimal medium (22 mM Na$_2$HPO$_4$·2H$_2$O, 22 mM KH$_2$PO$_4$, 100 mM NaCl) containing no carbon or nitrogen sources. Incubation of the plates is performed in the dark at 20°C. To prevent the media from drying out during the incubation, plates are placed in plastic bags containing moist filter paper. In addition to field tests, germination and growth of plants in nonsterile soil is traditionally performed

[3] J. W. Kloepper and C. J. Beauchamp, *Can. J. Microbiol.* **38**, 1219 (1992).
[4] X. Latour, T. Corberand, G. Laguerre, F. Allard, and P. Lemanceau, *Appl. Environ. Microbiol.* **62**, 2449 (1996).
[5] P. E. Maloney, A. H. C. van Bruggen, and S. Hu, *Microb. Ecol.* **34**, 109 (1997).
[6] E. A. S. Rattray, J. I. Prosser, L. A. Glover, and K. Killham, *Appl. Environ. Microbiol.* **61**, 2950 (1995).
[7] M. Simons, A. J. van der Bij, I. Brand, L. A. de Weger, C. A. Wijffelman, and B. J. Lugtenberg, *Mol. Plant–Microbe Interact.* **9**, 600 (1996).
[8] B. Assmus, P. Hutzler, G. Kirchhof, R. Amann, J. R. Lawrence, and A. Hartmann, *Appl. Environ. Microbiol.* **61**, 1013 (1995).
[9] L. A. de Weger, L. C. Dekkers, A. van der Bij, and B. J. Lugtenberg, *Mol. Plant–Microbe Interact.* **7**, 32 (1993).
[10] L. Kragelund, and O. Nybroe, *FEMS Microbiol. Ecol.* **20**, 41 (1996).

in microcosms placed in growth chambers or greenhouses. The type and age of the plants under analysis will determine the volume of soil and the watering system to be used. We have mainly worked with roots of barley (*Hordeum vulgaris*) and corn (*Zea mays*) plants growing during a maximum period of 15 and 7 days, respectively, in 50 ml plastic tubes filled with 50 g of a 1 : 1 mixture of sand and sandy-loam soil. Addition of sand to the soil microcosms facilitates bacterial extraction and preparation of root samples for visualization by microscopy. To avoid soil drying and watering during the incubation period, soil moisture is adjusted to 15% (w/w) with sterile water, the tubes are placed in a beaker containing moist filter paper, and the whole system is placed in a transparent plastic bag. Seeds coated with bacteria are sown about 1 cm below the soil surface, and the plants are left to germinate and grow at 18° (barley) or 25° (corn) in a growth chamber with a 12 hr light: 12 h dark photoperiod.

We have developed a system for direct on-line microscope examination of the root surface of young barley seedlings. Rhizosphere chambers (length 55 mm × width 20 mm × depth 3 mm) are constructed by sticking a silicone rubber gasket on top of a microscope slide and mounting a cover slip (24 mm × 60 mm) on top of the silicone layer. The chambers are filled with vermiculite or soil soaked with 500 μl of water. Barley seeds coated with bacteria are placed at about 5 mm below the soil or vermiculite surface, and the whole system is placed inside a 50 ml centrifuge tube. To avoid soil drying, moist filter paper is placed inside the tubes. To ensure the localization of the root system in the proximity of the cover slip, the tubes containing these growth chambers are placed on a rack forming a vertical angle of approximately 45°. Seeds are allowed to germinate in the dark at 20° during a maximum period of 3 days.

The Mouse Gut

It is an object of discussion whether the bacterial population of the animal intestine can be considered a biofilm or not. The viscous nature of the intestinal mucus keeps the bacteria growing there in a more or less fixed position, but it is definitely not directly comparable to an exopolysaccharide matrix. However, some similarities between the flow-chamber biofilms and the intestinal populations make it obvious to compare the two: (1) the high density of the bacteria, which literally form a "layer" on the inside of the animal epithelium, and (2) the fact that the intestinal bacteria, just like the biofilm in the flow chamber, continuously receive a flow of nutrients, and continuously are diluted/washed out of the system with this flow. A recent comparative study from our group discusses the similarities between the two systems.[11] Since conventional animals are not easily colonized with new ingested

[11] T. R. Licht, T. Tolker-Nielsen, K. Holmstrøm, K. A. Krogfelt, and S. Molin, *Env. Microbiol.* **1**, 23 (1999).

bacteria,[12] animals treated with antibiotics are often used as models for bacterial colonization. We chose streptomycin-treated mice[13,14] as models for *in situ* colonization studies involving *Escherichia coli*,[15-17] *Salmonella typhimurium*,[18] and *Klebsiella pneumoniae*.[19]

In all cases, the bacteria in question were simply given to the animals *per os*, and numbers of bacteria excreted with the feces were followed by plating. At the end of the experiment, animals were sacrificed, and the intestinal area of interest was prepared for *in situ* studies as described later.

In a simpler intestinal model system bacterial growth may be followed *in vitro* in mucus extracted from the mouse gut. The idea of simply inoculating crude, nonsterile intestinal extracts with gram-negative streptomycin-resistant bacteria, and subsequently following the proliferation of inoculated bacteria by plating on selective plates, was originally introduced by P. S. Cohen and co-workers.[20,21] This technique gives stable results and allows the evaluation of different intestinal extracts (mucus and contents) as substrates for bacterial growth. We have expanded the applications of the method to also include hybridization studies of individual bacterial cells as described below.[11,18] Determination of colony forming units of a single bacterial species is possible since the gram-positive bacteria of the intestinal microbiota are mostly anaerobic and will not grow on aerobically incubated plates, whereas the indigenous aerobic, gram-negative bacteria (mainly *E. coli*), will not grow in the presence of streptomycin. The mouse cecum is chosen as the organ used for extraction of intestinal contents and mucus, because (i) the physiology of the cecum makes it easier to manually separate these two fractions than is the case for the rest of the intestine, and (ii) the cecal microbiota is believed to be representative for the rest of the bacterial microbiota in the large intestine.[22] The procedure for preparing cecal mucus and contents is described below.

Mice are sacrificed and the bowels opened. The cecum is cut free from the large and small intestines with a scalpel and saved on ice. It is important to avoid any

[12] D. van der Waij, J. M. Berghuis-De Vries, and J. E. C. Lekkerkerk-van der Wees, *J. Hyg.* **69,** 405 (1971).

[13] D. J. Hentges, J. U. Que, S. W. Casey, and A. J. Stein, *Microecol. Ther.* **14,** 53 (1984).

[14] J. U. Que, S. W. Casey, and D. J. Hentges, *Infect. Immun.* **53,** 116 (1986).

[15] L. K. Poulsen, T. R. Licht, C. Rang, K. A. Krogfelt, and S. Molin, *J. Bacteriol.* **177,** 5840 (1995).

[16] T. R. Licht, B. B. Christensen, K. A. Krogfelt, and S. Molin, *Microbiology* **145,** 2615 (1999).

[17] L. K. Poulsen, F. Lan, C. S. Kristensen, P. Hobolth, S. Molin, and K. A. Krogfelt, *Infect. Immun.* **62,** 5191 (1994).

[18] T. R. Licht, K. A. Krogfelt, P. S. Cohen, L. K. Poulsen, J. Urbance, and S. Molin, *Infect. Immun.* **64,** 3811 (1996).

[19] S. Favre-Bonté, T. R. Licht, C. Forestier, and K. A. Krogfelt, *Infect. Immun.* **67,** 6152 (1999).

[20] B. A. McCormick, B. A. D. Stocker, D. C. Laux, and P. S. Cohen, *Infect. Immun.* **56,** 2209 (1988).

[21] E. A. Wadolkowski, D. C. Laux, and P. S. Cohen, *Infect. Immun.* **56,** 1030 (1988).

[22] B. S. Drasar and P. A. Barrow, *in* "Intestinal Microbiology." American Society for Microbiology, Washington, D.C, 1985.

unnecessary manipulations of the cecum, which may cause mixing of the mucus layer on the inside of the cecal epithelium with the cecal contents. Carefully cut open the cecum, e.g., with sharp scissors or a scalpel. Transfer the cecal contents, which "fall out," to an Eppendorf tube, and keep cold. Wash the cecal epithelium with a physiologic salt solution or any buffer as desired, until no dark material from the contents sticks. Scrape the yellow/white mucus layer from the epithelium with a rubber policeman, and transfer to an Eppendorf tube. The intestinal extracts should be used as soon after isolation as possible. The yield from this procedure is usually around 0.5 ml of mucus and 0.5 ml of contents per mouse. Mice given streptomycin in drinking water have enlarged ceca, which makes the procedure easier to carry out.

Microscopy and Image Analysis

Fluorescence Microscopy

Microscopy is obviously essential for *in situ* investigations. The epifluorescence microscope is used for the analysis of single cells spread out in single layers on microscope slides. Microscopy should be performed with a research quality epifluorescence microscope such as the Zeiss Axioplan (Carl Zeiss AG, Oberkochen, Germany). Fluorescence signals are captured with a charge coupled device (CCD) camera, capable of recording signals quantitatively. We use two different systems, a Photometrics CH250 or a Princeton Instruments PentaMAX (Roper Scientific, Bogart, GA). These cameras are able either to collect data at frame rates of up to 25 frames per second for live imaging, or to add occasional faint signals for quantitative imaging (slow scan mode). Light intensities are collected in quantitative mode, which can subsequently be subjected to Cellstat image analysis (see below). The epifluorescence microscope is mainly useful for two-dimensional samples, since the instrument is only able to record one single focal plane; any out-of-focus regions are seen as blurs in the image. Since most biofilm samples are typically up to 100 μm thick, another instrument is required to discriminate and capture sharp images from within such objects: the scanning confocal laser microscope.

The confocal microscope is based on a conventional epifluorescence microscope, but with additional features. The confocal microscope is equipped with a pinhole aperture placed in the focal point of the light path from a single (focal) plane in the sample. Light from all other planes will have different focal points, and the images from these planes are in practice eliminated by the pinhole. By moving the sample up or down, images from a range of focal planes can be collected, and, using a computer, a 3D representation can be constructed. Furthermore, a laser beam is used for excitation of fluorochromes, eliminating the need for detection of entire frames of images at a time: the laser line is scanning the sample, and the emitted fluorescent light is detected by a photomultiplier tube (PMT).

We use the Leica TCS4D confocal microscope (Leica Lasertechnik GmbH, Heidelberg, Germany) for all biofilm microscopy. The microscope is equipped with an Ar-Kr laser carrying three laser lines, which emits excitation light with wavelengths of 488, 568, and 647 nm. This is appropriate for a number of fluorophores, including fluorescein, rhodamine, CY2, CY3, CY5, and GFP, just to mention a few. To excite DAPI or GfpUV an additional UV laser is used [Coherent Enterprise 651 (Coherent, Inc., Santa Clara, CA); an Ar laser with a primary laser line in the range 351 to 364 nm]. The various excitation wavelengths are attenuated by an acousto-coupled transmission filter (AOTF). Excitation beam splitters separate the excitation light from the emission signals, and emission beam splitters separate the emission wavelengths and direct them to the appropriate PMTs.

Using this tool, basic structural analysis has been performed on thin biofilms. However, several problems are connected with the use of the confocal microscope, in particular with thick specimens. Some of the potential problems are as follows: (1) all layers are illuminated as long as the laser scans the object, regardless of which layer is scanned, which may lead to bleaching problems; (2) similarly, the intense laser light may be phototoxic to live specimens, leading to tissue or cell damage during prolonged scans (e.g., when scanning several layers); (3) deep layers (i.e., more than 30–40 μm) are typically distorted if recording of such structures is at all possible; (4) the lasers of the conventional confocal microscope are limited to only emitting light of certain, narrow band wavelengths; and (5) the pinhole aperture, by default, limits the amount of light reaching the detector.

The multiphoton confocal microscope in theory alleviates all these limitations. The multi- or two-photon systems consist of a tunable laser, typically with light emission wavelengths in the range of 800–1000 nm. The light is emitted in very brief (pico- to femtosecond) pulses with relatively low energy corresponding to the relatively long wavelengths. The light is focused through the microscope optics; in the focal point (within the object), the laser light intensity is very high, and simultaneous delivery of two photons to a fluorophore occurs frequently. These excitation energy "packets" of longer wavelength photons are capable of exciting the fluorophores, even though they normally require shorter wavelength excitation light. Since the density of photons is high enough only in the focal plane, this results in excitation and subsequent light emission from only that single plane. This eliminates the need for a pinhole in the detection path.

Furthermore, the long wavelengths cause less damage to the samples, since only objects in the focal plane are excited. This means that there will be no noise from objects in front of or behind the focal plane. Also, no bleaching or quenching effects are notable, since only a single point is exposed transiently to the high-intensity pulses. Side effects are that the physics of multiphoton technology allows simultaneous excitation of several fluorophores with a single wavelength of light, which may create problems when using more than one fluorophore in the same sample. UV excitation is facilitated with the standard configuration without the costly and technically complicated UV-laser equipment. One last but very important feature of multiphoton technology is penetration capability. Standard confocal microscopy only permits clear imaging of specimens with depths of up to 50 μm. In contrast, the multiphoton microscope can be used for clear imaging of samples, which are thicker than 150 μm. At present, multiphoton systems are somewhat difficult to operate compared to standard confocal microscopy, and the equipment is expensive.

Microscopic techniques in biofilm research has been described in several publications.[23–25]

The Cellstat Image Analysis Program

The UNIX Cellstat program was developed as a tool for analyzing fluorescence data from bacteria labeled with fluorophore-tagged oligonucleotide probes. Cellstat uses gray-scale images that are captured quantitatively using a CCD (charge coupled device) camera mounted on an epifluorescence microscope. Cellstat performs a number of advanced image processing algorithms on the data to detect and quantify the signals from individual cells. The first operation is to compensate for uneven illumination of the object image; this is done using a top-hat algorithm, which results in an image with a black background (rather than levels of gray) with bright objects. The preprocessing then determines the intensity distribution, and on the basis of this an initial threshold gray value is defined for identification of objects compared to background or noise. The image is scanned and potential objects marked. Each object is subsequently further analyzed. The extent of objects is determined as the pixels having a given fraction (typically 0.2) of the maximum intensity of the object. Subsequently, object size and orientation are determined. The objects are finally tested against exclusion criteria (minimum and maximum dimensions, form factor, etc.), and accepted cells are listed in spreadsheet format, with intensity, cumulative pixel intensity (equivalent to the amount of fluorophores), size, projected volume (based on rod-shaped cell morphology), position, and orientation. Data from each image are subjected to basic statistics, and these data are included in the output file. Cellstat has no user interface but operates entirely from the command line. A processed image is produced and can be used for verification of the results. Cellstat is designed to operate batchwise without operator intervention on multiple images. Cellstat has become available for Windows 95/98 and Windows NT 4. Further details and examples are available on the Cellstat home page: http://www.im.dtu.dk/cellstat.

[23] R. J. Palmer, Jr., and C. Sternberg, *Curr. Opin. Biotech.* **10,** 263 (1999).
[24] J. R. Lawrence and T. Neu, *Meth. Enzymol.* **310,** 131 (1999).
[25] D. Phipps, G. Rodriguez, and H. Ridgway, *Meth. Enzymol.* **310,** 178 (1999).

Recovery of Bacteria from Biofilm Systems

Recovery of Cells from Flow Chambers

In principle, cells can be recovered from flow chambers in two ways: As dispersed cells or as cells with preserved surroundings, i.e., as intact three-dimensional structures. When single cells are to be examined, three easily separable fractions are available. First of all, cells can be collected from the effluent medium. This contains detached cells, either as truly single cells or as flocs. The second fraction comprises the loosely attached cells. This fraction can be harvested by passing an air bubble through the flow channel and collecting the effluent on recommencing of the liquid flow. The remaining, firmly attached cells can only be recovered by physical means. In practice, the flow cell is disassembled and cells are recovered from the glass by sonication. Sonication is also employed for dissociation of flocs and clumps.

Recovery of Bacteria from Rhizosphere Samples

To obtain information about the *in situ* physiology of inoculant cells in soil and rhizosphere samples, a number of methods based on cell extraction and subsequent microscopic observation are currently available, e.g., using redox or fluorescent dyes, fluorescent antibodies or rRNA probes, and Gfp-tagged cells. Microscopic visualization and quantification of bacteria extracted from soil or rhizosphere samples have several limitations: (1) possible changes in the physiological state of the cells during the extraction procedure, (2) low efficiency of the extraction method, especially from soils rich in organic matter or clay, (3) the fact that bacterial cells may be masked by soil particles and plant material and (4) autofluorescence of the soil and plant debris particles. Changes in the physiological state of the cells can be minimized by using precooled solutions and keeping the samples on ice during the extraction procedure. When soils are used as substrata for plant growth, a high content of sand facilitates the extraction of the cells and increases the purity of the samples. Severe masking and autofluorescence can be reduced by dilution of the sample and sedimentation of coarse particles, by using a polyphenol binding agent (i.e., polyvinylpolypyrrolidone, PVPP), and by gradient centrifugation. The protocol given below includes all these options, but nevertheless, a fraction of the cells will be lost during the sedimentation and centrifugation steps. Even when using sufficiently diluted soil, some masking will occur and autofluorescent soil and plant material particles will appear. Extraction of bacteria from the root system (the rhizoplane) and/or the soil adhering to the roots (the rhizosphere) is performed essentially as previously described.[26] Seedlings are removed from the soil or agar-plate system, and the roots separated from the seeds and placed into

---

[26] A. Unge, R. Tombolini, L. Mølbak, and J. Jansson, *Appl. Environ Microbiol.* **65**, 813 (1999).

2–5 ml of 0.9 % cold sodium chloride containing 16 mg of acid washed PVPP prepared as previously described. [27] After mixing in a vortex for 3 min, the larger soil particles are allowed to settle in the tube (2 min). Next, 700 μl of the upper phase is transferred to a new tube, from which serial dilutions are made for plating on media containing the necessary antibiotics for the selection of the corresponding strain in order to enumerate the number of bacteria. When necessary, bacteria are also counted on LB plates as sterility control. The remaining suspension is centrifuged for 4 min at 4° and 200 g, and the supernatant is carefully transferred to a microcentrifuge tube containing 700 μl Nycodenz (Nycomed Pharma, Norway) with a density of 1.3 g/ml. The discontinuous Nycodenz/cell suspension gradient is centrifuged for 30 min at 12,000g and 4°, the upper 500 μl of the gradient is discarded, and the lower 500 μl, containing the soil bacteria fraction, is transferred to a new microcentrifuge tube containing 1.5 ml of PBS buffer (130 mM NaCl, 10 mM NaPO$_4$, pH 7.2).

Ribosome Contents in Growing and Resting Bacteria Monitored in Fixed Cell Preparations

Since the ribosome is a phylogenetic marker of great significance,[28] techniques for *in situ* rRNA hybridization have been used extensively for identification of specific organisms in complex environments.[29] In addition, the concentration of rRNA increases in a linear manner with protein synthesis, and hence with bacterial growth rate.[30–32] Generally, this linear correlation is true for bacterial growth rates above 0.5 doublings per hour. Therefore, *in situ* rRNA hybridization also yields information about the physiological activity of the cell. The overall strategy of quantitative *in situ* hybridization is to fix cells in their particular environmental setting, hybridize with specific fluorescence labeled oligonucleotide probes (15–25 nucleotides) targeted against specific domains of 16S or 23S rRNA, and subsequently quantify the fluorescence intensity in the cells by image processing. The resulting signal intensities must then be compared with a standard curve obtained for the organism under investigation, based on hybridization to cells growing under defined conditions yielding different growth rates.

[27] W. E. Holben, J. K. Jansson, B. K. Chelm, and J. M. Tiedje, *Appl. Environ. Microbiol.* **54,** 703 (1988).

[28] C. R. Woese, *Microbiol. Rev.* **51,** 221 (1987).

[29] R. I. Amann, L. Krumholz, and D. A. Stahl, *J. Bacteriol.* **172,** 762 (1990).

[30] O. Maaløe and N. O. Kjeldgaard, *in* "Control of Macromolecular Synthesis." W. A. Benjamin, Inc., New York, 1966.

[31] H. Bremer and P. P. Dennis, *in* "*Escherichia coli* and *Salmonella typhimurium:* Cellular and Molecular Biology" (F. C. Neidhardt, ed.), p. 1553. ASM Press, Washington, D.C., 1996.

[32] F. C. Neidhardt, J. L. Ingraham, and M. Schaechter, *In* "Physiology of the Bacterial Cell: A Molecular Approach." Sinauer Associates Inc., Sunderland, MA, 1990.

The standard curve is essential if quantitative information concerning growth activity is sought, and in most cases the more informative data will be obtained from chemostat cultures in which the growth rate of the organism is determined by the nutrient dilution rate. One important advantage of chemostats is that low levels of growth activity can be obtained, which help define the lower limits of a reliable correlation between ribosome contents and growth rate. Below these limits the uncertainties of the correlation prevent meaningful estimates of growth rate, because distinctions between slow growth and the stationary phase of the cells will no longer be significant. It may also be important to obtain information about ribosome decay under conditions of nutrient starvation and other stress conditions, if such conditions are expected in the environment under investigation.

Fluorescent in Situ Hybridization

The protocol described in this section for fluorescent *in situ* rRNA hybridization (FISH) is generally applicable to cells extracted from a broad range of environmental samples.

Cell fixation is performed essentially as previously described[33] in 3% paraformaldehyde by mixing 250 μl of cell suspension and 750 μl of fixative solution (or 4% paraformaldehyde[1]). After mixing by vortexing for 1 min, the cell suspension is incubated at 4° for at least 5 min. The standard incubation time is 1 hr, but this time may be increased to 24 hr without consequences. After fixation, cells are spun down at 7000g for 5 min, the supernatant is decanted carefully, and the cells are suspended in 900 μl of PBS buffer and 100 μl 0.1% Nonidet P-40. After mixing by vortexing for 1 min, cells are harvested by centrifugation at 7000g for 5 min, the supernatant is decanted carefully, and the cells suspended in 10 to 20 μl storage buffer (50% ethanol, 10 mM Tris pH 7.5, 0.1% Nonidet P-40). Fixed cells can be stored at $-20°$C for several months until use, but for quantitative hybridization, fixed cells should be used within 1 month (50% degeneration has been observed after a period of 2 months).[33] 16S rRNA hybridization of fixed cells is performed using fluorescence labeled oligonucleotide probes.[33] We recommend the use of rRNA probes labeled with the carbocyanine fluorescent dyes, CY3 (red) or CY5 (far red), because they usually give the highest fluorescence intensity. Hybridization is performed in 6-well (8–10 mm diameter) heavy Teflon slides coated with poly-L-lysine. Coating is performed by transferring slides washed in acid alcohol (1% HCl in 70% ethanol) to a 100 ml coplin jar containing 0.01% (w/v) poly-L-lysine. After 5 min, slides are dried for 1 hr at 60° or overnight at room temperature. One to three μl of fixed cells is immobilized on a coated slide and air dried. The cell suspension should be distributed homogeneously over the surface of the well with the help of a pipette tip. Cells are

---

[33] L. K. Poulsen, G. Ballard, and D. A. Stahl, *Appl. Environ. Microbiol.* **59,** 1354 (1993).

dehydrated by transferring the slide to a 100 ml coplin jar containing a series of ethanol solutions (50%, 80%, and 96% ethanol; 3 min each). After the slides are dried at room temperature, 10 μl of hybridization solution (hybridization buffer plus 16S rRNA probe) are applied to each well. To ensure that all the wells receive the same concentration of the probe, the hybridization buffer [30% formamide, 0.9 M NaCl, 0.1 M Tris pH 7.5, 0.1% sodium dodecyl sulfate (SDS)] and the probe (final concentration of approximately 2.5 μg/ml) are mixed for all the wells. Slides are incubated at 37° in the dark for at least 16 hr in a formamide saturated atmosphere (e.g., in a sealed large test tube containing tissue paper wetted with the hybridization buffer). To remove unspecifically bound probe, two washing steps are performed.[1] Finally, the slides are rinsed in distilled water and air dried in the dark. Image capture and analysis using the Cellstat program[34] is performed as described above.

Application of FISH to Bacterial Cells Extracted from Rhizosphere Samples

Application of FISH techniques to analyze the activity of bacterial cells from rhizosphere samples is limited at the moment. In addition to the limitations described above for the detection of rhizobacteria by epifluorescence microscopy, the major reason is the low physiological activity of many soil and rhizosphere microorganisms, resulting in weak or even undetectable hybridization signals due to low ribosomal contents of slow-growing and starved cells. Determinations of exact growth rates of bacteria in rhizosphere systems through the use of quantitative FISH will therefore be impossible in most cases. Instead of correlating fluorescence intensities obtained from rhizosphere extracted cells with standard curves obtained for cells growing at defined growth rates, we recommend that they be compared with those measured in cells kept under starvation conditions and/or growing at rates below the limit of detection using quantitative FISH (0.5 hr^{-1}).

In some instances, e.g., in nutrient rich soil systems or sterile systems, FISH methods have a potential for providing estimates of the active fraction of soil and rhizosphere bacteria. Hybridization protocols for determinations of bacterial numbers in soil without an extraction step have already been developed.[35,36] The FISH protocol given above has been shown to be suitable for quantification of ribosomal contents and cell volumes of *Pseudomonas putida* cells in rhizospheres of both agar (Fig. 1) and soil-grown barley seedlings.[37]

[34] S. Møller, C. S. Kristensen, L. K. Poulsen, J. M. Carstensen, and S. Molin, *Appl. Environ. Microbiol.* **61,** 741 (1995).

[35] H. Christensen and L. K. Poulsen, *Soil Biol. Biochem.* **26,** 1093 (1994).

[36] H. Christensen, M. Hansen, and J. Sørensen, *Appl. Environ. Microbiol.* **65,** 1753 (1999).

[37] C. Ramos, L. Mølbak, and S. Molin, *Appl. Environ. Microbiol.* **66,** 801 (2000).

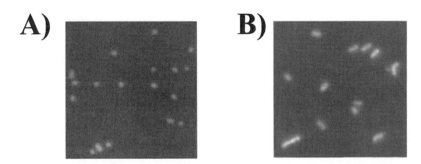

FIG. 1. Epifluorescence micrographs of fixed and hybridized *Pseudomonas putida* cells extracted from the root system of barley seedlings. Sterile barley seeds were coated with stationary-grown *P. putida* cells at approximately 10^2 CFU/seed, placed in agar plates, and incubated at 20°. After 1 day, bacteria were extracted from the root system and hybridized with a CY3-labeled ribosomal RNA probe specific for the *P. putida* subgroup A. An axioplan epifluorescence microscope (Carl Zeiss) was used to visualize the hybridizations. The microscope was equipped with a 100 W mercury lamp, and filter set no. XF40 (Omega Optical, Brattleboro, VT) was used to visualize the red fluorescence. A slow-scan charge-coupled device camera CH250 (Photometrics, Tucson, AZ) equipped with a KAF 1400 chip (pixel size 6.8 by 6.8 μm) was used for capturing digitized images. (A) Epifluorescence micrograph of fixed and hybridized cells extracted from the root system of 1-day-old barley seedlings; (B) epifluorescence micrograph of the starved cells used as inoculum. Magnification × 630.

Ribosomal RNA Precursor Molecules

The number of ribosomes in a cell is not always correlated with the growth rate, since a substantial fraction of the ribosomes is inactive during slow growth.[38,39] If bacteria are exposed to a nutritional downshift, e.g., by removing amino acids from the medium, or by changing the carbon source from glucose to acetate, an inhibition of rRNA transcription follows immediately (stringent response).[40] However, a breakdown of already functional ribosomes does not occur instantaneously, and the number of ribosomes in a cell right after a nutritional downshift therefore cannot be used as a measure of growth.[41,42] Similarly, the ribosomal concentration immediately after a nutritional upshift does not reflect the bacterial growth activity, as the bacteria need time to synthesize new ribosomes. Therefore, more precise methods are needed to determine the cellular physiological state, i.e., the true growth rate, at such points of measurement.

[38] A. L. Koch, *J. Theor. Biol.* **28,** 203 (1970).
[39] A. L. Koch, *Adv. Microb. Physiol.* **6,** 147 (1971).
[40] M. Cashel, D. R. Gentry, V. J. Hernandez, and D. Vinella, *in* "*Escherichia coli* and *Salmonella typhimurium:* Cellular and Molecular Biology" (F. C. Neidhardt, ed.), p. 1458. ASM Press, Washington, D.C., 1996.
[41] F. Ben-Hamida and D. Schlessinger, *Biochim. Biophys. Acta* **119,** 183 (1966).
[42] R. Kaplan and D. Apiron, *J. Biol. Chem.* **250,** 1854 (1975).

A)

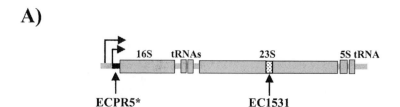

ECPR5* EC1531

B)

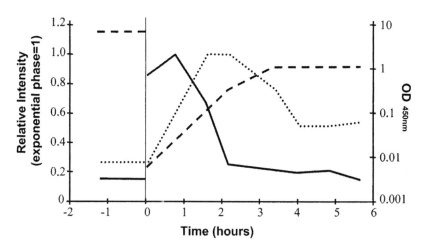

Time (hours)

FIG. 2. Precursor-16S rRNA levels in *E. coli*. (A) Total ribosomal RNA transcript from *E. coli rrn* operons. The position of two different *E. coli* specific oligonucleotide probes is indicated. ECPR5* (black box) hybridizes to precursor 16S rRNA, and EC1531 (dotted box) hybridizes to the mature 23S rRNA. (B) Growth curve for *E. coli* and relative hybridization signals from each of the two probes during up-shift, exponential growth, and run-out at 37° in L-broth. At time 0 the culture was diluted 1000-fold in L-broth preheated to 37°. Relative fluorescence intensity measurements for ECPR5* (solid line) and EC1531 (dotted line) are affiliated to the left axis, setting the maximal intensities measured for each of the probes during exponential phase to 1. Measurements of optical density (dashed line) are affiliated with the right axis. Redrawn from Fig. 4.8 in Molecular ecology of biofilms, In "Biofilms II" (ed. James Bryers), Wiley-Liss, New York.

A method for determination of rRNA synthesis rates in bacteria has been presented.[43] It benifited from the fact that the rRNA molecules (16S, 23S, and 5S) are transcribed together as one long precursor transcript, which is posttranscriptionally processed to the final rRNA products. In this processing reaction, small precursor sequences outside the mature RNA domains (Fig. 2) are quite rapidly

[43] G. A. Cangelosi and W. H. Brabant, *J. Bacteriol.* **179,** 4457 (1997).

removed by nuclease activities. Even though the lifetime and cellular concentrations in precursor-RNA molecules are significantly lower than in mature rRNA, they are present in growing cells in amounts sufficient for allowing their detection by quantitative hybridization. In cells that have stopped growing, the transcription of rRNA is blocked, resulting in a rapid disappearance of precursor-RNA molecules. The concentration of precursor rRNA molecules in single bacteria can be determined by fluorescent *in situ* hybridization as described above for mature rRNA molecules. However, detection of the much lower amount of target molecules requires a better signal/noise ratio, and it is therefore absolutely essential to use fluorophores with very high fluorescence intensity (e.g., CY3), and to lower the amount of fluorescent background material in the samples as much as possible. It should be stressed that determination of levels of precursor rRNA may yield controversial results, which only make sense when compared with further data about the physiology of the bacteria under investigation. This was documented in a study of *E. coli* in intestinal contents, where the cellular amounts of pre-rRNA were found to be surprisingly high because of a growth inhibitory effect similar to that caused by chloramphenicol.[11]

In Situ Estimations of Ribosome Contents

In Situ Investigations of Flow-Chamber Biofilms

To recover structurally intact samples for *in situ* determinations of ribosome contents, it is necessary to embed the cells in a matrix, which retains the three-dimensional relations of the cells constituting the biofilm community. We routinely use acrylamide embedding, which is a gentle, yet efficient method of preserving biofilm structure. In brief, the medium in flow channels is replaced with an acrylamide solution, and after solidification it is a relatively simple procedure to disassemble the flow chamber and recover the acrylamide-embedded biofilm sample. This procedure, including *in situ* rRNA hybridization, was previously described in detail.[1] Acrylamide embedded biofilm samples are easy to handle because of the stability of the gel stabs. Samples are embedded and the solidified samples are transferred to coated cover slips for labeling. In principle, embedded samples are labeled as suspended cells. When fluorescence-labeled oligonucleotide probes are used, labeling can be carried out even in relatively thick samples (we have successfully labeled samples with a thickness of more than 100 μm). However, not all probes are suitable for use with acrylamide embedded samples, e.g., it is not possible to use immunolabeling with antibodies since these molecules cannot penetrate the polyacrylamide matrix. After labeling and washing, samples can be mounted for microscopy and examined with the confocal microscope.

Quantitative measurements of fluorescence intensities in such samples have not yet been achieved, since the three-dimensional nature of the sample and virtual slicing by the confocal microscope makes it difficult to determine signal properties. The problem arises when some cells may occur in more than one "slice," perhaps even in an ill-defined angle. To be able to measure quantitatively, we need to be able to determine the total signal (not more or less) from each cell. Using advanced image processing and correction we expect this to be possible in the future.

Procedure for rRNA Hybridization of Cells on Root Samples

Plants are carefully removed from the rhizosphere system, and the roots are washed by shaking in sterile PBS buffer. Root pieces (10–20 mm) are transferred to a fixation buffer (3% paraformaldehyde in PBS) and fixed for 4–12 hr at 4°. After washing briefly in PBS buffer, samples are transferred to hybridization buffer (as described above) at 37° for 30 min to ensure that the root samples are saturated with formamide before hybridization. Root pieces of approximately 5 mm are placed in the wells of a poly-L-lysine coated slide, and 50 μl of hybridization solution (hybridization buffer + rRNA probe) is added to each root sample. Slides are incubated for 3–12 hr at 37° in the dark in a formamide-saturated atmosphere. To remove unspecifically bound probe, two washing steps are performed.[1] Root pieces are placed on a glass slide, a drop of antifade solution is added, and a cover slip is mounted on top using silicone glue (3M, St. Paul, MN). The root sample is then ready for microscopic inspection.

The effectiveness of this method depends mainly on the ability of the bacterial cells to remain attached to the root surface after the fixation and hybridization. The age of the plant is another limiting factor for visualization of hybridized cells in the rhizoplane. In the case of *P. putida* cells colonizing the rhizosphere of barley plants, a large number of cells can usually be detected on the root surface of young seedlings. However, as the plants grow older, the bacterial cells are more difficult to see, since bigger areas of the root surface are not covered by bacteria, and root hairs develop covering almost the entire root surface, thus obstructing visualization. Although autofluorescence from the root epidermal cells is always recorded during visualization of root preparations by confocal microscopy, it is an advantage of the visualization of the root cells together with the root colonizing bacteria. We recommend the use of filters increasing the detection of autofluorescence from the root surface at a different wavelength than that used to record the fluorescence emitted from the bacteria (Fig. 3).

Preparation and Fixation of Intestinal Tissue Sections for Hybridization

The viscous mucus layer covering the inside of the animal intestine is often important in colonization studies of intestinal bacteria. Since the mucus is not part of the eukaryotic tissue and does not have a fixed structure, it can be quite laborious to produce nice histological cross sections showing the spatial distribution of bacteria within the mucus. However, in spite of the difficulties, it is possible to

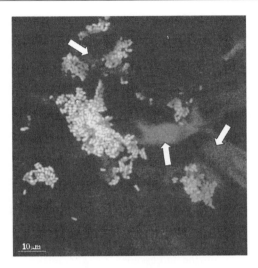

FIG. 3. *In situ* visualization of *Pseudomonas putida* cells in the root hair zone of a 3-day-old barley seedling by hybridization with a fluorescently labeled ribosomal-RNA probe. *P. putida* JB156 cells (light gray) were hybridized with a CY3-labeled probe specific for the *P. putida* 16S RNA (probe PP986-CY3). Microscopic observation and image acquisition were performed on a Leica TCS4D confocal microscope (Leica Lasertechnik, GmbH, Heidelberg, Germany) equipped with an argon–neon laser and two detectors for simultaneous monitoring of green and red fluorescence. Multichannel SFP (simulated fluorescence projection) images were generated using IMARIS (Bitplane AG, Zürich, Switzerland) software running on an Indigo 2 workstation (Silicon Graphics, Mountain View, CA). Images were processed for display using PhotoShop (Adobe, Mountain View, CA). The green autofluorescence exhibited by the root material was used to visualize root hairs (arrows). Magnification × 630.

preserve the mucus in either frozen intestinal cross sections[17] or paraffin embedded tissue sections.[18] Both types of sections can be made according to standard histological methods, and the sections are subsequently immobilized on coated microscope slides. Usually, sections are made from the large intestine (colon), since the epithelial tissue as well as the intestinal contents are more stable here than in the cecum. Furthermore, the majority of the gram-negative bacteria that have been subject to our investigations colonize the colonic area in high numbers.

The intestinal tissue may be fixed prior to and/or after sectioning. Typically, cryosections are fixed in paraformaldehyde (3%) after sectioning, whereas the tissue is fixed in formalin (10%) before paraffin embedding. Dehydration of the sections with ethanol or xylene often causes shrinking of the mucus layer and should be avoided when possible. For cryosections, a dehydration step can be left out completely, while paraffin embedded sections must be deparaffinated with xylene followed by 96% ethanol prior to hybridization. However, it is usually easier to preserve the epithelial cell layer nicely when using paraffin embedded tissue. Paraffin embedded specimens furthermore preserve rRNA extremely well and can be kept at 1–4° for several months. Thin intestinal sections (5–10 μm) can easily be examined with two-dimensional fluorescence microscopy, whereas thicker sections (20–30 μm) require confocal microscopy to exclude light from the areas outside of the focal plane.

Histological staining, e.g., with hematoxylin, makes the tissue highly fluorescent and cannot be used in combination with fluorescent *in situ* hybridization. Instead, the tissue can be stained after hybridization and recording of the pictures of the hybridized bacteria. Before hybridization, a circle is drawn on the slide around each section with a fatty pen (DAKO, PAP pen), in order to keep the hybridization solution in place.

Quantification of bacterial rRNA is preferably done in smears from fluent samples made as described earlier, because the bacteria are spread out and can be visualized as evenly distributed single cells, which makes it easier for the Cellstat software to recognize them. Hybridization of histological sections primarily provides information about the spatial distribution of selected bacteria in the intestine, but it is possible to measure the fluorescence "by eye" and thereby get an idea about the physiological state of the bacteria in the sections.

The Use of Reporter Genes for On-Line, Nondestructive Activity Estimations

Genetic tagging has alleviated the need for fixation and physical dismantling of the biofilm community. Tagging with constitutively expressed *lacZ* or *luxAB* markers facilitates detection of the labeled bacteria within a complex community, but these markers require addition of exogenous compounds, i.e., a chromogenic β-galactosidase substrate or a long-chained aliphatic alcohol, respectively. Insertion of the entire *lux* operon, however, permits detection without any substrate addition, since the aliphatic alcohol is synthesized intracellularly by operon encoded enzymes. The marker protein Gfp (green fluorescent protein), encoded by a gene originating from the jellyfish *Aequorea victoria,* is self-contained and requires only minute concentrations of oxygen to fold and become fluorescent. Gfp has become the preferred marker in biofilm research because of this virtue. Native Gfp is excited by long-wave UV light (396 nm), with a secondary excitation peak in the blue band (476 nm), and emits green light at 509 nm. Newer variants have been optimized to have excitation maxima around 470 nm, with little excitation in the UV band and markedly better quantum efficiency. These mutants often have shorter maturation times, typically $T_{1/2}$ of 10–30 min compared to the 2 hr for wild-type *Aequorea* Gfp.

Wild-type Gfp is degraded at a very low rate (half-life of several hours to days under physiological conditions). Novel variants have been created that have considerably shorter half-lives, ranging from a few minutes to several hours. The degradation of the reporter protein is carried out by C-terminal tail specific proteases as a consequence of engineering the proper signal peptide sequences into the Gfp protein.[44] Transfer of these variant Gfp genes to a number of different bacterial species has shown that the instability of the protein is fairly universal, although the actual half-lives may vary from one organism to another. Using Gfp in biological experimentation facilitates self-contained marker systems, ideal for biofilm research, and the destabilized mutants (dsGfp) further make investigations of gene regulation *in situ* feasible.

Monitoring Growth Activity

Investigations of localized biological activity in biofilms have been described, most of which involve more or less destructive methods ranging from addition of presumably harmless substrates, over the use of microelectrodes, to complete dissociation of biofilms. Gfp has been utilized for studies of biological activity in biofilms. Biological activity, in the form of growth activity (cell proliferation) or specific gene expression, can be monitored *in situ* using the new Gfp variants. Growth activity is closely linked to the synthesis of ribosomes, and the *rrn* ribosomal promoters are highly regulated. It is, therefore, useful to construct *rrn* promoter fusions to ds*Gfp* in order to monitor growth rates *in situ*. Specifically, we have used the P1 promoter from the *E. coli rrnB* operon fused to a variant *gfp* gene, encoding a dsGfp, Gfp(AAV), with a relatively short half-life. This cassette has been introduced into the chromosome of several bacterial species, in which the expression of green fluorescence as a function of growth activity has been analyzed. It must be ensured that expression is appropriately effective since the turnover of the protein both requires constant synthesis and synthesis to a protein level of mature nondegraded protein sufficient for detection. This can be modulated to some extent by the choice of Gfp variant and promoter. In the case of the *rrnBP1* promoter, expression levels during exponential growth in batch experiments are normally high, and the use of a Gfp variant with a short half-life (Gfp(AAV)) is optimal for detection. Chemostat experiments can be used to determine "cutoff" growth rates, i.e., the growth rate where Gfp fluorescence becomes indistinguishable from the background fluorescence. With *P. putida* RI::*rrnBP1*-dsGfp(AAV), this was achieved at growth rates below 0.6 hr^{-1}. Another consideration when using Gfp in biofilms is to ensure that oxygen levels are adequate for maturation of the protein. We have determined the minimum concentration of oxygen still allowing full maturation of the Gfp fluorophore and found this to be less than 0.1 ppm.[45] Thus, even in microaerobic environments, Gfp is a relevant marker for cell identity and activity. For *in situ* estimations of bacterial growth activity in the plant rhizosphere, a similar *rrn–gfp* fusion based on a different *gfp* allele, *gfp*[AGA]^{1,37} (Fig. 4), which encodes an unstable Gfp with a half-life longer than

[44] J. B. Andersen, C. Sternberg, L. K. Poulsen, S. P. Bjorn, M. Givskov, and S. Molin, *Appl. Environ. Microbiol.* **64**, 2240 (1998).
[45] M. C. Hansen, R. J. Palmer, Jr., C. Udsen, D. C. White, and S. Molin, unpublished.

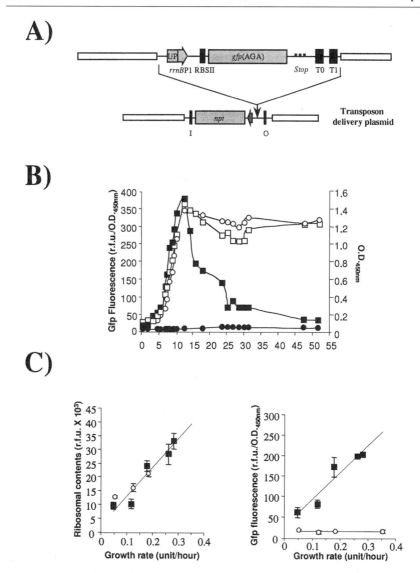

FIG. 4. Cell growth regulation of Gfp expression in a *P. putida* strain carrying a chromosomal insertion of a Gfp-based growth activity reporter. *P. putida* SM1700[37] carries a chromosomal insertion of a *gfp* allele, *gfp*[AGA],[1,37] which encodes an unstable Gfp, fused to the ribosomal *E. coli* promoter *rrnB*P1. (A) Schematic representation of the cloning strategy used for the construction of the transposon delivery plasmid containing the *rrnB*P1–*gfp*[AGA] cassette. *rrnB*P1, growth rate regulated *E. coli* ribosomal promoter; UP, 21-bp region responsible of the strong stimulation of transcription of the *E. coli* *rrnB* P1 and P2 promoters; RBSII, synthetic ribosomal binding site; *Stop*, stop codon in all three reading frames; To and T1, termination sites; I and O, 19-bp transposon inverted repeats; *npt*, gene encoding

that of *gfp*[AVA], is useful as an on-line monitor system, and in particular it may disclose hot spots of microbial activity. This reporter system allows determination of the spatial distribution of active cells introduced in the rhizosphere of plants grown under sterile conditions.[37] However, the slow growth rates supported by root exudates from plants growing under nonsterile conditions are not compatible with the method of monitoring fast growth in connection with the use of these Gfp-fluorescent cells. Therefore, the lack of apparent Gfp expression should not be taken as an indication of starvation; the promoter could be active (and the cells growing) but at levels not allowing Gfp detection.

We have not yet completed our studies on using fluorescent marker genes for *in situ* studies of intestinal samples. Preliminary data show that Gfp is a tool that can be used in the gut as well as in the above described systems. However, we have to bear in mind that it is important to choose marker proteins with a very high fluorescence intensity, since intestinal samples often contains autofluorescent objects (nondigested plant material, etc.) emitting light in the blue-green area.

Monitoring Specific Gene Activities

We have previously reported the design of a toolbox of general cloning vectors allowing the combination of any promoter of interest with the most convenient marker gene for a given application.[1] The resulting monitor cassettes can subsequently be introduced into the chromosome of virtually any gram-negative bacterium. To ensure that the insertion of the monitor cassette into the chromosome of the strain does not interfere with any essential function, the tagged strains must be tested in batch culture experiments prior to their use *in situ*. It is also recommended to determine the insertion point of the cassette by sequencing. The promoter fusions should also be tested to comply with the expected behavior before their use. Additionally, to ensure that the insertion does not alter the *in situ* behavior of the strain, the parental and marked strains should be tested together.

---

kanamycin resistance. (B) Growth phase-dependent expression of the *rrn*BP1 promoter in *P. putida* SM1700 growing in batch cultures. *P. putida* SM1700 (squares) and *P. putida* JB156 (wild-type, circles) were grown at 20° on minimal media containing 8 mM sodium citrate as the sole carbon source. At the indicated period of time the O.D.$_{450nm}$ of the cultures (open symbols) and the intensity of the green fluorescence emitted by the cells (closed symbols) were measured. (C) Variation in ribosomal contents (left panel) and ribosomal synthesis (right panel) as a function of growth rate in *P. putida* SM1700 (squares) and *P. putida* JB156 (circles) growing in chemostats at 20°. Ribosomal contents and ribosomal synthesis were estimated by quantitative hybridization with fluorescently labeled ribosomal RNA probes and by measuring the Gfp fluorescence emitted by the cells, respectively. Each point represents an average based on three independent measurements. The error bars indicate standard deviations from the averages.

Application of Growth Rate Estimation Tools to Environmental Investigations

Essentially all methods and tools described in this article require detailed knowledge about the targeted organism obtained from pure cultures, genomic sequence information, controllable growth conditions, possibilities of introducing engineered strains, and sometimes even specific equipment for placing the system under investigation directly under the microscope. In short, most or all these methods are only relevant for laboratory model systems. This does not, however, exclude the possibility that some may also be applied to more natural systems—possibly after modifications of the methodology.

As long as the targeted organisms belong to well-known species, for which information about their performance and activity under controllable conditions is available, it may be possible to expand the analysis, e.g., from a laboratory plant or animal system to field conditions. A special case in this direction concerns deliberate release scenarios, in which engineered strains of bacteria (usually well-known laboratory strains) are introduced in the field with specific purposes in mind. In such cases there are excellent possibilities for assessing their function not only with respect to the purpose of the release, but also with respect to general growth-related activities, such as those described here. In natural ecosystems, characterized by an enormous biodiversity of organisms and a huge heterogeneity of microbial communities and their activities, it is, however, only FISH that today carries the potential for growth activity determinations. In the future, the assessment of microbial activity in natural scenarios will have to deal with microbial communities as "complex multicellular organisms" in which we do not need to understand all the individual cellular activities in order to develop an understanding of the total community. The consideration of microbial communities as multicellular units as much as assemblies of distinct individuals will require the development of molecular tools allowing the study of the interrelationships between the cells as much as monitoring activities within the individual cells. This challenge now confronts microbiologists, who will have to interact with scientists from several other disciplines to approach the inherent problems.

[3] Use of a Continuous Culture System Linked to a Modified Robbins Device or Flow Cell to Study Attachment of Bacteria to Surfaces

By MICHAEL R. MILLAR, CHRISTOPHER J. LINTON, and ANDREA SHERRIFF

Introduction

In medical practice *Staphylococcus epidermidis* is one of the most frequent causes of biomedical device-associated infection,[1] and as such is a major cause of hospital-acquired infection (HAI). In the United Kingdom the prevalence of HAI is 5–10% in hospitalized patients,[2] with an annual cost to the National Health Service of £1000 million.[3] *Staphylococcus epidermidis* is the commonest cause of intravascular catheter-associated bacteremia. The additional hospital costs associated with hospital-acquired bacteremia has been estimated in the United Kingdom to be in excess of £5000/episode.[4]

Bacteria in foreign body associated biofilms are relatively resistant to the action of host defense factors or antibiotics,[5] and much of the morbidity and cost associated with foreign body associated infection arises from the relative ineffectiveness of current treatment strategies for these infections.[6] The attachment of bacteria to surfaces is a prerequisite for the formation of biofilm. Many factors are important in influencing the rate of attachment of bacteria to surfaces, including the substratum,[7] bacterial characteristics,[8–11] shear forces,[12] and salt concentrations.[13]

[1] A. G. Gristina, C. D. Hobgood, and E. Barth, *Zentralblatt Bakteriol. Mikrobiol. Hygiene.* **1** Suppl. 16, 143 (1987).

[2] M. Emmerson, A. M. Enstone, J. E. Griffin, M. Kelsey, and M. C. Smyths, *J. Hosp. Infect.* **32,** 175 (1996).

[3] R. M. Plowman, N. Graves, and J. A. Roberts, "Hospital Acquired Infection," p. 15. BSC Print Ltd., London, 1997.

[4] R. Plowman, N. Graves, M. Griffin, J. A. Roberts, A. V. Swan, B. Cookson, and L. Taylor, "The Socio-Economic Burden of Hospital Acquired Infection." Public Health Laboratory Service, 1999.

[5] M. R. W. Brown, D. G. Allison, and P. Gilbert, *J. Antimicrobial Chemother.* **22,** 777 (1998).

[6] R. S. Baltimore and M. Mitchell, *J. Infect. Dis.* **141,** 238 (1980).

[7] M. Fletcher and G. I. Loeb, *Appl. Environment. Microbiol.* **37,** 67 (1979).

[8] A. H. Hogt, J. Dankert, and J. Feijen, *FEMS Microbiol. Lett.* **18,** 211 (1983).

[9] S. McEldowney and M. Fletcher, *Appl. Environ. Microbiol.* **52,** 460 (1986).

[10] E. Evans, M. R. W. Brown, and P. Gilbert, *Microbiol.* **140,** 153 (1994).

[11] A. Ljungh and T. Wadstrom, *Microbiol. Immunol.* **39,** 753 (1995).

[12] H. J. Busscher and H. C. van der Mei, *Methods Enzymol.* **253,** 445 (1995).

[13] C. J. Linton, A. Sherriff, and M. R. Millar, *J. Appl. Microbiol.* **86,** 194 (1999).

There is increasing information on the genetic and phenotypic characteristics that determine biofilm formation by *Staphylococcus epidermidis*.[14–22]

Many biomedical devices are exposed to conditions of shear stress, which is an important determinant of the attachment of microorganisms to artificial[23] and natural surfaces.[24] Until recently most studies of bacterial adhesion did not take account of conditions of flow.[12] The parallel plate flow chamber has become increasingly used to allow direct observation of microbial interactions with surfaces under conditions of flow and has been used to study the physicochemical determinants of microbial adhesion.[25] The parallel plate system does not usually allow direct comparison of different materials using the same microbial suspension. Also, direct observation is inevitably confined to small selected areas of the surface and may not give results representative of the whole surface. Indirect measurement of bacterial attachment such as charge transfer has been used to study attachment of bacteria to a semiconducting surface in a parallel plate flow chamber.[26] Substratum topography[27] and surface physicochemical properties have a major impact on bacterial attachment so that surface heterogeneities can give rise to large variations in measurements at different points.

Principles and Objectives of Methods

The objectives of the method described were to do the following:

1. Define bacterial culture conditions of media composition, growth rate, pH, temperature, and oxygenation
2. Use a growth rate below μ_{max}
3. Limit surface conditioning effects of growth media components
4. Limit interexperimental variation
5. Directly compare surfaces using the same bacterial suspension
6. Use a bacterial strain widely used in previous studies
7. Limit variations in results associated with surface heterogeneities

[14] E. Muller, J. Hubner, N. Gutierrez, S. Takeda, D. A. Goldmann, and G. B. Pier, *Infect. Immun.* **61,** 551 (1993).

[15] D. Mack, M. Nedelmann, A. Krokotsch, A. Schwarzkopf, J. Heesemann, and R. Laufs, *Infect. Immun.* **62,** 3244 (1994).

[16] M. E. Rupp, N. Sloot, H. G. Meyer, and H. J. S. Gatermann, *J. Infect. Dis.* **172,** 1509 (1995).

[17] L. Baldassarri, G. Donnelli, A. Gelosia, M. C. Voglino, and A. W. Simpson, *Infect. Immun.* **64,** 3410 (1996).

[18] C. Heilmann, C. Gerke, F. Perdreau-Remington, and F. Gotz, *Infect. Immun.* **64,** 277 (1996).

[19] C. Heilmann, O. Schweitzer, C. Gerke, N. Vanittanakom, D. Mack, and F. Gotz, *Mol. Microbiol.* **20,** 1083 (1996).

[20] D. Mack, W. Fischer, A. Krokotsch, K. Leopold, R. Hartmann, H. Egge, and R. Laufs, *J. Bacteriol.* **178,** 175 (1996).

[21] G. J. C. Veenstra, F. F. M. Cremers, H. van Dijk, and A. Fleer, *J. Bacteriol.* **178,** 537 (1996).

Continuous Culture and Modified Robbins Device

In a chemostat, culture medium is added to the vessel while a constant volume of culture is maintained by removal of medium at the same rate at which it is added to the vessel. At "steady state" the biomass is limited by utilization of medium components such as carbon or energy sources.[28] The addition of fresh medium ensures that the cells do not pass into stationary phase. The number of cells, concentration of products and substrates, and growth rate remain constant. Disadvantages of a chemostat system include the accumulation of biofilm material over time, which may make a variable and unpredictable contribution to the culture, and also the selection of mutants with different adhesive properties. Continuous culture with cyclic on–off flow has been used to select for adhesive bacteria.[29]

Staphylococcus epidermidis strain ATCC 35984–RP62A has been widely used in previous studies of biofilm formation and attachment[30] and was used in the development of this system. An overnight batch culture of *Staphylococcus epidermidis* RP62A (ATCC 35984) is harvested to provide cells that are stored on plastic beads at −70° in a 10% glycerol broth. An individual bead is cultured on 5% horse blood agar overnight at 37°, and this culture is used to provide the inoculum at the start of each continuous culture run. A schematic diagram of the experimental system is shown in Fig. 1.

Iso-sensitest broth (CM473, Oxoid Ltd., Basingstoke, Hants.) is relatively defined by comparison with Mueller Hinton or tryptone soy broth and supports the growth of *Staphylococcus epidermidis* RP62A. The growth medium is diluted to 1/10 recommended strength both to reduce the impact of media conditioning on bacteria/surface interactions and to reduce frothing. The medium is sterilized using a 0.2 μm filter to avoid variations associated with autoclaving large batches of medium. The continuous culture system used is an Inceltech LH.SGI Discovery Bioreactor series 100. The operating volume of the chemostat vessel with

[22] M. Hussain, M. Herrmann, C. von Eiff, F. Perdreau-Remington, and G. Peters, *Infect. Immun.* **65,** 519 (1997).

[23] K. W. Millsap, R. Bos, H. C. van der Mei, and H. J. Busscher, *Antonie Van Leeuwenhoek* **75,** 351 (1999).

[24] N. Mohammed, M. A. Teeters, J. M. Patti, M. Hook, and J. M. Ross, *Infect. Immun.* **67,** 589 (1999).

[25] R. Bos, H. C. van der Mei, and H. J. Busscher, *FEMS Microbiol. Rev.* **23,** 179 (1999).

[26] A. T. Poortinga, R. Bos, and H. J. Busscher, *J. Microbiol. Methods* **38,** 183 (1999).

[27] T. R. Scheuerman, A. K. Camper, and M. A. Hamilton, *J. Colloid Interface Sci.* **208,** 23 (1998).

[28] D. W. Tempest, *in* "Methods in Microbiology" (J. R. Norris and D. W. Robbins, eds.), Vol. VI, pp. 259–276. Academic Press, London, 1970.

[29] G. A. Murgel, L. W. Lion, C. Acheson, M. L. Shuler, D. Emerson, and W. C. Ghiorse, *Appl. Environ. Microbiol.* **57,** 1987 (1991).

[30] G. D. Christensen, L. Baldassarri, and W. A. Simpson, *Methods Enzymol.* **253,** 477 (1995).

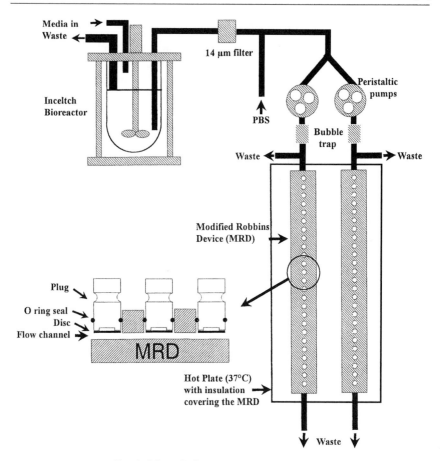

FIG. 1. Schematic diagram of experimental setup.

mixing at 400 rpm is adjusted to 1.25 liter. The incubation temperature is $37°$ and the pH maintained at 7.4. Air is pumped through a bacterial filter directly into the vessel. The chemostat is inoculated and operated without dilution for the first 5 hr and then under continuous culture conditions with a dilution rate of 0.14 for a minimum of 4 days before adhesion experiments. At this dilution rate the oxygen level stabilizes within the range of 80–90% with an optical density of approximately $0.7_{540\ nm}$ and viable count of 4×10^8 colony-forming units (cfu) ml^{-1}. Over time biofilm material accumulates on the vessel walls and other surfaces within the continuous culture system. Experiments are carried out within a week of initiating the continuous culture to avoid the impact of variable removal of accumulated biofilm on adhesion results. Longer culture times are

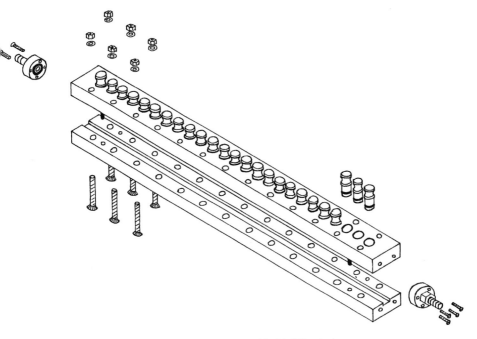

FIG. 2. Schematic diagram of modified Robbins device.

associated with increasing variability of test results and variable accumulation and stripping of biofilm material from the surfaces within the bioreactor. An additional reason for limiting the time that the chemostat is run to provide cells for experiments is to limit the possibility of selection of mutants. Care is needed to avoid bubbles being taken into the tubing, bubble trap, and modified Robbins device.

The modified Robbins device (MRD) is a laminar flow colonization chamber developed by McCoy et al.[31] An MRD is designed so that it can be dismantled totally for cleaning (Fig. 2). The main body of the device is in two pieces bolted together. A thin layer of sterile silicon grease is applied between the two parts to seal the device. The inlet and outlet end plates are designed with an O-ring seal. The flow channel is polished. The sample ports are large enough to accept 11 mm diameter glass cover slips. Constructed of acrylic/Perspex, this device has 25 evenly spaced sample ports that allow sample discs to lie flush with the inner surface of the lumen. The discs are held in place with removable watertight O-ring sealed plugs. The dimensions of the flow channel are 40 × 9 × 2 mm.

[31] W. F. McCoy, J. D. Bryers, J. Robbins, and J. W. Costerton, *Can. J. Microbiol.* **27,** 910 (1981).

Enumeration of Attached Bacteria

Epifluorescent microscopy is employed to allow examination of surfaces that do not transmit light. Bacteria are stained with acridine orange (AO), which binds to nucleic acids and fluoresces when excited with light at a wavelength of 436 or 490 nm. The nucleic acid–AO complex fluoresces greenish-yellow or orange/red depending on experimental conditions (dye concentration, pH, fixative) and the viability of test cells. A staining procedure has been optimized for *Staphylococcus epidermidis* on different test surfaces. To remove operator bias in determining which area to count, a special disc holder has been made so that 10 specific locations on all the test disc are determined by the microscope stage micrometer. Ten fields ($\times 1000$ magnification) are routinely counted for each disc. Using this technique, large variations in the distribution of attached cells over the test surfaces are observed. Even cleaned and polished surfaces show large variations in the distribution of bacteria across the surface.

An alternative approach to microscopy is to remove the bacteria from the test surfaces and quantify the number by viable counting.[32] Two removal techniques, sonication and vortexing, have been evaluated. The test discs are removed from the MRD and placed in glass bottles in a buffer. These bottles are placed in a sonicated water bath for different lengths of time, the resulting suspensions are cultured, and the number of colony forming units (CFU) counted. The discs are then stained with AO to determine the numbers of bacteria remaining on the surface. Generally sonication for 15 to 30 min removes the most bacteria from the test surfaces. These results have also been compared with vortexing the discs for 2 min in a buffer containing Triton X-100. The optimum sonication time (to obtain the highest viable count and the lowest number of attached bacteria by microscopy) is dependent on the test material. Vortexing reliably removes bacteria attached to the disc surfaces and gives results that are independent of the test material. Buffer solutions are chilled at 4° to prevent microbial growth.

Standard Operating Procedures

Continuous Culture

1. Clean chemostat and wash tubing through with analytical grade (Elgastat) water.
2. Calibrate pH electrode.
3. Autoclave chemostat and tubing.
4. Filter 20 liter of 1/10th strength Iso-senstest broth (Oxoid Ltd., Basingstoke, Hants.) through a 0.2 μm bacterial capsule filter (Sartorius).
5. Fill chemostat vessel with 1.15 liter of media and switch all the modules on.

---

[32] A. A. Miles and S. S. Misra, *J. Hygiene* **38**, 732 (1938).

6. Leave overnight to check for sterility.

7. Calibrate the oxygen and pH electrodes and check that the temperature is 37°
and stirrer at 400 rpm.

8. Inoculate with a suspension of *Staphylococcus epidermidis* (RP62A) and
leave to batch culture for 5 h.

9. Switch on medium in pump (156 ml/hr). Under these conditions after 4 days
the culture will be growing at pH 7.4 with an oxygen level in the range of 80–90%
and a dilution rate of 0.14.

Modified Robbins Device

For studies of early attachment it is very important to ensure that the lumen
and other surfaces within the MRD are clean. Failure to adequately remove material associated with surfaces from previous experiments results in unpredictable
variations in the degree of attachment.

The MRD is disinfected with 1% w/v Vircon (Antec International Ltd.,
Sudbury, Suffolk). A peristaltic pump is used to pump the culture through the
MRDs and is calibrated before each run; a bubble trap is placed between the MRD
and the pump to stop bubbles passing through the MRD and to reduce the effect
of sedimentation. An in-line 14 μm filter is also used to remove large clumps of
cells. The MRD is placed on a hot plate and insulated so that the temperature of
the MRD is maintained at 37°.

1. The MRD cannot be autoclaved and so is removed from the system before
the chemostat and tubing are autoclaved.

2. Also prior to autoclaving the in line 14 μm filter holder is dismantled and
cleaned and the filter replaced with a new one.

3. The chemostat and attached tubing are autoclaved.

4. Dismantle MRDs 1 and 2 completely, cleaning the inlet and outlet end
plates, flow channel, and plugs.

5. Rebuild the MRDs, sealing the two halves with a small amount of sterile
silicon grease.

6. Lightly grease the O rings of the end plates and plugs.

7. Attach MRDs 1 and 2 into the system, fill with Vircon (Antec International
Ltd.), and leave overnight.

8. Flush the system with 500 ml of analytical grade water.

9. During the flushing process calibrate the flow rate of the pumps for each
MRD to 60 ml/hr.

10. Pump the system dry and leave until start of run.

The MRDs are flushed with 0.2 μm filtered analytical grade water (using
the side-arm demonstrated in Fig. 1) and then loaded with 11 mm discs using

aseptic technique. Glass discs (11 mm diameter, No. 1½, Chance Propper Ltd., Smethwick, Warley, England) can be used as a common comparator and are placed in every other sample port, 1, 3, 5, . . . 25. Test material discs are placed in the remaining positions 2, 4, 6, . . . 24, and all the discs are placed in the MRD so that they are flush with the upper interior wall of the flow channel (avoiding large particle deposition). Bacterial culture is pumped from the bioreactor through the MRD at a rate of 60–300 ml hr^{-1}, producing a wall shear stress well within the range producing laminar flow. There is a small amount of depletion of the bioreactor volume when flow rates to the MRD exceed the bioreactor fill rate (156 ml hr^{-1}). After 2 hr the culture fluid is replaced with phosphate buffered saline, pH 7.4 (PBS, Oxoid BR14a), using the side arm, which is perfused through the MRD for 45 min at the same pump rate as used for the culture fluid. Test and glass comparator discs are removed from a raised outflow end and then sequentially toward the inflow end to avoid movement of air bubbles through the device.

Evaluating the Numbers of Bacteria Attached to the Discs

1. Clamp the inlet to the MRDs and start removing the plugs from the outlet end (position 25) so as not to allow air bubbles to travel over the remaining discs.

2. Using aseptic technique, remove the disc from each plug and place it in a sterile 5 ml bottle containing 1 ml of chilled wash buffer (0.075 mol $liter^{-1}$ phosphate buffer, pH 7.9, with 0.1% v/v Triton X-100). Place the bottle in the rack on an ice block to keep cool during the removal of all the discs.

3. After all the discs have been removed store all on the rack at 4°.

4. The discs are processed in batches of six, and the bottles are kept on ice blocks at all times.

5. Vortex each bottle vigorously for 1 min, cool on an ice block for 5 min (while the other bottles in the batch are vortexed), and then repeat the vortex step before diluting the resulting suspension for viable counting.

6. The numbers of bacteria in the suspension are determined by viable counting.[32] The efficiency of the washing procedure is monitored by staining washed discs with 0.01% w/v acridine orange for 5 min and then viewing the discs by epifluorescent microscopy.

Test Materials and Adhesion Conditions

The test materials used are glass, siliconized coated glass (using the method of Sambrook, Fritsch and Maniatis, *Molecular Cloning, a Laboratory Manual*, page E2, using 2% w/v dimethyldichlorosilane in 1,1,1-trichloroethane), PTFE, stainless steel, and titanium. The influence of increasing sodium chloride, calcium, magnesium, or EDTA concentration is determined by adding concentrated solutions directly to the culture vessel immediately before turning on the pump to the MRD. Glass discs are conditioned with plasma proteins by exposing the discs in a MRD to 10% pooled citrated human donor plasma in PBS pumped at a flow rate of 30 ml hr^{-1} for 1 hr.

Statistical Analysis and Mathematical Modelling

The results of using this system to compare initial adhesion of *Staphylococcus epidermidis* RP62A to different test surfaces have been previously reported.[13]

Statistical Methodology

If it is assumed that the number of bacteria adhering to the material is linearly related to the position along the MRD, then a separate linear regression of the number of bacteria adhering (Y) versus MRD position (x) can be postulated for each material ($i = 1,2$).[13,33]

$$Y_i = \bar{y}_i + b_i(x - \bar{x}_i)$$

If the slopes of the regression lines (b_1, b_2) differ, there is an indication that an interaction is present between the test material and the rate of adhesion along the MRD.

In this case (when only two materials are being tested), one can use a simple t-test for the difference between the two slopes (b_1 and b_2) to test the hypothesis that the fitted regression lines are identical:

$$t = \frac{(b_1 - b_2)}{SE(b_1 - b_2)}$$

on ($n_1 + n_2 - 4$) degrees of freedom, where

$$SE(b_1 - b_2) = \sqrt{s^2 \left\{ \frac{1}{\Sigma_1(x - \bar{x}_1)^2} + \frac{1}{\Sigma_2(x - \bar{x}_2)^2} \right\}}$$

and

$$s^2 = \frac{\Sigma_1(y - Y_1)^2 + \Sigma_2(y - Y_2)^2}{n_1 + n_2 - 4}$$

which is the pooled estimate of the residual variance (as in a two-sample t-test).

If the slopes do not differ, the lines are parallel and one can then ask whether the lines differ in height above the x-axis (MRD position) or whether the two lines coincide. This is considered to be an analysis of covariance (ANCOVA). Two parallel lines may be fitted with common slope b. The equations of the two parallel

---

[33] P. Armitage, "Statistical Methods in Medical Research." Blackwell Scientific, Oxford, 1987.

lines are

$$Y_1 = \bar{y}_1 + b(x - \bar{x}_1)$$
$$Y_2 = \bar{y}_2 + b(x - \bar{x}_2)$$

The difference between the values of Y at a given x is therefore

$$d = Y_1 - Y_2$$
$$= \bar{y}_1 - \bar{y}_2 - b(\bar{x}_1 - \bar{x}_2)$$

As only two materials are being compared at any one time, one can use a t-test (with the standard error of d) to test the null hypothesis that the regression lines coincide, $E(d) = 0$,

$$t = \frac{d}{SE(d)} \quad \text{on } (n_1 + n_2 - 3) \text{ degrees of freedom}$$

The standard error of d can be calculated as

$$SE(d) = \sqrt{S_p^2 = \left\{ \frac{1}{n_1} + \frac{1}{n_2} + \frac{(\bar{x}_1 - \bar{x}_2)^2}{\Sigma(x - \bar{x}_1)^2 + \Sigma(x - \bar{x}_2)^2} \right\}}$$

where S_p^2 can be estimated by the residual mean square error about the parallel lines.

Comparisons of different regression slopes or analysis of covariance techniques are both implemented in standard statistical packages where calculations of the appropriate variances are carried out automatically.

Mathematical Modeling Approach

Using the previous statistical approach it was assumed that bacterial depletion was a linear function of MRD position. There was no attempt to describe the behavior of the process, only to determine if the rate of depletion through the MRD differed between materials.

An alternative approach is to directly simulate the flow of bacteria through the MRD. Complex systems that cannot easily be translated into mathematical equations and solved analytically can be simulated. If a physical or biological process has been modeled adequately, computer simulations offer a quicker and less expensive method of experimentation and addressing "what-if" questions.

The Model

As bacteria flow through the MRD they may be lost from the interacting stream by mechanisms including sedimentation, they may stick to a disc made of material 1 or material 2, or they may continue flowing through the MRD.

A simple mathematical model for this physical process is formulated as a pair of recurrence relations.[34] Equation (1) describes the number of bacteria adhering to material 1 (glass) and Eq. (2) describes the number of bacteria adhering to material 2 (siliconized glass). In the experiment the odd numbered discs are made of material 1 and the even numbered discs are made of material 2.

$$N(p) = N(0)(1 - \alpha)^p (1 - P\beta_1)^{(p-1)/2} (1 - P\beta_2)^{(p-1)/2} P\beta_1 \qquad (1)$$

$$N(p) = N(0)(1 - \alpha)^p (1 - P\beta_1)^{p/2} (1 - P\beta_2)^{(p/2)-1} P\beta_1 \qquad (2)$$

$N(p)$ is the number of bacteria adhering to the disc at position p along the MRD, $N(0)$ is the initial number of bacteria in the culture, α is the rate of loss to the interacting stream along the MRD sampled from a uniform distribution: $\alpha \sim U(0,1)$, and P is a random variable sampled from a normal distribution: $P \sim N(\mu, \sigma^2)$. It is the inherent adhesion of the bacterial culture. $\beta_1, \beta_2 = (0,1)$ are constants associated with the adhesion of material 1 and material 2, respectively.

There are several assumptions made within this model. We assume that the adhesion of the bacterial culture follows a Gaussian (normal or bell-shaped) distribution. The rate at which bacteria are lost from the interacting stream, α, is assumed to be constant at each position p throughout the MRD. Finally, the adhesion of the bacteria is assumed to be independent of their position along the MRD.

The Simulation

The simulation of the recurrence relations (1) and (2) was implemented in Microsoft Excel (v7) using the @RISK (3.5.2, Palisade Corporation) add-on demonstration software. The Monte Carlo sampling technique was used to sample from the two probability distributions (the normal distribution and the uniform distribution) within the model. This technique simply generates pseudorandom numbers to sample from a predefined probability distribution.[35-37]

Two separate simulations have been run. The first simulates bacterial flow along the MRD with a flow rate of 60 ml/hr, the second with a flow rate of 300 ml/hr. In both simulations, glass is compared with siliconized glass. For illustration purposes, the initial parameters in this problem are estimated empirically from the

[34] R. E. Mickens, "Difference Equations: Theory and Applications." Van Nostrand Reinhold, New York, 1990.

[35] N. T. J. Bailey, "The Elements of Stochastic Processes with Applications to the Natural Sciences." Wiley, New York, 1964.

[36] C. L. Chiang, "An Introduction to Stochastic Processes and Their Applications." Robert E. Krieger Publishing Company, Huntington, New York (1980).

[37] W. H. Press, B. P. Flannery, S. A. Teukolsky, and W. T. Vetterling, "Numerical Recipes in Pascal." Cambridge University Press, UK (1992).

TABLE I

PARAMETER ESTIMATES FOR THE DISTRIBUTION OF ADHESION OF THE BACTERIAL CULTURE
AND THE SEDIMENTATION RATE

	Flow rate 60 ml/hr		Flow rate 300 ml/hr	
	Glass	Siliconized glass	Glass	Siliconized glass
Truncated normal dist[n]				
μ	0.0038	0.002	0.0038	0.001
σ^2	0.001	0.001	0.001	0.0001
	$\beta_1 = 1$	$\beta_2 = 0.53$	$\beta_1 = 1$	$\beta_2 = 0.26$
Uniform distribution				
α(min, max)	(0.01, 0.05)	(0.01, 0.05)	(0.01, 0.05)	(0.01, 0.05)

data, and the model is run to explore how well the predictions from the model agreed with the observed data.

The data from the two experiments (glass vs siliconized glass at 60 ml/hr and 300 ml/hr) are presented in Figs. 3a and b (reproduced from ref. 13). At a flow rate of 60 ml/hr there appears to be a gradual depletion in bacteria adhering to the discs as they move down the MRD for both materials, whereas at 300 ml/hr this depletion was only apparent for glass. Clearly the number of bacteria adhering to both materials is reduced by higher flow rates. This effect is particularly marked for siliconized glass at 300 ml/hr.

The parameter estimates used for the simulations are presented in Table I. Of interest is the relative adhesion between the two materials for both flow rates; therefore the sedimentation rate, α, was held constant. Figures 4a and b are results of the simulation of the system at a flow rate of 60 ml/hr. At 60 ml/hr the model appears to simulate the observed data well. From this model, glass appears to be almost twice ($\beta_1 = 1$) as adhesive as siliconized glass ($\beta_2 = 0.53$). Under the assumption that the bacterial culture used for the second experiment had the same distribution of adhesion as that used for the first, we have modeled the system under a flow rate of 300 ml/hr using identical parameters as before (Figs. 5a and b). The simulation overestimates the number of bacteria adhering to the discs along the MRD for both materials. On examination of the data, it was observed that although there appears to be a difference between the numbers of bacteria adhering to glass at different flow rates, the greatest difference appears to be between the numbers of bacteria adhering to siliconized glass. This suggests that glass is relatively unaffected by flow rate, but siliconized glass was not. In order to improve the fit of the model to the observed data, the adhesion of the siliconized glass must be reduced to $\beta_2 = 0.26$. This greatly improves the fit of the model (Figs. 6a and b).

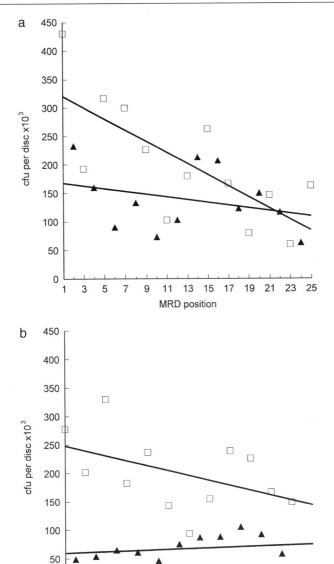

FIG. 3. Initial adhesion (2 hr) of *Staphylococcus epidermidis* RP62A to glass and siliconized glass at a flow rate of: (a) 60 ml/hr; (b) 300 ml/hr. — □ —, glass; — ▲ —, siliconized glass.

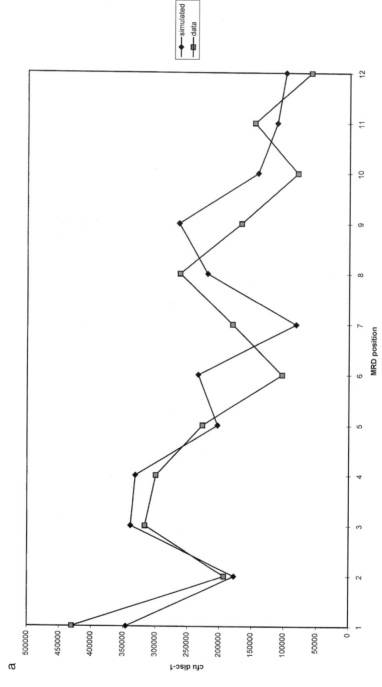

FIG. 4. (a) Adhesion of *S. epidermidis* to glass at 60 ml/hr. (b) Adhesion of *S. epidermidis* to siliconized glass at 60 ml/hr.

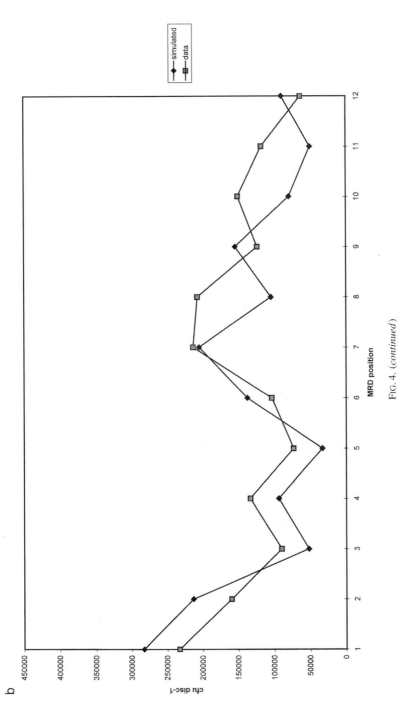

FIG. 4. (*continued*)

57

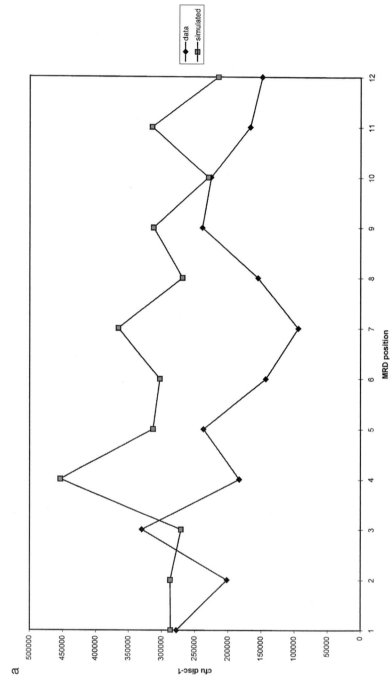

FIG. 5. (a) Adhesion of *S. epidermidis* to glass at 300 ml/hr. (b) Adhesion of *S. epidermidis* to siliconized glass at 300 ml/hr.

58

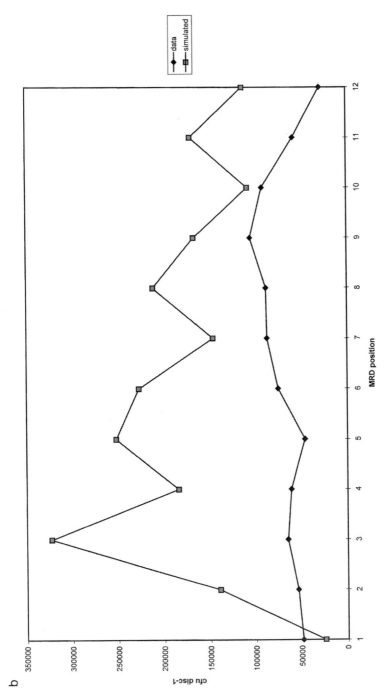

FIG. 5. (*continued*)

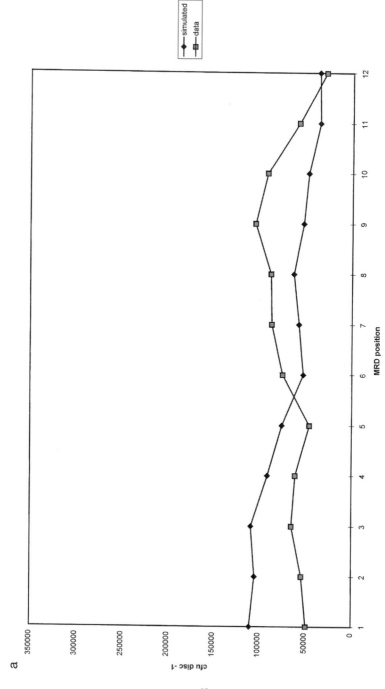

FIG. 6. (a) Adhesion of *S. epidermidis* to siliconized glass at 300 ml/hr. (b) Adhesion of *S. epidermidis* to glass at 300 ml/hr.

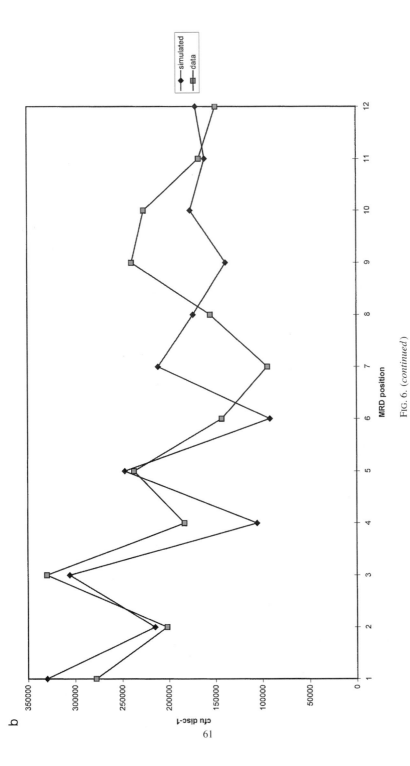

FIG. 6. (*continued*)

61

Using this modeling approach it appears that glass is more adhesive than siliconized glass under both sets of conditions (60 ml/hr and 300 ml/hr). However, the relative difference in adhesion between the two materials increases when the flow rate increases from 60 ml/hr to 300 ml/hr.

Concluding Remarks

We describe the use of a continuous culture system linked to a modified Robbins device to compare the adhesion of bacteria grown under defined conditions to different surfaces. The chemostat allows the numbers, growth conditions, and growth rates of the bacteria used in adhesion experiments to be controlled. Use of a chemostat allows conditions that promote adhesion and/or biofilm formation to be selected. For example, *Helicobacter pylori* forms a biofilm in a defined medium with a high carbon/nitrogen ratio.[38] We have used this system to study the impact of shear stress and flow rate, changing ionic environment, and substrate properties on adhesion of *Staphylococcus epidermidis* RP 62A. In principle the methods described could be used to select for bacterial phenotypes or genotypes with preferential adhesion to different surfaces using mono- or polymicrobial cultures. The MRD has been used predominantly for biofilm studies rather than for adhesion studies. In experiments, particularly at low flow rates, there was a trend toward higher numbers adhering to the discs at the in-flow end of the MRD than at the outflow end, probably reflecting depletion of adherent bacteria in the interacting stream.

We have described a mathematical simulation model of the experimental system. Other representations may yield better agreement with experimental results. We have demonstrated how statistical modeling tools can aid in the understanding of complex systems, which cannot always be formulated and solved analytically. Further work has to be undertaken before this model can be established as the most appropriate representation of the experimental system. Rigorous methods of parameter optimization are required to estimate optimal values for the parameters of the distributions, rather than estimating parameters empirically (although this is a good starting point).[37] Large numbers of iterations must be run in order to obtain an average performance of the model, from which one can predict the statistical error (variance) in this result and hence the precision of the model's predictions.[35,36] A sensitivity analysis should be performed to determine how sensitive the model is to small changes in the parameter estimates, and thus which factors are important and which have relatively little influence on the model outcome.

The MDR linked to a chemostat can be used to directly compare bacterial adhesion to potential biomaterials. Experimental results can be simulated using a mathematical modeling approach.

[38] R. M. Stark, G. J. Gerwig, R. S. Pitman, L. F. Potts, N. A. Williams, J. Greenman, I. P. Weinzweig, T. R. Hirst, and M. R. Millar, *Letts. Appl. Microbiol.* **28,** 121 (1999).

[4] Direct Biofilm Monitoring by a Capacitance Measurement Probe in Continuous Culture Chemostats

By Jana Jass, J. Gary O'Neill, and James T. Walker

Introduction

Traditional techniques have often involved indirect methods for monitoring of surface biofilms and biofouling. These include direct contact plates, pour plates, and spread plates to assess total viable bacterial numbers and identification of species to be enumerated. These techniques are limited in that they require long incubation periods to allow most organisms to grow, especially those found in oligotrophic environments. A qualified person is required to identify the microbial populations on the plates and to interpret results. In addition, the only organisms detected are those that can be grown on plates; thus, viable but nonculturable organisms,[1] those in a dormant state or dead cells contributing to biofouling, are not detected. Therefore, there is a requirement for improving detection methods and simplifying microbial biofilm detection.[2]

A primary requirement of most biofilm determination techniques is that access to the site where the fouling occurs is required to assess the microbiological contamination of a surface. There are a number of side-stream devices that are used and they provide easy access to the biofilm for collection and analysis.[3] One of the most prolifically used engineering devices for monitoring biofilms on industrial surfaces is the Robbins device.[4] Over the years it has been copied and modified for both in-laboratory simulation models and small-scale pilot studies.[5] These types of devices provide convenient mechanisms for presenting a substrata for biofilm development within systems being investigated. However, even these systems require the removal of substratum from the device for analyses, which involves the disruption of the biofilm integrity for such quantitation using total viable counts.

Ideally, methods are required that will allow direct on-line and in-line sensors to monitor bacterial accumulation and biofouling of a surface.[6] Such technology has the capability to allow downstream data to be collected and remotely assessed so that decisions can be made on the basis of real-time results from the developing

---

[1] M. R. Barer and C. R. Harwood, *Adv. Microbiol. Physiol.* **41,** 93 (1999).
[2] M. Sara, A. P. Ison, and M. D. Lilly, *J. Biotechnol.* **51,** 157 (1996).
[3] J. W. Hopton and E. Hill, "Industrial Microbiological Testing." Blackwell Scientific, Oxford, 1987.
[4] W. F. McCoy and J. W. Costerton, *Dev. Indust. Microbiol.* **23,** 551 (1982).
[5] C. L. Schultz, M. R. Pezzutti, D. Silor, and R. White, *J. Indust. Microbiol.* **15,** 243 (1995).
[6] D. E. Nivens, R. J. Palmer, and D. C. White, *J. Indust. Microbiol.* **15,** 263 (1995).

biofilm. Thus, process systems could be automated so that an alarm would be signaled when specific measurements are obtained by in-line instrumentation. This is much preferred to waiting for problems to arise through systems failures before biofouling tests are initiated.[7] Such equipment must be rapid, able to measure and differentiate between planktonic and biofilm cells, simple to operate, accurate, reliable, reusable, and robust.[8]

Many of the sensors and methods automated for on-line biomass determinations rely on optical measuring principles[9,10] or exploit filtration characteristics,[11,12] density changes,[13] or dielectric properties of suspended cells, such as capacitance.[14-17] The device presented here is responsive to viable cells adherent on a surface by measuring the capacitance under an electric field.[15,16] This system was primarily developed for yeast cell systems; however, it is also able to detect biomass produced by bacterial cells in the form of biofilms on a platinum electrode. This enables on-line monitoring and studying of biofouling within a laboratory model system with the potential to be used in natural systems.

Laboratory Model Systems

To study natural systems in the laboratory, the researcher must first identify which environmental features would be required to include in his model. For potable water systems, it is important to control environmental factors, including temperature, dissolved oxygen, pH, and nutrient sources such as carbon and trace elements, so that it will closely resemble the natural system. In addition, the microbial community must be selected from the system being modeled, since these organisms will be best suited to survive the conditions set and will more closely resemble the natural system. The most suitable method is to use the continuous culture system where many parameters can be controlled and held constant during biofilm development.

---

[7] H.-C. Flemming and G. Schaule, in "Microbial Deterioration of Materials" (W. Sand, E. Heitz, and H.-C. Fleming, eds.), p. 121. Springer Verlag, Heidelberg, 1996.

[8] M. Harris and D. Kell, Biosensors 1, 17 (1985).

[9] J. Nielsen, K. Nikolajsen, S. Benthin, and J. Villadsen, Analyt. Chim Acta 237, 165 (1990).

[10] S. Iijima, S. Yamashita, K. Matusunaga, H. Miura, and M. Morikawa, J. Chem. Technol. Biotechnol. 40, 203 (1987).

[11] E. Nestaas and D. I. C. Wang, Biotechnol. Bioeng. 25, 1981 (1983).

[12] D. C. Thomas, V. K. Chittur, J. W. Cagney, and L. H. C, Biotechnol. Bioeng. 27, 729 (1985).

[13] B. C. Blake-Coleman, D. J. Clarke, M. R. Calder, and S. C. Moody, Biotechnol. Bioeng. 28, 1241 (1986).

[14] G. D. Austin, R. W. J. Watson, and T. D'Amore, Biotechnol. Bioeng. 43, 337 (1994).

[15] R. Matanguihan, K. Konstantinov, and T. Yoshida, Bioprocess Engineering 11, 213 (1994).

[16] K. Mishima, A. Mimura, Y. Takahara, K. Asami, and T. Hanai, J. Ferm. Bioeng. 72, 291 (1991).

[17] R. Fehrenbach, M. Comberbach, and J. O. Petre, J. Biotechnol. 23, 303 (1992).

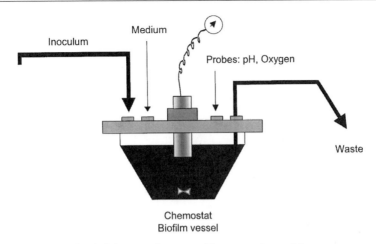

Inoculum
Medium
Probes: pH, Oxygen
Waste
Chemostat
Biofilm vessel

FIG. 1. Laboratory chemostat—biomass monitor model.

Continuous Culture Laboratory Model

The laboratory model was a continuous culture chemostat apparatus where the environmental conditions could be controlled.[18,19] It consisted of a glass vessel and a titanium top plate with glass or silicone rubber attachments to minimize exogenous metals or carbon sources being leached into the water medium (Fig. 1). The titanium top plate was modified and adapted to accept the 25 mm diameter probe. The culture was stirred which provided not only mixing of nutrients and bacteria but also an active flow which would be present in most water systems.

The inoculum, selected from the modelled environment, included organisms such as[20] *Pseudomonas* spp., *Flavobacterium* spp., *Acinetobacter* spp., and *Methylobacterium* spp. These were obtained by filtering a volume of water from a water system[21] and preparing a microbial concentrate. This microbial concentrate was added to inoculate the chemostat and maintain a suitable population of organisms for biofilm formation.

To demostrate this procedure an example model system is described. The selected inoculum was initially diluted with soft water at pH of 7.1 and hardness

[18] J. T. Walker, A. B. Dowsett, P. J. L. Dennis, and C. W. Keevil, *Internat. Biodeterior.* **27,** 121 (1991).
[19] J. Rogers and C. W. Keevil, *in* "Protozoal Parasites in Water" (K. C. Thompson and C. Fricker, eds.), p. 209. Royal Society for Chemistry London, 1995.
[20] J. Rogers, A. B. Dowsett, P. J. Dennis, J. V. Lee, and C. W. Keevil, *Appl. Environ. Microbiol.* **60,** 1585 (1994).
[21] J. S. Colbourne, R. M. Trew, and P. J. Dennis, *J. Appl. Bacteriol.* **65,** 299 (1988).

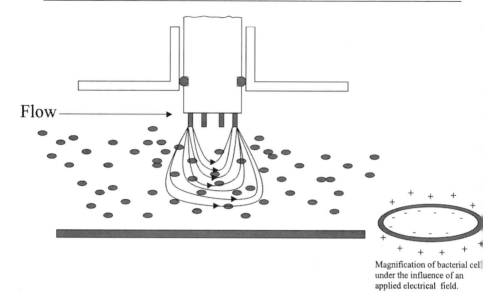

FIG. 2. Schematic of biomass probe.

Magnification of bacterial cell under the influence of an applied electrical field.

of <20 ppm CaCO₃. This was followed by the addition of R2A nutrient broth that was maintained at 25° and at a dilution of 0.05 hr^{-1}. The chemostat system itself was at ambient room temperature. It is possible to introduce additional monitoring probes for pH and dissolved oxygen. Not only can the pH be monitored, but it can be controlled by the addition of an acid or base.

Biomass Monitoring Probe

The biomass monitor 214M was designed for the on-line detection of viable bacteria, yeasts, fungi, animal cells, and other cellular biomass and can be purchased from Aber Instruments Ltd. The probe consists of a 25-mm (5-inch) black plastic core into which four platinum electrodes, of 2 mm diameter and 8 mm in height, are inserted (Fig. 2). The signal from the probe was amplified using a head amplifier and then transferred to the biomass monitor. The instrument was set on the "lowrange" and operated over a frequency of 200 kHz to 10 MHz. The conductivity range was 1 to 36 MS/cm with analog outputs (0–10 V or 4–20 mA) for frequency, conductance, temperature, offset capacitance, and biomass. The instrument could be reused by operating an antifouling clean pulse to remove any material from the probes, enabling the instrument to be returned to a baseline monitoring range. Calibration was carried out using 20 mM KCl before undertaking any experimental procedures.

Microbial Analysis

Microscopy and Image Analysis

The bioprobe was assessed periodically using noncontact differential interference microscopy. A Nikon Labophot 2 microscope with episcopic DIC (reflected light), powered by a 100 W halogen lamp, was used to examine the biofouling on the platinum probes fluorescence without the need for fluorescent dyes. The bioprobe was fixed to the specimen stage and the platinum probes examined using noncontact M Plan Apo lenses ($\times$40), allowing a working distance of up to 1 cm. Neutral density filters were used where appropriate to suppress high background fluorescence. Images were relayed to 35 mm transparencies (Nikon F801) and saved as computer files (*.tif) for image analysis (Optimus, Datacell, U.K.) of percentage coverage[22].

Quantitation of Planktonic Bacteria and Biofouling. Total viable counts (TVC) of the heterotrophic bacteria in the planktonic (liquid) phase were assessed by serially diluting (1/10 dilutions) the liquid culture in phosphate buffered saline (PBS) and spreading, in duplicate, 0.1 ml onto appropriate (i.e., R2A, nutrient agar) agar plates. The plates were incubated at 37° for 48 hr.[23]

The extent of biofouling within the system was assessed by traditional quantitation methods for comparison and standardization of the probe. Glass coupons immersed in the culture vessel were removed and washed to dislodge any adherent planktonic cells by gently rinsing the coupons in PBS. The biofilm was removed by scraping the surface with a scalpel blade into 1 ml PBS, and the cells were suspended by vortexing for 30 sec. The biofilm samples were then serially diluted and cultured on appropriate (i.e., R2A, nutrient agar) plates as described above.

Monitoring of Biofilm Growth

To help describe typical results and their interpretation, our model system of bacterial growth in soft potable water is used. The probe was first autoclaved and introduced aseptically into the chemostat biofilm vessel through a 25 mm port machined into the titanium top plate. It was allowed to stabilize for a few hours prior to initiating the experiment.

Baseline Calibration and Lower Detection Limit. The instrument must be calibrated to have a constant baseline, and the lower limits of detection should be determined prior to experimental setup. Lower bacterial densities of approximately 10^4 cfu/ml are not detected in our model system even though it was evident from the microscopic analysis of the probe that the surface was fouled up to 75%. Figure 3 demonstrates a recognizable measurable deflection in the capacitance

[22] S. B. Surman, J. T. Walker, D. T. Goddard, L. H. G. Morton, C. W. Keevil, A. Skinner, and J. Kurtz, *J. Microbiol. Meth.* **25**, 57 (1996).

[23] D. J. Reasoner and E. E. Geldreich, *Appl. Environ. Microbiol.* **49**, 1 (1985).

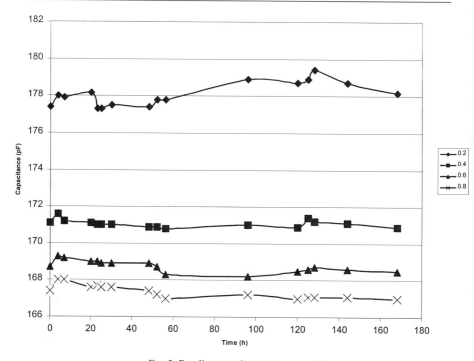

FIG. 3. Baseline capacitance measurements.

with increasing time when biofilm formation is below the measurable limit. The results will vary with the changing frequency. For example, at 0.2 kHz, the capacitance will vary from 177.2 to 179.4 pF over a 170-hr time period. Increasing the frequency will decrease the capacitance start point; however, it may also decrease the variance in the capacitance. For example, at 0.4 kHz the range is 0.8 pF (170.8–171.6) and at 0.6 kHz the range is 0.7 pF (168.2–168.9). The capacitance is measured over the frequency range 0.2–0.8 kHz to determine which frequency will be the most sensitive in detecting a change in culture volume.

Interpretation of Results. A typical response from the probe is demonstrated in Fig. 4. After the probe is introduced into the biofilm vessel (Fig. 4, point A) and is allowed to stabilize overnight, the clean cycle on the probe is activated (Fig. 4, point B). After the clean cycle, the recorded data stabilized and no obvious variations are observed. This steady state must be reached before continuing. The experiment is started by the addition of nutrient medium to increase bacterial growth. Only following the addition of the R2A (Fig. 4, point C) did the recorded results start to increase at all frequencies.

The increases in capacitance for the different frequencies listed in Table I correspond with the increase in biomass to 7.45×10^8 cfu/ml in the liquid culture

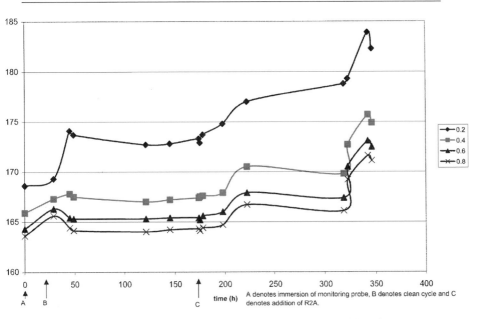

FIG. 4. Measurement of capacitance in response to increased microbial numbers.

resulting from the addition of nutrient. The biofilm formation on surfaces within this environment increased to 1.28×10^8 cfu/cm^2, recovered from the coupons immersed in the culture vessel. The best response was at the lowest frequency; however, it was also the frequency with the greatest variation in the baseline detection.

Instrument Limitations. The biomass monitor device described here may not be sensitive enough to be used as a biofouling monitor in process water. In natural environments, the numbers of bacteria are generally less than 10^4 cfu/ml and the monitor is currently not capable of detecting those numbers of bacteria. However,

TABLE I

FREQUENCY EMPLOYED RESULTS IN DIFFERENT RESPONSES
TO BIOFILM FORMATION

Frequency (kHz)	Capacitance (pF)	Change in capacitance (pF)
0.2	168.6–179.3	10.7
0.4	165.9–175.7	9.8
0.6	164.3–172.5	8.2
0.8	163–171.6	8.6

it is a good research tool for monitoring fouling in fermentation systems and long-term studies where a certain level of biofilm formation is desired. It also has potential usage in "dirty" process waters where the numbers of bacteria are greater and within the limits of the biomass probe.

The ability to monitor on-line biomass in a nondestructive manner is an extremely valuable asset to bioprocess technologists, as it removes the necessity for predictive controls by providing instantaneous results.[24] Particular usefulness may be in industrial settings where biofouling can be monitored remotely with automated immediate responsive control measures.

Acknowledgments

The authors thank Yorkshire Water for financially supporting this work, and in particular Dr. C. W. Keevil (CAMR) for guidance and advice.

[24] D. E. Nivens, J. Q. Chambers, T. R. Anderson, A. Tunlid, J. Smit, and D. C. White, *J. Microbiol. Meth.* **17,** 199 (1993).

[5] Use of Continuous Culture Bioreactors for the Study of Pathogens Such as *Campylobacter jejuni* and *Escherichia coli* O157 in Biofilms

By Clive M. Buswell, Helen S. Nicholl, and James T. Walker

Introduction and Background

Bacterial growth is a contentious issue, where a multitude of different modes of growth from individual laboratory species in monoculture to multiple species cohabiting in the environment have to be considered. Consideration of these varying complexities is required during experimental design. Decisions have to be made as to which growth phase is to be studied, whether in mono- or multiculture, batch, semicontinuous, or continuous culture, and whether a complex, minimal, or defined medium is used as the growth medium.

Water has been implicated as a reservoir for the transmission from the environment of many major human pathogens, including *Campylobacter jejuni*,[1,2]

[1] M. B. Skirrow, *Lancet* **336,** 921 (1990).
[2] S. A. Leach, *Rev. Med. Microbiol.* **8,** 113 (1997).

Legionella pneumophila,[3] and *Escherichia coli* O157.[4-6] Often, however, the offending pathogen cannot be cultured from the water source implicated because (i) too much time is taken to trace the source after the reported infection; (ii) the strains are present in too low numbers; and (iii) the strains are viable but nonculturable.[7]

Studying biofilms in the environment can be difficult because of their heterogenous nature.[8] Hence, laboratory models can be used to provide more reproducible and defined conditions. In this way a number of different models have been developed to investigate biofilm growth. These include (i) flow cells[9]; (ii) channel reactors[10]; (iii) Robbins devices[11]; (iv) rototorque[12]; (v) packed bed reactors[13]; (vi) airlift reactors[14]; (vii) constant depth film fermentors[15]; and (viii) chemostat bioreactors.[16]

There is a vast range of experimental systems with which to develop biofilms, and the model used depends on the nature of the problem that one is investigating. Flow cells using glass are useful when the investigator wants to follow attachment using light microscopy combined with fluorescent dyes for species specificity or viability assays. Flow regimes in pipe work can be modeled using the Robbins device, which can be flexibly applied to the simulation of different modeling scenarios such as water systems or urinary catheters. Where shear rate is to be examined, the rototorque proves an efficient tool for the purpose. In a number of studies the continuous culture bioreactor has been used because of the flexibility in providing a situation where the planktonic and biofilm phases are cultured together. Further, a standardized inoculum can be supplied to a series of "experimental vessels" in which culture conditions can be carefully controlled and the resultant effects compared. Finally, when generating biofilms involving microbial pathogens, health and safety implications arise and in some cases facilities for operator protection are required.

[3] P. M. Arnow, D. Weil, and M. F. Para, *J. Infect. Dis.* **152,** 145 (1985).

[4] M. Alary and D. Nadeau, *Can. J. Public Health* **81,** 268 (1990).

[5] P. M. Arnow, T. Chou, D. Weil, E. N. Shapiro, and C. Kretzschmar, *J. Infect. Dis.* **146,** 460 (1982).

[6] H. E. Bryant, M. A. Athar, and C. H. Pai, *J. Infect. Dis.* **160,** 858 (1989).

[7] M. R. Barer and C. R. Harwood, *Adv. Microb. Physiol.* **41,** 93 (1999).

[8] J. W. Costerton, Z. Lewandowski, D. DeBeer, D. Caldwell, D. Korber, and G. James, *J. Bacteriol.* **176,** 2137 (1994).

[9] D. E. Caldwell and J. R. Lawrence, *in* "CRC Handbook of Laboratory Model Systems for Microbial Ecosystems" (J. W. T. Wimpenny, ed.), p. 117. CRC Press, Boca Raton, FL, 1988.

[10] D. DeBeer, P. Stoodley, and Z. Lewandowski, *Biotechnol. Bioeng.* **53,** 151 (1997).

[11] W. F. McCoy, J. D. Bryers, J. Robbins, and J. W. Costerton, *Can. J. Microbiol.* **27,** 910 (1981).

[12] M. M. Gabriel, D. G. Aheam, K. Y. Chan, and A. S. Patel, *J. Cataract. Refract. Surg.* **24,** 124 (1998).

[13] F. A. MacLeod, S. R. Guiot, and J. W. Costerton, *Appl. Environ. Microbiol.* **56,** 1598 (1990).

[14] A. Gjaltema, P. Arts, M. van Loosdrecht, J. Kueun, and J. Heijen, *Biotechnol. Bioeng.* 194 (1994).

[15] R. Coombe, A. Tatevossian, and J. Wimpenny, *in* "Surface and Colloid Phenoma in the Oral Cavity: Methodological Aspects" (R. Frankm and S. Leach, eds.), p. 239. IRL Press, London, 1981.

[16] D. J. Bradshaw, P. D. Marsh, K. M. Schilling, and D. Cummins, *J. Appl. Bacteriol.* **80,** 124 (1996).

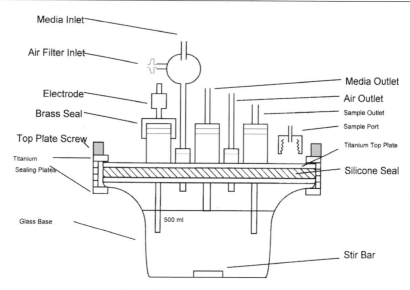

Media Inlet

Air Filter Inlet

Electrode

Brass Seal

Top Plate Screw

Titanium Sealing Plate

Glass Base

500 ml

Media Outlet

Air Outlet

Sample Outlet

Sample Port

Titanium Top Plate

Silicone Seal

Stir Bar

FIG. 1. Schematic of biofilm bioreactor.

General Description of Model Continuous Culture Bioreactor

The basic design, operating conditions, and development of continuous culture models in the study of oral biofilms were described in an earlier volume of this series.[17,18] However, the continuous culture laboratory model bioreactor has also been used in other studies to obtain defined, controlled, and reproducible experimental conditions, particularly with respect to biofilms in the aquatic environment.[19–21] In our model, a titanium head plate is used in which the number of ports available has been maximized to allow the insertion of as many monitoring/control sensors and surfaces for biofilm development as possible. The diameter of the head plate is such that it fits comfortably onto a 1 liter volume desiccator jar that can be bought off the shelf. We find that these vessels, being of a short, squat design, maximize the surface to volume ratio of the system and therefore the surface area available for the development of biofilm (Fig. 1). The models are designed to investigate the influence of parameters, such as temperature, water type/content,

[17] D. J. Bradshaw and P. D. Marsh, *Methods Enzymol.* **310,** 279 (1999).

[18] J. M. Vroom, K. J. DeGrauw, H. C. Gerritsen, D. J. Bradshaw, P. D. Marsh, G. K. Watson, J. J. Birmingham, and C. Allison, *Appl. Environ. Microbiol.* **65,** 3502 (1999).

[19] J. T. Walker, A. B. Dowsett, P. J. L. Dennis, and C. W. Keevil, *Int. Biodeterior.* **27,** 121 (1991).

[20] J. Rogers, A. B. Dowsett, P. J. Dennis, J. V. Lee, and C. W. Keevil, *Appl. Environ. Microbiol.* **60,** 1585 (1994).

[21] P. J. Robinson, J. T. Walker, C. W. Keevil, and J. Cole, *FEMS Microbiol. Lett.* **129,** 183 (1995).

and plumbing material, on the survival and growth of pathogens such as *Campylobacter jejuni, Legionella pneumophila,* and *Escherichia coli* O157 in biofilms. Some experiments have been undertaken within an ACDP category III containment cabinet (CAMR Engineering) in ACDP category III containment laboratory suites to ensure operator protection due to the low infectious dose or route of infection of the pathogen being studied.[22] The model system consists of two or more glass vessels linked in series. The first vessel (seed vessel) contains 500 ml of culture and is mixed using a magnetic stirrer at 250 rpm (LH Fermentation, U.K.; Anglicon Microprocessor, Brighton). Sterile water is supplied at a chosen flow rate to achieve the desired dilution rate, which is routinely between 0.025 and 0.05 hr^{-1}, although much higher rates can be set if required. The limit of the flow rate is the speed of the peristaltic pump and tube diameter. The water supplied to the seed vessel is taken from a number of sources and filter-sterilized (0.2 μm, Pall Filtration). Membrane filtration of the water has been shown to result in sterile water that is less chemically altered than autoclaved water.[23,24] The vessel volume is maintained by the removal of culture through an overflow weir set to the desired volume.

The culture volume in the second vessel (experimental vessel), where the biofilm development takes place, is typically at least 500 ml and sometimes a 1.0 liter working volume. This enables an increased number of suspended surfaces to be immersed in the water for development of biofilm. The vessel is inoculated by supplying effluent from the seed vessel. Then, by a combination of supplying effluent and fresh sterile water to this vessel, a consistent population is developed. The parameters in this vessel are then controlled to maintain conditions for the particular experimental regime being undertaken. Examples of the parameters studied using this system include temperature, water quality (including carbon loading), and different oxygen tensions. Temperatures used range between 4° and 40°. The lower temperatures represent a cold water mains supply, whereas those around 20° represent supplies within buildings or within pipes during a hot summer spell. The higher temperatures represent the lower range of temperatures that would be found within a hot water system. To maintain temperature below 20° required a chiller unit (Grant LTD 6, UK).

Inoculum Preparation and Seeding of Model

The inoculum used depends on the study being undertaken. For some studies the total native autochthonous population is required; in others a defined population

[22] N. Banatvala, M. M. Debeukelaer, P. M. Griffin, T. J. Barrett, K. D. Greene, J. H. Green, and J. G. Wells, *Pediatr. Infect. Dis. J.* **15,** 1008 (1996).

[23] J. S. Colbourne, R. M. Trew, and P. J. Dennis, *J. Appl. Bacteriol.* **65,** 299 (1988).

[24] C. M. Buswell, Y. M. Herlihy, L. M. Lawrence, J. T. McGuiggan, P. D. Marsh, C. W. Keevil, and S. A. Leach, *Appl. Environ. Microbiol.* **64,** 733 (1998).

is used (we use an eight-membered consortium of gram negative and gram positive bacteria).[29] The autochthonous population of a given water source is obtained by filtering 20 liters of water through a polycarbonate 142 mm diameter membrane filter (0.2 μm, Pall Filtration). The filter is then placed in a sterile sample container and shaken at ambient temperature on an orbital shaker to suspend the bacteria. For the preparation of the defined population, samples (0.1 ml) of the microbial suspension are taken after 1 hr and plated onto R2A minimal recovery agar[25] and incubated for 48 hr at 37°. Selection of representative colonies is based on morphological characteristics and subcultured onto R2A for a further 48 hr at 37°. Subsequently, colonies of the eight different microbial types are stored on beads at −19°. The bacterial inoculum is pooled together and used as inoculum or, alternatively, the filter is shaken overnight and the entire contents of the sample container used as inoculum.

Case Studies

The continuous culture system can be adapted or controlled to study the effects of a variety of changes in parameters, as has already been discussed. Two case studies are presented below that provide details of how the model and procedures can be adapted according to the design of the experiment.

Survival of E. coli O157

The defined eight-membered consortium is added to vessel 1, grown in batch for 48 hr, and then operated under continuous culture for 1 week for culture stabilization before the biofouling materials are introduced. A human isolate of E. coli O157, strain PS14, was obtained from the Central Public Health Laboratory (Colindale, London). This is incubated overnight in TSB broth culture in an orbital shaker incubator (Gallenkamp, U.K.) at 100 rpm at 37°. Immediately after the plumbing materials are placed into the test vessel, the seed vessel is inoculated with 10 ml of E. coli O157 broth suspension to simulate a worst-case scenario of run-off from a field contaminated with cattle feces into a private water well.

Standard plumbing tube materials such as copper, polybutylene plastic, and stainless steel are used, along with control glass surfaces, as the substrata on which the biofilms are developed. The plumbing materials are sectioned into 1 cm^2 coupons (CAMR Engineering). Four coupons of each material are suspended by titanium wire, one above the other, from a silicone rubber bung and introduced aseptically into the vessel. Three of the coupons are used for microbiological enumeration and the fourth for microscopy assessment of the biofilm coverage.

[25] D. J. Reasoner and E. E. Geldreich, *Appl. Environ. Microbiol.* **49**, 1 (1985).

Water samples are removed from the seed and test vessels aseptically through a sample port at day 7 postseed. Each coupon assembly for biofilm analysis is removed and replaced with a sterile silicone rubber bung. The water and biofilm coupon samples are placed inside double sterile plastic bags that have been sprayed with disinfectant [5% Hycolin, (v/v), Solmedia, Essex] before removal to a separate ACDP category III containment cabinet for microbiological analyses.

Each coupon is immersed gently in 10 ml of nonflowing sterile phosphate buffered saline (PBS) to remove nonadherent bacteria and then placed in 10 ml sterile PBS in which the coupon surfaces are scraped with a sterile implement to remove the attached bacteria. The samples are homogenized by vortexing for 15 sec.

Serial dilutions are prepared from the water and biofilm samples and homogenized by vortexing for 15 sec in PBS. Aliquots (0.1 ml) are then plated on R2A medium to quantify heterotrophic plate count (48 h at $37°C$) and chromogenic E. coli medium (CECM) (Oxoid, U.K.) agar for enumeration of the E. coli O157 (incubated for 24 hr at $37°$).

Identification of E. coli O157 and Assessment of Verotoxin Production

E. coli O157 is identified using morphological characteristics and the serotype confirmed with a latex agglutination test kit (Oxoid). Toxin production is also assessed using a VTEC reversed passive latex agglutination (RPLA) E. coli verotoxin test kit (Oxoid).

Results of Studies Involving E. coli O157

The results exhibited for E. coli O157 demonstrate that when grown in mixed culture, this water-borne human pathogen develops as part of a biofilm consortium (Table I). There were differences between the bacterial numbers recovered from

TABLE I
E. coli O157 (Human Isolate PS 14) Survival in Potable
Water Biofilms Formed on Different Materials at
Temperatures Typical of Cold and Warm Water Supplies[a]

Temperature	Glass	Copper	Stainless steel	Polybutylene
10	+	+	++	+++
20	+	+	++	+++
40	+	+/−	+	+++

[a] +/− denotes little or no colonization; + denotes light colonization; ++ denotes good colonization; and +++ denotes heavy colonization.

the biofilm, and this was dependent on the type of substratum on which the biofilm formed. Of the metallic plumbing materials, copper was fouled less than stainless steel, and the plastic polybutylene tubing was fouled more than either of the metals. Glass was used as a control and was found to have a similar fouling level to copper.

Persistence of *Campylobacter jejuni* in Water from Different Sources

Two systems are run in parallel for this series of experiments. The persistence of two *C. jejuni* strains is investigated in water from different sources (potable tap water, bore hole source, and poultry farm supply). The total native microbiota is recovered from these waters as described above and used as the inoculum in the two models. The seed vessels are maintained in batch for 21 days to allow formation of biofilm. Each vessel is supplied with fresh filter-sterilized water from the same original source that is fed continuously into the each vessel by a peristaltic pump. The flow rate of fresh water into the vessel is maintained at 12.5 ml hr^{-1} (dilution rate $= 0.025$ hr^{-1}, mean generation time $= 27.7$ hr) and the temperature at ambient.

The progress of biofilm development was determined by removing glass coupons and examining them by light microscopy and using a selection of oligonucleotide probes.[26] The viable cell count and autochthonous population in the seed vessel stabilise 30 days post inoculation and are then used to supply inocula to the experimental vessels.

Inoculation of the experimental vessels is achieved by flowing the effluent from the seed vessel to the experimental vessel for 7 days and then to the waste bottle. A continuous flow of fresh filter sterilized water is also then pumped to the experimental vessel. The vessel conditions are maintained aerobically at ambient temperature (nominally 20°).

Once established, the experimental vessels are allowed to equilibrate for 14 days before being challenged with cells of *C. jejuni*. These are harvested from overnight batch cultures. The cell densities of the washed suspensions are adjusted so that 10 ml of inoculum achieves a cell number in the experimental vessel of 2×10^7 cfu $> $ nl^{-1} (total count).

The persistence of the challenge strain in every case is determined by performing colony counts on serial 100-fold dilutions of the planktonic phase in sterile water. Each dilution (100 μl) is spread onto Skirrow selective plates and incubated microaerobically.[1,27,28] Persistence in the biofilms is also monitored

[26] R. I. Amann, W. Ludwig, and K. H. Schleifer, *Microbiol. Rev.* **59**, 143 (1995).
[27] M. B. Skirrow, *Brit. Med. J.* **2**, 9 (1977).
[28] R. M. Baird, J. E. L. Corry, and G. D. Curtis, *Internat. J. Food Microbiol.* **5**, 268 (1987).

TABLE II
SUMMARY OF RESULTS: USE OF GROUP-SPECIFIC OLIGONUCLEOTIDE PROBES AND PRESENCE
AND PERSISTENCE OF *C. jejuni* IN BORE-HOLE AND POULTRY HOUSE MAIN SUPPLY

Water source	Major differences between groups within the autochthonous populations	*C. jejuni* persistence
Bore-hole water	Beta group ~35% High GC ~20% Gamma ~20%	2 to 4 times longer in poultry water than in bore hole source, when analyzed by culture
Poultry house mains supply	Beta group ~17% High GC ~50% Gamma ~6%	Still detectable using rRNA probe long after culture negative

by fluorescence microscopy on glass coupons removed from the cultures using *Campylobacter*-specific rRNA probes.[24]

Results of Studies Involving *C. jejuni*

Persistence of *C. jejuni* was considerably longer in the poultry house water than in the bore hole source water (Table II). These results suggest that water quality is an important factor in the survival of this organism. Other experiments have been undertaken to study the effects of water quality[29] and other parameters that could be examined using this system. The presence of *C. jejuni* was detected in the biofilm using specific rRNA probes long after they were culture negative (Table II), perhaps indicating a viable but nonculturable status.[7] Analysis of the total autochthonous population with group-specific oligonucleotide probes found, not surprisingly, that the proportions of certain microbial groups differed in the two water types, although the total numbers remained very much the same. No specific association between *C. jejuni* and any one group could be detected. Alternative techniques, however, that retain the structure and morphology of the biofilm may be more appropriate for this type of study (see below).

Limitations of the Continuous Culture Bioreactor

The continuous culture bioreactor can be used for modeling aquatic systems as it can simulate natural aquatic biofilm establishment. The biofilm is allowed to develop and subsequently progress through to steady state, though this could be

[29] C. Buswell, Y. Herlihy, C. Keevil, P. Marsh, and S. Leach, *J. Appl. Microbiol.* **85,** 161S (1999).

said to be contradictory when biofilms are themselves known to be of a heterogeneous nature. Each experiment can be designed to simulate initial development to investigate early attachment, or to study established colonization. With the design as it is, materials have to be removed for examination. Removal could affect the biofilm and system in a number of ways: (i) disturbing the attached biofilm as the tile assembly is moved through the water; (ii) causing loss of biofilm as the tiles pass from the water to the air interface[30]; (iii) causing collapse of the biofilm as the water phase evaporates; and (iv) perturbing the homeostasis of the bioreactor when the biofilm population has been removed.

In studies of the persistence of pathogens in aquatic systems, however, detailed analysis of the intact spatial arrangement of the biofilm is not necessarily so important. For detailed structural analysis an on-line flow cell would be more suitable.

The bioreactor vessel itself has advantages in (i) generating a planktonic population that is stabilized and reproducible; (ii) generating a biofilm from the planktonic cells; (iii) allowing the biofilm and planktonic phases to be studied in the same environment; and (iv) enabling many physical and chemical changes to be made to determine the effects on the population under controlled conditions.

The model bioreactor demonstrated is adaptable and an important tool in the study of aquatic biofilms and the persistence and survival of pathogens within them.

Acknowledgments

The authors thank the Department of Education, Northern Ireland, the Department of Health, London, and the International Copper Association Ltd., New York, for financial assistance in this work. Drs. C. W. Keevil and A. Maule are thanked for their scientific expertise and contributions to the *E. coli* O157 sections of the work.

[30] H. J. Busscher, R. Bos, and H. C. van der Mei, *FEMS Microbiol. Lett.* **128**, 229 (1995).

[6] An Open Channel Flow Chamber for Characterizing Biofilm Formation on Biomaterial Surfaces

By YUEHUEI H. AN, JONATHAN B. MCGLOHORN, BRIAN K. BEDNARSKI, KYLIE L. MARTIN, and RICHARD J. FRIEDMAN

Following attachment to a surface, bacteria produce slime composed usually of polysaccharides and other extracellular material. As more and more bacteria adhere, a biofilm is formed. As a working community, the bacteria on the surface of the biofilm capture inorganic and organic molecules from the bulk liquid and make them available as nutrients to the cells within the biofilm. The bacteria within a biofilm work together to remove waste and toxins from the biofilm. As a unit, biofilm is quite successful; antibacterial agents, antibiotics, phagocytic white blood cells, and other chemical biocides are much less effective against bacteria within a biofilm as compared to planktonic bacteria. Biofilm-forming bacteria constitute over 75% of all infectious organisms associated with surgical implants. As a result of biofilm formation, approximately 1–3% of all implant patients experience severe infection following surgery. Biofilm may develop on an implant as late as 10 years following surgery, and the resulting infection quickly wreaks havoc. Infections often require the replacement of implants and can result in the loss of a limb or even the patient's life.[1]

Many methods are currently available to cultivate and study biofilms *in vitro.*[2] Perhaps the simplest was that developed by Christensen *et al.*[3] which tests for the presence of biofilm *in vitro*. A biofilm was considered present if staining showed a lining along the inner surface of the tube. Prosser *et al.*[4] developed another method for producing biofilms. *Escherichia coli* cells were incubated overnight on Mueller–Hinton agar plates, suspended, and dispensed onto catheter disks. The disks were incubated, washed, transferred to petri dishes containing nutrient broth, and incubated a second time, producing thick biofilms upon the catheter disks.

To gain more accurate models of the dynamics of biofilms, flow models have been developed and used with great success. The most popular of these flow models is the Robbins device.[5,6] It allows for multiple samples to be tested simultaneously.

[1] Y. H. An and R. J. Friedman, *J. Hosp. Infect.* **33**, 93 (1996).

[2] C. H. Sissons, L. Wong, and Y. H. An, *in* " Handbook of Bacterial Adhesion—Principles, Methods, and Applications " (Y. H. An and R. J. Friedman, eds.), p. 133. Humana Press, Totowa, NJ, 2000.

[3] G. D. Christensen, L. P. Barker, T. P. Mawhinney, L. M. Baddour, and W. A. Simpson, *Infect. Immun.* **58**, 2906 (1990).

[4] B. L. Prosser, D. Taylor, B. A. Dix, and R. Cleeland, *Antimicrob. Agents Chemother.* **31**, 1502 (1987).

[5] W. F. McCoy, J. D. Bryers, J. Robbins, and J. W. Costerton, *Can. J. Microbiol.* **27**, 910 (1981).

[6] A. Kharazmi, B. Giwercman, and N. Hoiby, *Methods Enzymol.* **310**, 207 (1999).

The Robbins flow chamber allows for any sample(s) to be removed and analyzed independently of other samples without upsetting the balance of the system. This enables different samples to be cultured for varying lengths of time. The samples are mounted on removable plugs that fit flush with the inside surface of a pipe containing fluid. The difficulties with the Robbins device are its high cost, lengthy assembly time, the development of boundary layers, and the potential problem of tearing of the biofilm layer when samples are being removed. In an effort to eliminate these problems, a simple open channel-based flow chamber was developed and a preliminary test was carried out to verify its efficacy for biofilm formation.

Description of Open Channel Flow Chamber

An open channel flow chamber was constructed from acrylic plastic as shown in Figs. 1 and 2. The assembled chamber consists of two circulating chambers connected by a platform (central chamber) on which the samples are placed. The circulating chambers promote turbulent flow, thus randomly mixing the medium. However, since the flow must be laminar across the platform, a baffle is inserted into each circulating chamber to contain the turbulent flow. The flow is guided around the baffle, losing much of its turbulence before arrival at the platform. After passing over the platform, the flow exits through the second circulating chamber. The two identical circulating chambers allow for the flow to be reversed while maintaining uniform conditions.

The platform is the most critical component of the flow chamber. Not only does it hold the samples, but it also promotes uniform flow. More importantly, the platform also produces uniform shear stress upon the samples enabling the samples to be compared. The platform is constructed with six channels (Fig. 2). The channels are constructed by machining five evenly spaced slots along the length of a sheet of acrylic plastic, $6 \times 13 \times 0.5$ cm. The slots are sized to hold a

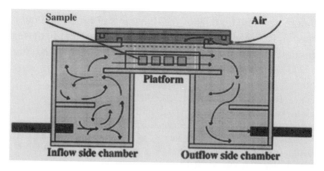

FIG. 1. Side view of the open channel flow chamber. Two circulating chambers are connected by a platform. A removable top is located above the platform so samples can be added or removed. The removable top allows ventilation to reach the system without contamination.

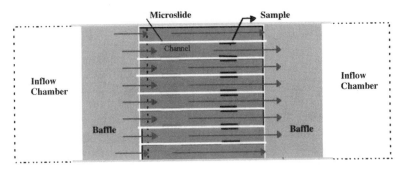

FIG. 2. Top view of platform. Arrows represent flow patterns over platform. Outer two channels are invalid for comparison because of slower flow velocity as a result of developed boundary layers.

microscope slide securely. Samples are attached to removable microslides. When all five microslides are in place, six channels are created. Only the center four channels are chosen to contain samples (see Fluid Dynamics section below).

After cutting, the individual parts of the chamber are sanded on a wheel grinder using 240-grit sandpaper to ensure a good fit. "Super glue" is used to fasten the parts together. Sealant (Household Glue and Seal, Silicon II, GE, Waterford, NY) applied along the edges and corners on both the inside and outside prevents leaks and provides support.

Fluid Dynamics

To examine the flow through the six channels, the Reynolds number (R) for parallel walls was calculated using Eq. (1):

$$R = Vd/v \qquad (1)$$

V is the mean velocity, d is the width of the channels, and v is the kinetic viscosity. In order for flow to be laminar, or layered, R should be less than 1000. The calculated value for R inside the channel is 145, indicating that flow is indeed laminar. To confirm flow patterns, a series of dye tests, using methylene blue, has been conducted. The tests showed that flow entering the platform is nonturbulent. As flow diverges among the channels, it is laminar and remains so throughout the length of each channel (Fig. 2).

Just as critical as laminar flow is the flow rate through each channel. For samples from different channels to be compared, they must have the same exposure to flow. Thus, the same amount of fluid must flow through each channel at the same flow rate. Hindering this requirement is the creation of a boundary layer along the walls of the platform. Boundary layers occur anywhere flow is laminar or only slightly turbulent. A boundary layer occurs due to a flow rate of zero at a surface containing

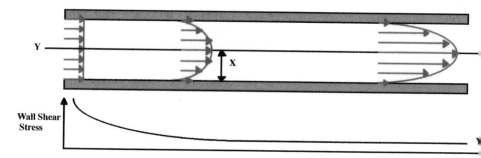

Fig. 3. Two-dimensional top view of a channel. Arrows indicate flow velocities. Because of the effects of boundary layers, the flow profile changes from uniform at the entrance of the channel to parabolic. The bottom diagram represents the shear stress experienced by the walls.

fluid flow (Fig. 3). As the distance, X, increases away from the surface, flow velocity (v) increases. As a result of the boundary layer developing along the walls, the two outer channels have low flow velocity and have been determined to be invalid for comparison purposes, so samples are placed only in the four inner channels. Boundary layers also develop on the bottom surface of the platform, but since all channels have equal exposure to this boundary layer, it has relatively little effect. Dye tests indicate that entering flow has equal velocity in these four channels. Since all channels are constructed with the same dimensions and materials, the flow velocities remain equal for all channels at the same distance from the entrance of the channel. However, the laminar nature of flow in the channels encourages the development of boundary layers. This has proved not to be a problem unless samples are placed at varying distances from the entrance of the channel, as with loading a channel with multiple samples. When this is done, each channel should be set up symmetrically with samples. Flow can be periodically reversed to expose each sample to the same flow conditions.

Circulating Flow System

Using a peristaltic pump, two 1000 ml Erlenmeyer flasks, 9.5 mm plastic tubing, four 3-way stopcocks, plastic inserts, and two #9 rubber stoppers, a circulating flow system was designed (see Figs. 1 and 4). Since plastic tubing inserted into the rubber stoppers deformed during autoclaving, plastic inserts were joined to the tubing and passed through the stoppers instead. The flow is directed clockwise by opening valves 1 and 2 to flask A and closing valves 3 and 4 to flask B. To reverse the flow, the pump is turned off, valves 1 and 2 to flask A are closed, valves 3 and 4 to flask B are opened, the direction of pump flow is reversed, and the pump restarted.

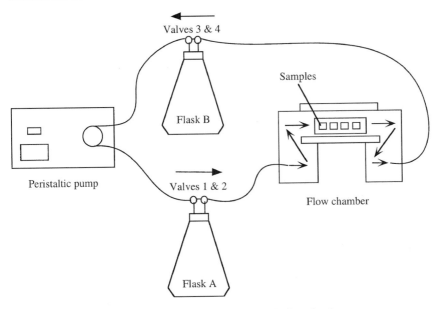

FIG. 4. Reversible flow system used with the flow chamber.

Experimentation

Bacterium, Growth Medium and Sample Surface

Staphylococcus epidermidis (RP-62A) was obtained from American Type Cell Culture (ATCC-35984). This strain was chosen because it has been characterized as a strong slime producer.[7] A recent study of aortofemoral graft infections reported that 18 of 34 (66%) patients had to be readmitted more than 4 months after initial surgery because of *Staphylococcus epidermidis* infection.[7]

One colony of RP-62A taken from an agar plate is added to a test tube containing 5 ml of tryptic soy broth. The tube is placed on a laboratory rotator (Glas-Col, Terre Haute, IN), rotated at a speed of 10 rpm, and incubated for 18 hr at 37°. Five milliliters of the overnight bacterial culture is added to 900 ml fresh culture medium. This mixture is used for daily replacement of the fluid in the circulating flask.

Growth medium used in the circulating system is prepared by adding 100 ml of tryptic soy broth to 900 ml of PBS. To this solution 70 g per liter of anhydrous dextrose is added. The mixture is autoclaved for 15 min at 122°C.

Titanium samples (10×15×1 mm) are grit-blasted with 25 μm aluminum oxide grit using a Micro Pencil Air Abrasive Unit (Crystal Mark, Glendale, CA).

---

[7] G. D. Christensen, W. A. Simpson, A. L. Bisno, and E. H. Beachey, *Infect. Immun.* **37**, 318 (1982).

The samples are treated using a standard passivation method (ASTM Standard F86-76, 1997), dried with nitrogen gas, and then attached to glass microscope slides using double-sided tape. The slides with adhered titanium plates are autoclaved for 15 min at 122° and dried for another 15 min. The slides containing the samples are placed into the slots on the platform, thus creating six channels, four of which are used for samples (see Fluid Dynamics above).

Experimental Procedure and Evaluation

The complete system is placed into an incubator and filled with 1600 ml of medium prewarmed to 37°. Flask A is filled with 900 ml of medium and valves 1 and 2 are opened; flask B is left empty. The system is allowed to circulate before addition of 10 ml of overnight bacterial culture. The system runs for 10 days at 37°. Each day flask A is replaced with a new flask containing 900 ml of new medium warmed to 37°. On the 10th day, the system is flushed with 4 liters of PBS. This is done by adding 1 liter of PBS (phosphate buffered saline) to flask A every 5 min, washing out most of the unattached bacteria. The titanium samples with attached biofilm are gathered using forceps and are prepared for evaluation.

Methods of Evaluation

Upon completion of the 10-day time period, the slides containing the samples are removed from the chamber and prepared for SEM and confocal microscopy as described below. Prior to preparation for the various analytical procedures, biofilm growth is readily apparent with the naked eye.

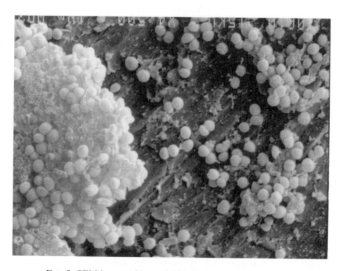

FIG. 5. SEM image of bacterial biofilm on titanium surface.

For scanning electron microscopy (SEM), samples are fixed in 2 ml of 3% buffered glutaraldehyde solution for 2 hr. Samples are washed three times with 0.2 M cacodylate buffer for a total of 15 min. The buffer is aspirated and 2 ml of 2% osmium tetroxide is added to postfix the samples. This is followed by dehydration using a graduated series of ethanol. The samples were stored in 100% ethanol. The samples are critical-point dried, mounted on metal stubs, and sputter-coated with gold for conductivity. The specimens are then viewed in a scanning electron microscope (JEOL 350, JEOL Ltd., Tokyo).

Confocal microscopy (MRC1000, Bio-Rad, Hercules, CA) has also been used to examine the growth of biofilm on the samples. Samples are removed from slides and placed in culture plate wells containing 2 ml of 70% ethanol for 30 min. The ethanol is aspirated and 2 ml of 0.05 mg/ml propidium iodide in PBS is added to each well. Following 5 min staining, each sample is washed twice with PBS and finally stored in 3 ml of PBS until viewing. To prevent drying of the biofilm while viewing, samples are placed into a small well containing PBS.

SEM provides a clear view of the extent to which biofilm has coated the titanium samples (Fig. 5). The SEM image demonstrated that the surfaces of the titanium specimens are randomly covered with scattered bacteria among numerous areas of biofilm growth. This confirms the initial visual observation of biofilm on the samples.

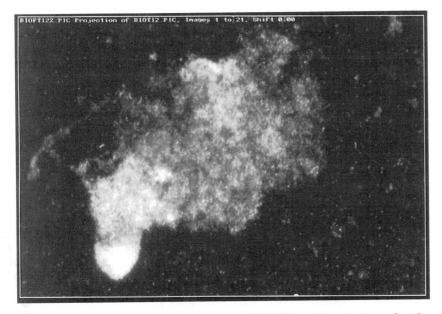

FIG. 6. Confocal laser scanning microscopy image of a biofilm patch on titanium surface (low power).

Confocal microscopy provides a unique and reliable means for measuring the thickness of the biofilm on the surface of the opaque metal. The bacteria again are seen to be covering the surface of the titanium specimens as indicated by numerous large patches of biofilm growth (Fig. 6). The thickness of random biofilm clumps is measured using a Z-series. The titanium samples yield biofilm growth with thicknesses ranging from 25 to 75 μm. Furthermore, confocal microscopy provides another means of viewing the extent to which the bacteria have colonized the surface of the metal.

With the results and images gained from the use of both SEM and confocal microscopic techniques, it has been concluded that the new continuous flow chamber is an efficient system for use in studying biofilm growth.

Conclusion

In the realm of bacterial adhesion research there are many flow chambers that have been designed to culture biofilm (Table I). Similar to the open channel flow chamber, both the tubular Robbins device[5,6] and the rotating annular reactor[8] enable individual samples to be removed and analyzed without disturbing the other specimens in the system. Furthermore, the rotating cylinder in the rotating annular reactor can be rotated at any speed, thus allowing control over the shear stress present in the chamber.[8] With their superior designs, these two devices have set the standard by which the open channel flow chamber should be measured. As mentioned above, the new system also has the ability to sustain multiple samples during each experimental run. In the Robbins device the samples are positioned such that they are flush with the wall of the chamber; thus, when the samples are removed there is a significant risk that the biofilm might be torn. In the new chamber, the samples are adhered to the surfaces of the parallel plates. In removing a sample, the entire surface is removed rather than just the sample itself and the biofilm is preserved. Another problem with both the Robbins device and the rotating annular reactor is their complexity. Assembly can be time consuming and expense can be significant. The new chamber overcomes these problems through a simple design based on readily available and inexpensive materials.

For the sole purpose of studying the growth of biofilm in real time, the microchambers designed by Van Pett et al.[9] and Peyton and Characklis[10] have been very successful. Furthermore they come in a range of complexities and cost. However, they have one major drawback: the limited materials that can be used in a real-time study. For studying the process of biofilm growth on transparent materials they are sufficient, but for studying other materials they are inadequate.

[8] M. G. Trulear and W. G. Characklis, *J. WPCF* **54**, 1288 (1982).

[9] K. Van Pett, D. J. Schurman, and R. L. Smith, *J. Orthop. Res.* **8**, 321 (1990).

[10] B. M. Peyton and W. G. Characklis, *Bioprocess Technol.* **20**, 187 (1995).

TABLE I
SELECTED FLOW CHAMBERS DESIGNED TO CULTURE BACTERIAL BIOFILM

Chambers	Author, year[a]	Estimated cost	Assembly time	Specimen number	Simplicity of flow system
Rotating annual reactor	M. G. Trulear et al., 1982[1]	Moderate	Moderate	Few(2–15)	Moderate
Robbins device	W. F. McCoy et al., 1981[2,3]	Expensive	Long	Many (15+)	Complicated
Microchambers	D. E. Caldwell et al., 1986[4]	Inexpensive	Short	One	Simple
	D. Kaplan et al., 1995[5]	Expensive	Moderate	One	Moderate
Parallel plate flow chambers (laminar flow)	K. Pedersen 1982[6]	Inexpensive	Moderate	Many (15+)	Moderate
	J. Ruel et al., 1995[7]	Moderate	Long	One	Moderate
	M. S. Zinn et al., 1999[8]	Expensive	Long	Five	Complicated
Open channel flow chamber (laminar flow)	Y. H. An et al., 2000 (This chapter)	Inexpensive	Short	Many (15+)	Simple

[a] Key for References: (1) Trulear, M. G., and Characklis, W. G., *J. WPCF* **54,** 1288 (1982); (2) McCoy, W. F., Bryers, J. D., Robbins, J., and Costerton, J. W., *Can. J. Microbiol.* **27,** 910 (1981); (3) Kharazmi, A., Giwercman, B., and Hoiby, N., *Methods Enzymol.* **310,** 207 (1999); (4) Caldwell, D. E., and Lawrence, J. R., *Microb. Ecol.* **12,** 299 (1986); (5) Kaplan, D. P. B., Sullivan, J., and Zimmerberg, J., *J. Microscopy* **181,** 286 (1995); (6) Pedersen, K., *Appl. Environ Microbiol.,* **43,** 6 (1982); (7) Ruel, J., Lemay, J., Dumas, G., *et al., Asaio J.* **41,** 876 (1995); (8) Zinn, M. S., Kirkegaard, R. D., Palmer, R. J., Jr., and White, D. C., *Methods Enzymol.* **310,** 224 (1999).

Having been designed for light microscopy, these chambers are limited to use with transparent materials. Most biomaterials are opaque and therefore cannot be examined using a traditional light microscope. Some of these chambers are very cost-effective and in fact very simple to prepare, but they are inadequate for the study of many biomaterial samples at one time.

It is this desire to study numerous specimens simultaneously that has popularized such chambers as the parallel plate chamber designed by Pedersen,[11] the laminar flow chamber by Zinn et al.,[12] the rotating annular reactor,[8] and the Robbins device.[5,6] All these designs provide for multiple inserts, giving the capability of testing many samples in one run under identical environmental conditions. The device reported by Zinn et al.[12] can hold five samples and facilitates real time observation. The Pedersen flow chamber[11] is most similar to the open channel

[11] K. Pedersen, *Appl. Environ. Microbiol.* **43,** 6 (1982).
[12] M. S. Zinn, R. D. Kirkegaard, R. J. Palmer, Jr., and D. C. White, *Methods Enzymol.* **310,** 224 (1999).

design. Both are based on a parallel plate design featuring controlled boundary layers and a laminar flow of liquid, but they differ greatly in several respects. The most important difference is in the ability to access samples without disturbing the system. In the open channel flow system, the specimens are more readily accessible and easier to remove without disturbing the remainder of the system. In the Pedersen system, removal of even one sample will require displacing the other samples being studied. In addition to eliminating this problem of sample handling, the new open channel flow device has a less complicated design and is easier to construct.

In conclusion, the open channel continuous flow chamber has met the standards set by the devices that have come before it. It has also proven to be an efficient means of producing and culturing biofilm on titanium surfaces. Plans have been made to use the chamber in another study comparing the growth of biofilm on various biomaterials.

[7] Biofilms in Flowing Systems

By T. Reg. Bott and Deon M. Grant

Introduction

Biofouling in flowing systems assumes particular importance in industrial operations involving water, e.g., cooling water circuits and paper manufacture. Many industries use cooling water as a utility for the removal of unwanted heat from liquid or gaseous products or by products or to control the temperature of chemical reactions. In most examples the bulk of the heat is recovered for reuse in suitable heat exchange between streams, but the final "trim" cooling is only achieved by the use of cooling water, to reduce the temperature of the process fluid to its desired temperature. The effectiveness of the cooling very much depends on the extent of the accumulation of deposits, including biofilms, on the heat exchanger surfaces. In some applications, however, for instance in electrical power generation, the cooling water plays a vital role in the efficiency of the turbine generator set. In order to recover the maximum energy from the steam passing through the turbine, it is essential to maintain the pressure in the steam condensers at as low a level as possible. The efficiency of the steam condensers in turn is related to the effectiveness of the cooling water system. The presence of a biofilm can seriously affect the performance of the steam condensers.

In paper manufacture the existence of biofilms in the system can seriously impair the efficiency of operation. Partial blockage of pipelines is a common

METHODS IN ENZYMOLOGY, VOL. 337
0076-6879/00 $35.00

problem and biofilm displaced from surfaces can affect the quality of the final product.

Furthermore, in many industrial water systems the presence of a biofilm can promote the corrosion of the metal surfaces on which it resides. The integrity of the surface can be affected by the microorganisms themselves, or the conditions that develop beneath the biofilm.

In summary, the presence of a biofilm in industrial water systems can lead to

1. Reduced flow or, for a given flow rate increased energy use to overcome the higher back pressure
2. Reduced heat transfer in coolers
3. Increased corrosion
4. Increased maintenance, including cleaning of equipment
5. Loss of production due to enforced reductions in throughput or cleaning operations
6. Necessary cleaning operations that can lead to supplementary problems such as the safe disposal of effluent

It will be apparent that all the effects involve extra cost and in turn affect the profitability of the enterprise.

The presence of biofilms in some industrial operations, however, is not a problem and is necessary for the process involved. For instance, the use of biofilms on surfaces can be a method for the treatment of effluents by the removal of toxic or other components, before discharge. The use of trickle-bed bioreactors for the removal of volatile organic compounds from a gaseous discharge is an example. In general, however, the velocities in the processes are very much lower than those experienced in cooling water circuits.

Basic Concepts

In order to thrive and develop, a biofilm requires nutrients including oxygen if it is aerobic, in addition to suitable temperature conditions. The availability of nutrients will depend on mass transfer to the surface. In static conditions it may be visualized that the availability of nutrients is dependent on diffusion (Brownian motion), but where the water is flowing the transfer of material from the bulk water can be assisted by the movement of the water itself. In conditions considered as streamline or laminar, where the velocities are relatively low, there is little influence on mass transfer; arrival of nutrients at the biofilm surface is, as with static conditions, due to diffusion from the bulk water. For diffusion to occur it is necessary to have a concentration "driving force," i.e., a concentration difference between the bulk and the biofilm. Consumption of nutrients by the developing biofilm ensures that the driving force is maintained. The situation is illustrated

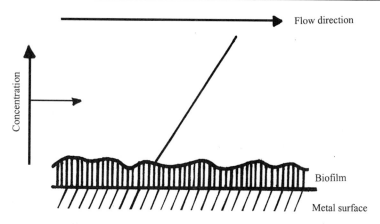

FIG. 1. Nutrient distribution under streamline conditions.

by Fig. 1. As the water flows through the system and nutrient is consumed, the nutrient concentration in the bulk water will fall so that the concentration driving force at the outlet will be less than at the inlet.

As velocity is increased the flow will eventually become turbulent. Under these conditions, although the bulk of the water is turbulent, i.e., well mixed, the flow near the wall remains laminar. The reason is that the viscous drag imposed by the wall reduces the velocity in that region. Nutrient flow across this viscous sublayer, as it is generally known, is by the same mechanism as for laminar flow. The laminar sublayer offers a resistance to mass transfer. The resistance to mass transfer in the bulk is very low since the eddy motion of the turbulence has the effect of keeping the components well mixed. Figure 2 illustrates the prevailing situation. Although a dotted line is shown on Fig. 2 to indicate the extent of the laminar sublayer, this boundary is not rigid, i.e., there is not a sudden change from turbulent to laminar flow conditions, but rather a more gradual change, albeit over a short distance. As the velocity is increased, the thickness of the laminar sublayer decreases and hence the resistance to mass transfer decreases. Whatever the velocity, however, a laminar sublayer remains at the wall.

The criterion for determining whether or not flow is turbulent or streamline is the so-called Reynolds number. The Reynolds number is a dimensionless ratio of the momentum forces to the viscous forces.

$$\text{Reynolds number (Re)} = \frac{Lv\rho}{\mu} \tag{1}$$

where v is the fluid velocity, ρ is the fluid density, μ is the fluid viscosity, and L is a characteristic dimension of the system related to the geometry; for flow through a tube, it is the inside diameter of the tube.

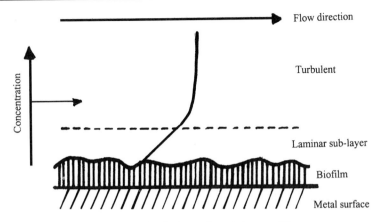

FIG. 2. Nutrient distribution under turbulent conditions.

The change from laminar to turbulent flow is considered to take place at Re = 2000, although this is not a precise demarcation; the change will be gradual in the range Re = 2000–4000. At the latter value the flow will be fully turbulent.

It is clear that for a given system, i.e., water, and a given geometry such as flow through a pipe, the crucial variable is flow velocity. In general terms in the turbulent region, the higher the velocity, the greater the mass transfer for a given bulk concentration.

Apart from the mass transport of nutrients, this has implications for the colonization of the surface, although the situation is somewhat more complex. Some microorganisms actively seek a surface through chemotaxis, but, at least in part, the colonization of a surface by microorganisms depends on enhanced mass transfer of cells across the laminar sublayer when the flow is turbulent. The principles of mass transfer apply.

Transfer across the laminar sublayer to the solid/liquid interface may also occur if the particles have sufficient momentum that they are able to "coast" through the viscous sublayer. The momentum is the product of the particle mass and its velocity. If the turbulence is such that a relatively high velocity is imparted to a particle, with a suitable trajectory, the particle will reach the solid surface. If the momentum is insufficient, the drag forces imposed on the particle by the fluid will prevent its reaching the surface, but it will continue to move with Brownian motion.

The velocity of a fluid across a surface produces a shear stress at the wall, or in simple terms a force in the direction of flow.

At a distance t from the surface where the fluid velocity is v, the shear stress τ is given by Eq. (2):

$$\tau = \mu \frac{v}{t} \qquad (2)$$

A is the area of the surface; μ is the fluid viscosity; τ is defined as the force F acting at the distance t from the surface, divided by area A, i.e.,

$$\tau = \frac{F}{A} \tag{3}$$

The magnitude of the force has implications for the morphology and the viability of any biofilm at the surface. It will be apparent from Eq. (2) that the greater the velocity, the greater the shear stress acting on the biofilm, with the distinct possibility that if the velocity is high enough, sloughing of the biofilm will occur. The extent of the removal of the biofilm will depend not only on the magnitude of the imposed velocity, but on the morphology of the biofilm. In response to the effects of increased velocity, the biofilm becomes more compact, i.e., more resistant to the removal forces.

The effects of velocity have implications for the control of biofilms in industrial equipment, not only in terms of mass transfer of biocides to the biofilm, but in terms of the application of removal forces and to a much lesser extent the effect it has on biofilm temperature.

The whole spectrum of microorganisms can exist in industrial water systems, but the biofilms that develop under flowing conditions are generally bacteria. Algae can exist where there is sunlight, but, in general, the associated water flow rate is very low, for example, in the basin of a large natural draft cooling tower. The effects of velocity therefore are negligible. Fungi can exist in industrial systems, but the opportunities for extensive growth are limited because the substrate, in general, is not consumable.

The discussion in the remainder of this contribution is exemplified by reference to biofilms consisting of bacteria, since these are common in industrial water systems.

Experimental Method

In research into the development of biofilms in flowing systems, it is essential that experiments be focused, so far as is possible, on simulating the industrial conditions to which the results of the research are to be applied. Laboratory studies therefore must be on a sufficiently large scale to provide reliable data with the maximum flexibility of operation. In general the aim of the work is to study mitigation techniques in preparation for plant-scale trials. The laboratory pilot trials reduce the extent of trial and error that would otherwise be necessary on the plant itself. In this way costs are reduced and the occurrence of problems resulting from inexperience and lack of fundamental knowledge is eliminated. A technique that has proved successful in this regard is based on the chemostat.

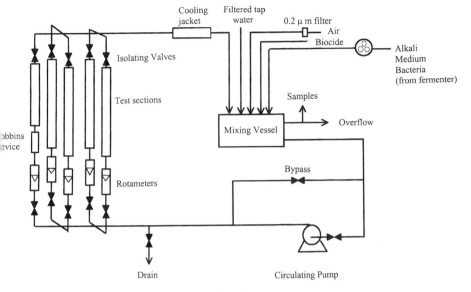

FIG. 3. Flow sheet of experimental apparatus.

A simple flow sheet is presented in Fig. 3. It is a feed and bleed system. The equipment can operate continuously for long periods up to several weeks, and this is essential if the necessary long-term data are to be obtained. The water used in the tests can originate from several sources; the choice depends to some extent on the tests to be carried out. Filtered town water, distilled water, or deionized water have all been used. Because of the large volumes of water used in the experiments, it will be necessary to ensure that the water supply is adequate. By far the simplest approach is the use of filtered tap water, which, if the pore size of the final filter is small, precludes contamination of the system. A carbon filter is necessary to remove contaminants and, in particular, the chlorine used in town water for disinfection. Furthermore, the use of tap water does, to a greater or lesser extent, simulate the water that is likely to be used in industrial equipment. The use of distilled or deionized water is likely to extend the conditioning time before the biofilm begins to develop.

A centrifugal pump is used to draw off the fouling solution from the mixing vessel and to recirculate the water through the system. After the pump, the flow is split so that it can flow through the vertical tubular test sections. The test sections are mounted vertically in order to eliminate the effects of gravity on the development of the biofilm in the tubes. The test sections are generally tubular in order to simulate industrial conditions, particularly tubular heat exchangers, but they could be made to have any cross-section to suit the requirements of the study. The tubes themselves may be made of any material. Glass, although not a usual industrial material of construction, is satisfactory for comparison purposes, allowing visible inspection of the biofilm to be made, and it does provide the opportunity to use a nonintrusive infrared monitoring device for continuous assessment of biofilm development.

The flow through the test sections is carefully controlled by a suitable valve and measured by use of variable orifice flow meters (rotameters). Tests at different flow velocities may be carried out, and with sufficient tubes replicate tests may be made.

Variables that can influence biofilm growth are carefully monitored and controlled and include microbial concentration, nutrient level, pH, and temperature. A dissolved oxygen probe indicates the saturation level of oxygen in the mixing vessel. The circulating water is made up of separate streams fed at the required rates into the mixing vessel, with each stream having a specified composition where appropriate. Filtered air is also sparged into the mixing vessel when aerobic bacteria are under test. The measurement and control systems associated with the mixing vessel are schematically illustrated in Fig. 4.

Nutrient Stream

Sterile nutrient, identical to that used for the continuous cultivation of microorganisms in the fermenter, is fed into the mixing vessel by a peristaltic pump. Sterile

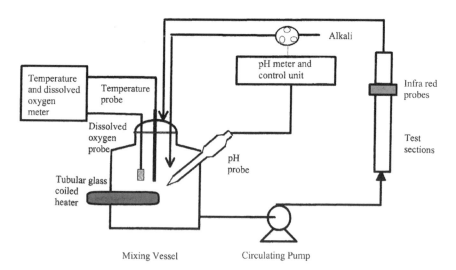

FIG. 4. The mixing vessel layout.

air is blown into the nutrient line through a glass medium break to prevent back growth of bacteria.

The carbon source in the nutrient stream is autoclaved separately to prevent browning reactions with other components of the medium.

Alkali Stream

To maintain the pH at the desired level, usually pH 7, potassium hydroxide (2.0 M) is automatically pumped by peristaltic pump into the mixing vessel in response to the pH monitor/controller.

Air Supply

The air from a compressor is filtered using a two-stage process. The first stage removes oil, water, and particulate matter down to 10 μm. The second stage consists of a carbon filter to remove volatile contaminants and odours, and a 0.2 μm pure Whatman PTFE filter is placed in the line to prevent microbial contamination.

Bacterial Supply Stream

A suspension of cells of known density is taken from the fermenter at a known rate, to provide the desired contamination level in the circulating water.

Biocide Stream

If a biocide is under test, the biocide is added to the mixing vessel at a rate designed to give the desired concentration in the circulating water.

Control Tests

It is usual when a biofouling control technique is being investigated, i.e., the use of biocides or biodispersants, to have a separate control circuit identical to the test system, but without the chemical additive. The two systems are fed from the same fermenter in order to allow a direct comparison between the biofilm accumulation with and without the additive.

Fermenter

Essential to the whole assessment procedure is the fermenter. A schematic figure of the layout of the fermenter circuit is given in Fig. 5.

During continuous fermentation, the pH is controlled and maintained at neutrality by automated addition of 2 M potassium hydroxide as described in relation to the mixing vessel. A constant working volume is maintained with a weir-type overflow. The temperature may be set at the optimum growth temperature for the bacterium under test, using a specially designed heating coil.

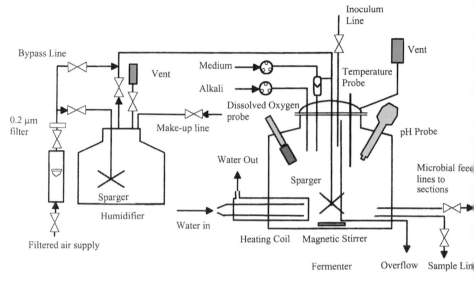

FIG. 5. The fermenter circuit.

Sterile air, produced as for the mixing vessel, is introduced through a sparger, via a sterile humidifier to prevent water loss from the fermenter.

Assessment of cell growth in the fermenter is obtained from viable cell counts, glucose level, and optical density.

Estimation of Biofilm Accumulation on the Test Sections

A simple technique that may be used for assessment of biofilm accumulation is by weighing the test sections before and after the trial. It is necessary to adopt a standard test procedure in relation to the draining of the tube before weighing. It is also important to reduce the flow through the test section slowly and carefully before removal from the circuit for weighing, in order to prevent biofilm disruption and removal from the tube. Weighing is a tedious technique, and by its very nature it does not allow a continuous assessment of biofilm accumulation. It may be the only way possible, however, where metals and other materials are used for the construction of the test sections.

A technique that has been developed at Birmingham University[1] is to use infrared absorbance, but this requires a transparent tube for the test section, e.g., glass. The technique is nonintrusive and continuous, and it is possible to measure the biofilm development over the length of the test section. The technique is particularly useful for comparing the growth of biofilms under different conditions.

[1] G. Bartlett, R. Santos, T. R. Bott, and D. Grant, *in* "Understanding Heat Exchanger Fouling and Its Mitigation" (T. R. Bott, ed.), p. 337. Begell House Inc., New York, 1999.

The device consists of an emitter and sensor for transmitted infrared radiation clamped onto the test section. The amount of radiation absorbed by the biofilm is a measure of biofilm accumulation. A correlation may be obtained by comparing the biofilm thickness obtained from the weighing method with the infrared absorbance. Biofilms grown under different flow conditions show different structures and densities so that it is necessary to correlate only data obtained under similar conditions. The absorbance readings are related to two layers of biofilm since the radiation is across a diameter.

The absorbance A, may be obtained from Eq. (4):

$$A = \log_{10} \left\{ \frac{I_o}{I_o - I_I} \right\} \qquad (4)$$

where I_o is the intensity of emitted radiation; I_I is the gain difference of intensity of infrared radiation.

An alternative device[2] that has a direct bearing on the effect of biofouling on heat transfer is to measure the changes in heat transfer over a period of time in response to the accumulation of biofilm. The method involves passing the contaminated water through an electrically heated tube. The increased tube wall accumulation of biofilm reduces the heat transfer, or for a given heat flux the wall temperature increases. The resistance to heat transfer is a direct measure of the biofilm thickness. In more precise terms it is possible to calculate the resistance to heat transfer from the electrical power input, the water temperature, and the wall temperature. If the assumption is made that the thermal conductivity of the biofilm is similar to that of water, it is possible to estimate the mean thickness of the biofilm on the tube wall.

The method would simply replace the glass test sections associated with the infrared device, and it does allow for the use of different materials of construction. The manufacture of the test sections to provide accurate measurements of wall temperature could be costly, particularly if a series is required to investigate simultaneously a number of different flow conditions.

A long established technique for biofilm accumulation is the use of the so-called Robbins device, in which removable studs are located in the test section. The method involves the removal of the individual studs; then, either by weighing or biofilm removal, the amount of biofilm deposited may be assessed. The technique may not give accurate data for several reasons:

1. The surface of the stud exposed to the microbial stream may not be flush with the test section, so that unknown hydrodynamic forces are imposed on the sample area, thereby modifying the accumulation (see Basic Concepts).

---

[2] J. M. Howarth, N. F. Glen, and A. M. Jenkins, in "Understanding Heat Exchanger Fouling and Its Mitigation" (T. R. Bott, ed.), p. 309. Begell House Inc., New York, 1999.

TABLE I
COMPOSITION OF MEDIUM USED FOR THE CULTURE OF *Pseudomonas fluorescens*

Mineral salts	g/liter	Trace elements	g/liter	Glucose[a]	g/liter
$NaH_2PO_4 \cdot 2H_2O$	1.01	$MnSO_4 \cdot 4H_2O$	13.3×10^{-3}	$C_6H_{12}O_6$	5
Na_2HPO_4	5.5	H_3BO_3	3.0×10^{-3}		
K_2SO_4	1.75	$ZnSO_4 \cdot 7H_2O$	2.0×10^{-3}		
$MgSO_4 \cdot 7H_2O$	0.1	$Na_2MoO_4 \cdot 2H_2O$	0.24×10^{-3}		
$Na_2EDTA \cdot 2H_2O$	0.83	$CuSO_4 \cdot 5H_2O$	0.025×10^{-3}		
NH_4Cl	3.82	$CoCl_2 \cdot 6H_2O$	0.024×10^{-3}		

[a] The glucose was sterilized separately from the remainder of the medium, in order to prevent browning reactions.

2. Loss (or even addition) of biofilm may occur during the removal of the stud from the apparatus.
3. The amount of biofilm on the stud is small so that inaccuracies in weighing may occur.

Example of the Method

In order to illustrate the procedure for the examination of biofilm growth in flowing systems, a typical experimental apparatus and methodology will be presented. It involves the growth of a monoculture of *Pseudomonas fluorescens* on glass test sections.[3] *Pseudomonas* species is a known slime former and is therefore very relevant to biofilm formation in cooling water systems.

Preparation of the Fermenter

A pure culture of *Pseudomonas fluorescens* from freeze-dried store was grown and stored short term at 4° on Oxoid nutrient agar slopes. The culture was streaked across agar plates and grown at a temperature of 25° for approximately 24 hr. Shake flasks containing 200 ml sterile mineral medium and glucose, the composition of which is given in Table I, were inoculated aseptically with the organisms developed on the agar plates.

After approximately 48 hr of batch growth, the flasks of bacteria, making an inoculum of 10% fermenter volume, were added to the presterilized fermenter, previously filled with sterile medium, which had been thoroughly aerated for roughly 24 hr. A glass fermenter of 5.5 liter capacity was used. The fermenter was steam sterilized at a pressure of 15 lb/in.^{2} and a temperature of 121° in an autoclave. The

[3] D. M. Grant, "Biofilm Control through Optimised Biocide Dosing," Ph.D. Thesis, University of Birmingham, 1998.

TABLE II
EXPERIMENTAL CONDITIONS OF THE FERMENTER

Parameters	
Volume	5.51 liter
Residence time	50 hr
Dilution rate	0.2 hr^{-1}
Medium input	1.86 ml/min
Dissolved oxygen	35–55% saturation at the fermenter temperature
Air flow rate	5.5 liter/min
pH	7
Temperature	27°
Density of cells	1×10^9 cells/ml

bacteria were allowed to grow as batch for approximately 48 hr and the exponential growth prolonged by gradual addition of fresh medium. The dilution rate was 0.02 hr^{-1}, which gave a residence time of 50 hr.

Once the fermenter had reached steady state, after approximately 8 days, the cell density remained constant and glucose concentration was reduced to between 2 and 4 mg/liter. The bacteria were grown under glucose limiting conditions ready for use in the biofouling experiments. The operating conditions of the fermenter are summarized in Table II.

The conditions were controlled as previously described. The pH and dissolved oxygen electrodes were calibrated prior to their use. The oxygen electrode gave percentage readings compared to saturation. For instance, a reading of 100% corresponded to the maximum saturation value of dissolved oxygen in distilled water. The dissolved oxygen probe was sterilized in place with the fermenter.

Calibration of the pH probe was performed using pH standard solutions at pH 4.0, 7.0, and 10.0. The pH probe was sterilized by separately autoclaving at 121° and 15 lb/in.^{2} for 30 min. To eliminate pH drift, samples of the fermentation broth were regularly taken and pH measured using a freshly calibrated pH probe.

Data on bacterial growth throughout the experiment in the fermenter were obtained from viable cell counts, glucose level, and optical density.

The Circulating System

The circulating system total volume was 12 liters, with a mixing vessel of 5 liters. Four vertical glass test sections were included in the system with diameters 23.5 and 9.0 mm. The operating conditions are given in Table III.

Before the start of an experiment, the equipment was steam sterilized at atmospheric pressure for at least 24 hr; the supply pressure of steam was 18 lb/in.^{2}.

TABLE III
OPERATING CONDITIONS OF THE APPARATUS

Test section no.	1	2	3	4
Flow velocity (m/sec)	1.3	1.3	0.5	0.5
Volumetric flow rate (liter/min)	4.96	4.96	12.57	12.57
Reynolds no.	11,700	11,700	11,750	11,750
Residence time	60 min			
Temperature	29–33°			
pH	7			
Filtered tap water flow rate	12 liters/hr			
Medium input	0.48 ml/min			
Bacterial input	1.83 ml/min			
Initial (glucose)	12 mg/liter			
Viable cell no. (cells/ml)	2×10^7 cells/ml			

The supply lines, i.e., silicone tubing for medium, bacteria, and alkali transport as well as the pH probe were sterilized separately in an autoclave. The pH probe was calibrated prior to sterilization by use of buffer solutions at pH values of 4, 7, and 10.

On cooling, the pH probe was inserted aseptically, the circuit was filled with filtered tap water, sterile air and nutrient solution were added, and pH was maintained at 7. The fluid velocities in the test sections were adjusted to the desired values and the infrared probes placed on the test sections and set to zero. The bacteria were added and glucose consumption and infrared signals were assessed and recorded as the experiment progressed.

In addition to the assessment of biofilm accumulation on the test sections, it was possible to carry out other tests on samples of biofilm to determine carbohydrate and protein levels.

After the experiment, the equipment was cleaned by circulating 12% chlorine solution produced from sodium hypochlorite. To ensure complete removal and inactivation of all biological matter, the sodium hypochlorite solution was added continuously to the apparatus over a period of 20 hr at levels sufficient to maintain a free chlorine level of at least 5 mg/liter. The addition of nutrients, air, and bacteria from the fermenter was terminated during the cleaning procedure.

Results of the Test

Figure 6 plots infrared absorbance against time for the two different velocities of 1.3 and 0.5 m/sec. The effect of velocity on biofilm accumulation is clearly demonstrated. The fall in biofilm accumulation on day 10 at 0.5 m/sec water velocity is due to the random sloughing of the biofilm. Biofilms formed at low

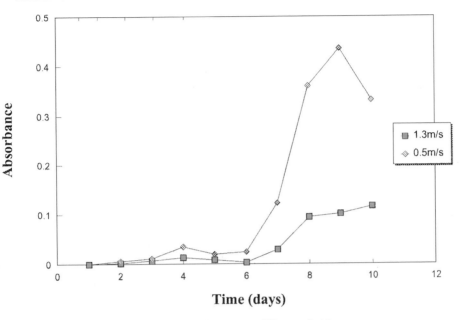

FIG. 6. Infrared absorbance at two different velocities.

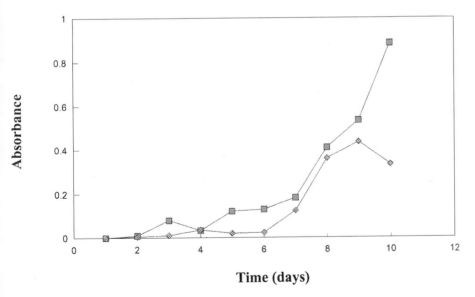

FIG. 7. Replication of data in Fig. 6.

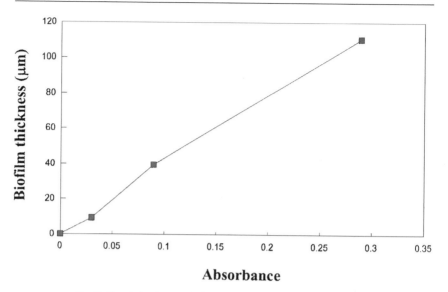

FIG. 8. Correlation between infrared absorbance and biofilm thickness.

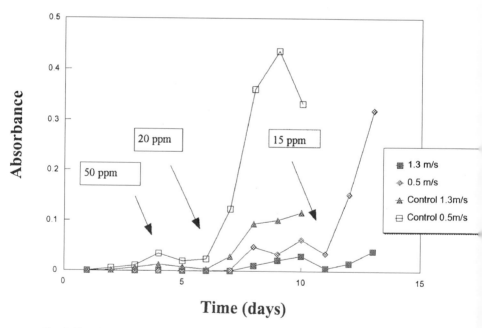

FIG. 9. The effect of reducing the concentration of a proprietary biocide on biofilm accumulation.

velocity are prone to extensive sloughing. Figure 7 compares replicate data. There is reasonable agreement except where the random sloughing at day 10 is apparent. Figure 8 provides a correlation between infrared absorbance and biofilm thickness. The weight change of the tube is used to determine the mean thickness of the biofilm for different absorbance determinations.

Figure 9 shows the effect of gradually reducing the concentration of a proprietary biocide on biofilm accumulation with time. In this particular example control appears to have been lost when the dose was reduced to 15 ppm.

Concluding Remarks

The method of assessing biofilm accumulation described in this chapter has been used extensively with success over the past 15 years to provide data on the establishment of biofilms in flowing systems and to investigate the effects of methods of control. A wide range of chemical additives (biocides and biodispersants) has been tested with the objective of determining the limiting concentration for effective control under operating conditions. Much of the work has been with proprietary formulations, but some data have been published.[4] The equipment has also been used to investigate the effectiveness of dosing programs.

Physical methods of control have also been studied, including the application of ultrasound,[5] the use of turbulence promotors inside tubes,[6] and the circulation of polymer fibers.[7] All have shown promise for biofilm control to a greater or lesser extent. The effects of surface on biofilm accumulation has also been studied by the technique of Mott and Bott.[8] A further series of tests has demonstrated that the combined use of chemical additives and physical methods of control can be very effective in preventing biofilm formation in flowing systems.[9]

A side stream from industrial installations has also been used in conjunction with the infrared device in place of the fermenter/circulating system, to provide data on the efficacy of biocides.[1]

[4] T. R. Bott and R. J. Taylor, in "Heat Transfer—Baltimore 1997" (M. S. El-Genk, ed.), p. 322. AIChE Symposium Series, Vol. 93, 1997.

[5] T. R. Bott, Heat Trans. Eng. 21, 43 (2000).

[6] A. Wills, T. R. Bott, and I. J. Gibbard, Can. J. Chem. Eng. 78, 61 (2000).

[7] T. R. Bott, M. J. Chen, and E. Kocianova, "Innovation in the Control of Biofouling in Cooling Water Systems," Proc. SAIChE 2000, South Africa, p. 168, 2000.

[8] I. E. C. Mott and T. R. Bott, in "Heat Transfer 1990" (G. Hetsroni, ed.), Proc. Ninth Int. Heat Trans. Conf. 5, 21 (1990).

[9] T. R. Bott and L. Trianqing, "Ozonation and AOPS in Water Treatment: Applications and Research," p. 34–41. Proc. Int. Regional Conf. Inst. Ozone Assn. Poitiers, 1998.

[8] Continuous Culture Models to Study Pathogens in Biofilms

By C. WILLIAM KEEVIL

A multistage continuous culture apparatus has been constructed to model mono-species biofilms and high species diversity biofilm consortia found in clinical practice, the natural environment, and the built environment. New artificial saliva and urine media were developed, and natural and treated potable waters obtained, for the defined, reproducible generation of the biofilms containing bacterial and protozoal pathogens on various hydroxylapatite, plastic, metal, and paint-covered substrata over many months. Following several modifications to the design, each chemostat vessel, linked in series or parallel, consists of a titanium top plate containing a variety of insertion ports for the monitoring electrodes, and addition/removal of gases, media, effluent, pH and redox titrants, antibacterial agents, and coupon substrata for biofilm generation. Each top plate is clamped to a 1 liter glass vessel, containing the culture medium, and mounted on a stirrer with a heater pad to facilitate external heating and stirring, controlling the shear rate across the immersed coupons. There are few internal parts. The growth rate is controlled by continuous addition of medium to the planktonic and sessile phases of the cultures. These model systems are ideal for generating biofilms either aerobically or anaerobically and containing microaerophilic or anaerobic species, even in highly aerated media.

Introduction

Microorganisms often survive on surfaces and within a film as an essential part of their life cycle. Biofilms are ubiquitous in nature and are formed at solid/liquid, oil/liquid, and air/liquid interfaces in response to hostile environments such as those with low nutrient status, adverse redox potential, or extreme temperature.[1] The physical barrier of the biofilm and the physiological interactions of complex microbial consortia may protect them not only against amoebal and protozoal grazing and disinfectants in the environment, but also macrophages, polymorphs, complement, and antibiotics *in vivo*. With advances in scanning confocal laser microscopy and episcopic differential interference contrast microscopy we are beginning to appreciate the open architecture or heterogeneous mosaic topography

---

[1] D. C. Ellwood, C. W. Keevil, P. D. Marsh, C. M. Brown, and J. N. Wardell, *Philos. Trans. R. Soc. Lond. B Biol. Sci.* **297,** 517 (1982).

of biofilms.[2,3] These open structures of "fronds," "streamers," or "mushrooms," frequently maturing into a sponge structure permeated with water channels for convective flow of nutrients and by-products, highlight the recent developments in molecular biology and microbial physiology showing the importance of the biofilm phenotype rather than a thick film barrier to provide a protective stress response in adverse environments.[4,5]

It is arguable whether any laboratory system can truly model the real world of biofilm formation and maturation, as this is a dynamic system subject to fluctuations of nutrient availability, growth rate, and shear rate, modified by predator–prey relationships and environmental parameters such as temperature, redox, and pH. Moreover, the physicochemistry of the substrata is altered by the complex nutrients of the local environment and the by-products of microbial metabolism adhering to the surfaces to be colonized. Nevertheless, laboratory studies form the basis of many of our observations on biofilm development, structure, microbial and chemical heterogeneity, and physiology. They offer the ability to dissect the effects of individual environmental and physicochemical parameters on biofilm structure and function, ideally in a reproducible fashion.[6] Moreover, laboratory biofilm systems can be pushed to investigate the extrema of environmental parameters that are probably never encountered *in situ*. Many model systems have been developed to study biofilms, but each can be argued to suffer particular flaws such as lack of reproducibility, defined physicochemical environments or control of the growth rate, and inability to operate for long periods. This paper will address the advantages and weaknesses of continuous culture systems for their application to study a range of clinical and environmental biofilms.

Chemostat Culture

The ease with which prokaryotes are cultivated in batch culture stems from their enormous plasticity in that they can adapt to sudden shifts in environment as encountered during the transitions from "lag" to "exponential" to "stationary" phases of batch growth. Indeed, bacteria are able to change themselves phenotypically to such an extent that it is impossible to define them chemically, structurally, or functionally without reference to the growth environment.[7] This cannot be

[2] J. Rogers, J. V. Lee, P. J. Dennis, and C. W. Keevil, *in* "Health Related Water Microbiology" (R. M. *et al.*, eds.), p. 192. IAWPRC, London, 1991.

[3] J. R. Lawrence, D. R. Korber, B. D. Hoyle, J. W. Costerton, and D. E. Caldwell, *J. Bacteriol.* **173,** 6558 (1991).

[4] J. L. Adams and R. J. McLean, *Appl. Environ. Microbiol.* **65,** 4285 (1999).

[5] W. L. Cochran, S. J. Suh, G. A. McFeters, and P. S. Stewart, *J. Appl. Microbiol.* **88,** 546 (2000).

[6] C. W. Keevil, *in* "The Life and Death of Biofilm" (J. T. W. Wimpenny *et al.*, eds.), p. 17. Bioline, Cardiff, 1995.

[7] D. Herbert, *Symp. Soc. Gen. Microbiol.* **11,** 391 (1961).

achieved easily in the "closed" system of batch culture and the only effective way of maintaining the chemical composition constant is to use the "open" flow system of continuous flow culture.[8]

The foundations of continuous culture theory were laid by Monod[9] in France and Novick and Szilard[10] in the United States. Subsequently, their mathematical analyses and rudimentary apparatus were refined by Herbert and colleagues[8] working at Porton Down in the United Kingdom.

Briefly, growth is considered to be an autocatalytic process that can be described by Eq. (1)

$$dx/dt = \mu x \tag{1}$$

where dx/dt is the growth rate of the cell population, x, and μ is the specific growth rate.

For exponential cell growth

$$x = x_0 e^{\mu t} \tag{2}$$

where x_0 is the initial cell population, and the doubling time t_d is represented by Eq. (3):

$$t_d = \ln 2/\mu \tag{3}$$

The dilution rate for continuous culture, D, is defined as the flow rate of fresh medium into the fermenter, f, divided by the fermenter volume, V:

$$D = f/V \tag{4}$$

During continuous culture,

Rate of change of organism = rate of growth − the rate of removal
density in the vessel   of the organism   of the organism

$$dx/dt = \mu x - Dx \tag{5}$$

where Dx is the rate of washout from the fermenter. By definition at steady state, the biomass and growth rate limiting substrate concentrations are constant, $dx/dt = 0$, and $ds/dt = 0$; therefore, $\mu = D$. As a consequence, the specific growth rate of an organism that is numerically equal to D, can be controlled by the substrate concentration s. At D_c, the critical dilution rate, $D > \mu$, and loss of cell population (washout) will occur. The maximum specific growth rate, μ_{max}, can therefore be determined by increasing the dilution rate until washout occurs.

[8] D. Herbert, in "Continuous Culture of Microorganisms," SCI Monograph 12, p. 21. Butterworths, London, 1961.

[9] J. Monod, Ann. Inst. Pasteur **78**, 807 (1950).

[10] A. Novick and L. Szilard, Proc. Nat. Acad. Sci. U.S.A. **36**, 708 (1950).

Consumption of the limiting substrate nutrient during continuous culture follows first order kinetics, defined by the Monod equation:

$$\mu = \mu_{max}s/s + K_s \tag{6}$$

where K_s is the limiting substrate constant and μ_{max} is the maximum specific growth rate. The biomass concentration and the growth limiting nutrient concentration can be determined from the following theoretical considerations:

$$ds/dt = D(S_R - s) - \mu X/Y \tag{7}$$

where S_R is the reservoir concentration of the growth limiting substrate, and Y is the biomass yield on that substrate. At steady state, $dx/dt = 0$ and $ds/dt = 0$; therefore

$$x^* = Y(S_R - K_sD)/\mu_{max} - D) \tag{8}$$

where x^* denotes the steady state value of x, the biomass concentration, and

$$s^* = K_sD/\mu_{max} - D \tag{9}$$

where s^* denotes the steady state value of s, the limiting substrate concentration.

Chemostat growth can be manipulated to obtain specific nutrient limitations and slow growth rates, as might be experienced in nature. This is particularly relevant to biofilm ecology, where doubling times of many hours or days have been reported. The chemostat quickly found favor with Czech[11] and Dutch scientists, as well as several notable workers in the United States[12] and proved a great asset to studies of the microbial physiology of monocultures of microorganisms and the ecology of complex microbial communities. All of these studies were of planktonically derived cells. Of note, Tempest[13] realized that useful information could be gained concerning either the selection of mutant organisms, present as a small proportion of the initial population, or phenotypic adaptation of the population from analysis of defined transient states as cultures adjusted from one steady state to another. The transition analysis utilized Eq. (10):

$$P_t/P_0 = e^{-Dt} \tag{10}$$

where D is the dilution rate, P_0 is the concentration of nutrient or product in the culture at zero time, and P_t is the concentration of P at time t. The analysis of the reverse transition utilized Eq. (11):

$$P_t/P_s = 1 - e^{-Dt} \tag{11}$$

[11] I. Malek, *J. Appl. Chem. Biotechnol.* **22,** 105 (1972).

[12] H. Veldkamp and H. W. Jannasch, *J. Appl. Chem. Biotechnol.* **22,** 105 (1972).

[13] D. W. Tempest, *in* "Essays in Microbiology" (J. R. Norris and M. H. Richmond, eds.), p. 1. Wiley, Chichester, 1978.

where P_s represents the final steady state concentration of the property and P_t is the concentration at time t. If a change in the production of a property following an alteration in environmental conditions is due to some phenotypic change, then the rate at which this will occur is defined by the exponential washout rate. A faster rate of change indicates turnover of the existing material, whereas a substantially lower rate of changeover could indicate mutant selection.

Clinical Models

Recognizing the need for reproducible biofilm generation systems to study defined clinical biofilms long term, and methods for their control, Keevil and colleagues[14] developed a versatile chemostat apparatus. This was initially used to study the complexity of dental plaque deposition and maturation of high species diversity consortia on dental acrylic tiles using continuous culture techniques. Dental plaque is found on healthy enamel, but it is also involved in the etiology of dental caries and periodontal disease.[15] The complex communities of plaque aerobes and anaerobes are closely associated, not only physiologically but also physically through specific coaggregation interactions. Stirred batch or even continuous culture systems cannot truly represent the intimate interactions involved. The apparatus developed attempted to reconcile the inherent faults of previous artifical mouths and stirred cultures by allowing the insertion of specific surfaces of dental interest in cultures of pooled plaque.

Plaque Model Design

The normal configuration of a 1 liter stainless steel fermenter head was redesigned, increasing the diameter from approximately 150 mm to 200 mm: this allowed the usual number of ports for housing electrodes, etc., together with an additional 8 ports around the periphery (19 mm external diameter, 12 mm internal diameter at the base) for the insertion and retrieval of coupons that were autoclaved and introduced aseptically during continuous culture (Fig. 1).

The head was initially constructed from autoclavable nylon but was subsequently replaced with an identical one made from stainless steel because the nylon head begins to warp after repeated autoclaving. The standard fermenter body was replaced with a wider, flanged, 1 liter Pyrex desiccator base or custom made vessel (200 mm external diameter, 150 mm internal diameter) to accommodate the modified head. The assembled fermenter contained a Teflon-coated magnetic stirrer bar and was mounted on a magnetic stirrer with a heater pad to facilitate external heating and stirring, controlling the shear rate across the immersed coupons

[14] C. W. Keevil, D. J. Bradshaw, A. B. Dowsett, and T. W. Feary, *J. Appl. Bacteriol.* **62**, 129 (1987).
[15] P. D. Marsh and C. W. Keevil, *in* "Microbial Metabolism in the Digestive Tract" (M. J. Hill, ed.), p. 155. CRC Press Inc., Boca Raton, FL, 1986.

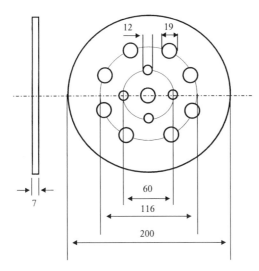

FIG. 1. Diagram of the chemostat biofilm model top plate showing dimensions (mm) and position of entry and exit ports (small ports for electodes and inlet/outlet pipes; large ports for coupon insertion/removal).

(Fig. 2). The chemostat was operated on its own or in series, with the first vessel providing a continuous seed into the second vessel, which also received its own supply of medium: F denotes the flow of medium into each chemostat vessel; G denotes flow of sparging gases, such as air or N_2/CO_2, for aerobic or anaerobic growth, respectively; C denotes flow of spent culture through a weir to another vessel or to waste; CHX denotes continuous or pulse addition of disinfectants such as chlorhexidine; and AT denotes insertion of plastic, epoxy, or metal coupons for cell attachment. The primary seed vessel ensures the reproducible maintenance of complex microbial consortia and supplies subsequent growth vessels in an open flow system for biofilm experimentation in defined environments. Coupons of known physicochemistry can be inserted and removed aseptically after hours or months, for microbiological and microscopy image analysis.

The growth rate was controlled by continuous addition of medium to the planktonic and sessile phases of the cultures. A trypticase/proteose peptone/yeast extract-based broth, supplemented with L-cysteine and glucose, provided the initial growth medium (BM medium). This was introduced into the 500 ml cultures at a constant rate of 25 ml hr^{-1} via a peristaltic pump to give a dilution rate of 0.05 hr^{-1} (equivalent to a mean generation time of 13.9 hr). The medium passed through an anti-grow-back tube, together with anaerobic sparging gas, to minimize grow-back of culture into the medium reservoir. The culture volume was maintained by removing the excess to a sterile, spent receiver (20 liter capacity) by an overflow weir

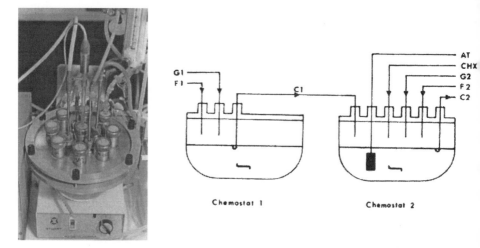

FIG. 2. Photograph of operating biofilm model chemostat vessel and diagram of two-stage assembly [C. W. Keevil, *in* "Recent Advances in Microbial Ecology" (T. H. *et al.*, eds.), p. 151. Japan Scientific Societies Press, Tokyo, 1989].

connected to a peristaltic pump. The pH of the culture was maintained automatically at pH 7.0 with 2 M NaOH using a pH controller linked to an autoclavable pH electrode. The surface of the culture was gassed with 50 ml min^{-1} 95% nitrogen: 5% carbon dioxide to maintain anaerobiosis. Surface gassing averted foaming and the necessity for control using antifoaming agents, which can affect the surface properties of microorganisms.

Plaque Inoculum and Coupon Recovery

Plaque was removed with a sterile dental probe from the tooth surface and gingival margin of 20 volunteers, suspended in prereduced BM medium (without glucose), and pooled before being used to inoculate the chemostat model. The coupons were made from a self-curing methyl methacrylate dental resin (product RR, De Trey Ltd., Weybridge, UK) which was poured into a glass petri dish to a depth of 1–2 mm and left overnight. Coupons, 20 × 10 mm, were cut from the cured matrix and hung from fine gauge stainless steel wire embedded in silicone rubber bungs that subsequently fitted into the fermenter ports (Fig. 3). Several coupons were suspended from the same bung when many biofilm samples were required. The combined coupons and bungs were inserted into the necks of glass bottles, the tops wrapped in aluminum foil, and the whole assembly autoclaved at 121° for 15 min. The foil was removed and the sterilized coupons were withdrawn from the bottle and introduced aseptically through the appropriate chemostat ports.

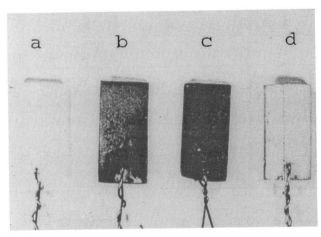

FIG. 3. Coupons stained with crystal violet to show plaque deposition. The dental acrylic tiles were stained (a) then were suspended from stainless steel wire prior to immersion and were incubated in the cultures for (b) 7 days or (c) 14 days before recovery for staining. The coupons were also scraped for microbiological analysis and were (d) subsequently stained to confirm complete recovery of the biofilms [C. W. Keevil, D. J. Bradshaw, A. B. Dowsett, and T. W. Feary, *J. Appl. Bacteriol.* **62**, 129 (1987)].

Data Obtained

Using this configuration of fermenter, substratum, and medium, the microbiological analysis of the planktonic and sessile biofilm phases indicated that a complex community of oral bacteria was established, and that there was reasonable agreement with samples taken many months apart.[14] The early pioneer species on the coupons included aerotolerant *Streptococcus sanguis* and *Neisseria* spp., as found *in vivo*.[15] Subsequently, anaerobic fusiforms, peptostreptococci, and *Veillonella* spp. became established, together with low numbers of lactobacilli, *Actinomyces* and *Bacteroides* spp., and *Streptococcus mutans,* the principal etiological agent of dental caries. Scanning electron microscopy showed that either side of the coupons contained a rough and smooth surface (because of the casting/curing procedure) and these initially favored the attachment of long fusiform bacteria, particularly on the rough surface. Cocci attached to those surfaces that were not heavily colonized by the fusiforms and eventually grew into and on the colonial sheets of the fusiforms.

Subsequent Modifications to the Plaque Model

Further work investigated incorporating 20% (w/w) hydroxylapatite in the acrylic coupons (to more closely model tooth enamel and increase the surface hydrophilicity), using dilution rates of 0.05 and 0.2 hr^{-1} (corresponding to mean

TABLE I

COMPOSITION OF THE PLAQUE BIOFILM FORMED ON ACRYLIC TILES
CONTAINING 20% (W/W) HYDROXYLAPATITE AFTER 7 DAYS[1] IMMERSION
IN CONTINUOUS CULTURES OF POOLED PLAQUE (D 0.05 HR^{-1})[a]

Medium	BMG	BMGM	BMM	Saliva
EPS streptococci	8	0.2	8.1	92
Low EPS streptococci	254	67	39	68
S. mutans	2	0.2	1.5	0.01
Peptostreptococci	345	445	1510	125
Lactobacilli	12	ND	0.02	0.01
Actinomyces spp.	ND	12	1.5	25
Veillonella spp.	40	6.5	ND	145
Bacteroides spp.	175	805	2675	1655
Fusiforms	190	1005	660	820
Neisseria spp.	5	3.5	ND	ND
Haemophilus spp.	ND	7.5	16	20
Treponema spp.	ND	D	D	D

[a] Results are expressed as $10^{-6} \times$ cfu cm^{-2} and are the mean of triplicate cultures with errors of <20%. ND denotes not detected; D denotes detection of spirochetes that could not be quantitated by the methods available [C. W. Keevil, in "Recent Advances in Microbial Ecology" (T. H. et al., eds.), p. 151. Japan Scientific Society Press, Tokyo, 1989].

generation times of 3.5 and 13.9 hr, respectively, which have been measured at various sites in the oral cavity for bacteria growing in young or mature plaque), and supplementing the BM medium with 0.1% (w/v) mucin (the major glycoprotein of saliva).[16] Addition of the mucin resulted in increases in the numbers of *Bacteroides* spp. and fusiforms, as well as *Actinomyces* spp. and spirochetes (Table I).[17] Withdrawal of glucose from the mucin-supplemented BM medium resulted in a decreased growth yield, the loss of *Neisseria* spp. from the planktonic and sessile phases, and decreased recovery of acidophilic streptococci and lactobacilli. Replacement of this medium with an artificial saliva containing 0.1% (w/v) mucin[16] enriched for *Bacteroides* spp., but lower numbers of the other genera commonly found in plaque were also maintained at either growth rate. The plaques formed were reminiscent of subgingival plaque and, of the species that were enriched, *Actinomyces* spp. have been implicated in root surface caries formation, whereas *Treponema, Bacteroides,* and *Haemophilus* spp. are suspected aetiological agents of the periodontal diseases. It proved possible to manipulate

[16] D. A. Glenister, K. E. Salamon, K. Smith, D. Beighton, and C. W. Keevil, *Microb. Ecol. Health Dis.* **1,** 31 (1988).

[17] C. W. Keevil, in "Recent Advances in Microbial Ecology" (T. H. et al., eds.), p. 151. Japan Scientific Societies Press, Tokyo, 1989.

TABLE II
INHIBITORY EFFECTS OF 0.1% (W/V) CHLORHEXIDINE GLUCONATE ON COMPLEX COMMUNITIES
OF PLAQUE BACTERIA GROWING IN CONTINUOUS CULTURE (D 0.2 HR^{-1}) IN ARTIFICAL SALIVA
MEDIUM AND ON 7-DAY-OLD BIOFILMS RECOVERED PRIOR TO BIOCIDE TREATMENT[a]

	Steady state culture			Biofilm		
	−CHX	+CHX	Kill factor	−CHX	+CHX	Kill factor
Gram negatives	140	0	>10^8	2300	2.4	10^3
Gram positives	7.5	0.0008	>10^4	420	14.5	30
Lactobacilli	0.002	0	>10^4	0.01	0.0007	16

[a] After 20 min treatment viable bacteria were recovered aerobically and anaerobically on blood agar in the presence (gram-negative plus lactobacilli count) and absence (total count) of vancomycin (2.5 mg liter^{-1}) or on Rogosa agar (lactobacilli count). Results are expressed as 10^{-6} × cfu ml^{-1} (culture) or cm^{-2} (biofilm) [C. W. Keevil, in "Recent Advances in Microbial Ecology" (T. H. et al., eds.), p. 151. Japan Scientific Society Press, Tokyo, 1989].

the complex microbiota and enrich species in the biofilm. For example, replacing glucose with sucrose enriched for *Eubacterium* spp., but not *Neisseria* spp,[18] while decreasing the growth pH from 7.0 to 4.0 (as found in caries-forming plaque), enriched for acid tolerant *S. mutans* and lactobacilli in the planktonic and sessile phases. The biofilms produced proved much more resistant to the effects of biocides than the corresponding dispersed cultures. For example, the minimal inhibitory concentration for the commonly used chlorhexidine gluconate (Hibitane, ICI) was >100-fold higher for the biofilm cells; only gram-positive bacteria, particularly streptococci, showed any resistance to its effects (Table II).[17] These studies with the two-stage biofilm model have subsequently been extended by Marsh and colleagues, using defined 9- and 10-species inocula to facilitate identification of the communities in different environments. Their work has highlighted the protective effects of aerobes such as *Neisseria* spp. in protecting oxygen-sensitive anaerobes in the planktonic and sessible phases of aerated saliva media.[19]

Urinary Catheter Infection

Brooks and Keevil[20] modified the chemostat biofilm model by replacing the stainless steel fermenter top, ports, and coupon wires with titanium. This material is increasingly being used in orthopedic and dental implants, both for its physiological properties and for its corrosion resistance. Its use here thereby avoided the

[18] D. J. Bradshaw, A. B. Dowsett, and C. W. Keevil, in "Recent Advances in Anaerobic Bacteriology" (S. P. B. et al., eds.), p. 327. Martinus Nijhoff, Dordrecht, 1987.
[19] D. J. Bradshaw, P. D. Marsh, C. Allison, and K. M. Schilling, *Microbiology* **142,** 623 (1996).
[20] T. A. K. Brooks and C. W. Keevil, *Lett. Appl. Microbiol.* **24,** 203 (1997).

leaching of Fe, Mn, or Cr from stainless steel into the culture medium; which might affect the growth of sensitive pathogens. To make the model physiologically relevant, a simple artificial urine medium (AUM) was devised to provide conditions similar to those found in human urine; this contained low concentrations of peptone, yeast extract, lactic, citric and uric acids, urea, creatinine, and various salts. AUM solidified with agar enabled the recovery of a wide range of urease-positive and -negative urinary pathogens. Liquid AUM supported growth at concentrations of up to 10^8 cfu ml^{-1}, as found in normal urine. Reproducible, steady-state growth also occurred over many generations in continuous culture. AUM was capable of forming crystals and encrustations resembling those found in natural urinary tract infections. The medium was found to be a suitable replacement for normal urine for use in a wide range of experiments modelling the growth and attachment of important urinary pathogens in the clinical environment. These included *Staphylococcus epidermidis, Pseudomonas aeruginosa,* and *Proteus mirabilis.* Staphylococcal biofilms on 1 cm^2 latex catheter coupons immersed in AUM produced carbonate–apatite crystals similar to those found in patients' infected urine. The pseudomonad produced far more alginic acid exopolymer in AUM than normally found with most growth media, and *P. mirabilis* formed biofilms with copious production of the characteristic struvite encrustations found *in vivo.*

Cystic Fibrosis Model, Antibiotic Resistance, and Serum Killing

Anwar and colleagues[21] have investigated the use of the chemostat biofilm model to study the physiology of carefully controlled sessile populations of *P. aeruginosa,* an opportunistic pathogen that produces copious alginic acid exopolymer in the lungs of cystic fibrosis patients. Mucoid *P. aeruginosa,* isolated from a patient with cystic fibrosis, was cultivated at a slow growth rate ($D = 0.05$ hr^{-1}) under iron limitation in a chemostat. Biofilm was allowed to form on acrylic tiles, similar to those described earlier. The kinetics of the biofilm formation and antibiotic resistance was then investigated. Planktonic cells were very sensitive to tobramycin. They were killed by exposure to 10 mg liter^{-1} tobramycin within 2 hr. Young biofilm cells of *P. aeruginosa* (day 2 of colonization) were found to be more resistant. Approximately 40% of the adherent cells remained viable after exposure to 10 mg liter^{-1} tobramycin for 5 hr. An increase in the concentration of tobramycin to 20 mg liter^{-1} resulted in an enhancement of the killing of young biofilm bacteria, and approximately 1.5% of them remained viable after exposure to this concentration of antibiotic for 5 hr. Old biofilm bacteria, examined at day 7, were the most resistant and 15% of the cells were found to be viable after they were exposed to 200 mg liter^{-1} of tobramycin for 5 hr. When either young or old biofilm cells of mucoid *P. aeruginosa* were scraped from the tiles to produce a

[21] H. Anwar, M. Dasgupta, K. Lam, and J. W. Costerton, *J. Antimicrob. Chemother.* **24,** 647 (1989).

planktonic cell suspension, they were sensitive to 5 mg liter^{-1} tobramycin. Further studies showed that young biofilm cells of mucoid *P. aeruginosa* could be effectively eradicated with chemostat-controlled doses of 500 mg of piperacillin liter^{-1} plus 5 mg of tobramycin liter^{-1}. However, old biofilm cells were very resistant to these antibiotics, and eradication of old biofilm cells was not achievable with the high doses used.[22] Aging biofilms (harvested on day 7) were also very resistant to the killing effect of whole blood and serum.[23]

In further studies, these workers[24] also showed the utility of the chemostat biofilm model to study biofilm formation and antibiotic resistance by *Staphylococcus aureus*. Planktonic and young biofilm cells were completely eradicated after exposure of these cells to drug levels representing one loading and two maintenance doses of tobramycin and cephalexin. A very different picture was observed when antibiotic exposure was initiated on day 21. Complete eradication of the old biofilm cells was not observed even when the antibiotic exposure was continued for an extra 6 days. Regrowth of the organism was observed when the antibiotic exposure was terminated. As a consequence of these chemostat studies, it was proposed that the establishment of aging biofilms may contribute to persistence of *P. aeruginosa* and *S. aureus* in biofilm-associated infections. More recent studies are using modifications of the principle of biofilm formation on coupons immersed in chemostat culture to study *Gardnerella vaginalis*[25] and respiratory pathogens such as *Streptococcus pneumoniae* and *Moraxella catarrhalis*.[26]

Environmental Model

It is now recognized that the natural and built environments provide a potential reservoir for many important respiratory and gastrointestinal pathogens, and that these must persist in the presence of complex autochthonous microbiota containing bacteria, yeasts, fungi and protozoal predators. Accordingly, Keevil and colleagues[27] modified their plaque biofilm apparatus to study these complex microbial communities in difficult environments. As before, a primary seed vessel ensured the reproducible maintenance of complex microbial consortia and supplies subsequent growth vessels in an open flow system for biofilm experimentation in defined environments. Tiles of known physicochemistry can be inserted and removed aseptically after hours or months, for microbiological and microscopy

[22] H. Anwar, J. L. Strap, K. Chen, and J. W. Costerton, *Antimicrob. Agents Chemother.* **36**, 1208 (1992).
[23] H. Anwar, J. L. Strap, and J. W. Costerton, *FEMS Microbiol. Lett.* **71**, 235 (1992).
[24] H. Anwar, J. L. Strap, and J. W. Costerton, *Antimicrob. Agents Chemother.* **36**, 890 (1992).
[25] F. Muli and J. K. Struthers, *Antimicrob. Agents Chemother.* **42**, 1428 (1998).
[26] R. K. Budhani and J. K. Struthers, *Antimicrob. Agents Chemother.* **42**, 2521 (1998).
[27] C. W. Keevil *et al.*, *in* "Advances in Pollution Control" (D. Wheeler, M. L. Richardson, and J. Bridges, eds.), Vol. II, p. 367. Pergamon Press, Oxford, 1989.

image analysis. By contrast to field studies, the model is cheap, reproducible, and easily manipulated. It can be used to model the effects of different growth rates, temperatures, environmental chemistry, and disinfectant concentration. The shear rate imposed on surfaces by water velocity can be determined by manipulation of the stirrer speed: typical water velocities of 0.2–3.0 m sec^{-1} have been investigated, the lower values maintaining the Reynolds number well below the transition zone for turbulent flow. Importantly, the model can be assembled in Class III containment cabinets to protect laboratory personnel during the study of more infectious pathogens persisting in biofilms, such as *Legionella pneumophila* and *Escherichia coli* O157.

Legionella Model

Initial experiments investigated growing sessile microbial consortia in natural and treated potable water, as prelude to colonization by *L. pneumophila,* the waterborne etiological agent of Legionnaires' disease and Pontiac fever. The continuous culture biofilm model system used the titanium materials in place of stainless steel, as described earlier, because of its corrosion resistance at higher temperatures and the fact that it does not leach other metal ions into solution, which would affect the water chemistry and perturb the microbial communities. The natural or potable waters were filter sterilized and used as the continuous culture growth medium. Membrane filtration of the water has been shown to result in water that is chemically unaltered. Two or three chemostats were linked in series to reproduce the conditions found in particular water systems. The first vessel was analogous to a reservoir or a storage or holding tank within a building; the second and third vessels represented a distribution system or a cold or hot water plumbing system within a building. The naturally occurring mixed microbiota for culture inocula were obtained by scraping biofilm from the surfaces of, or filter concentrating the water from, natural ground and surface waters, mains, distribution supplies, taps, or calorifers, depending on the environment to be modeled. The waters from these sources were filter-sterilized (0.2 μm, Pall Filtration) to provide the sole source of growth medium. They were designated as hard, moderately hard, and soft based on a hardness of 330, 150, and 34 ppm $CaCO_3$, respectively. The chemostats were held at various temperatures between 5° and 60°. The lower temperature of 5–10° represented a cold water mains supply; 20° would occur within buildings or within pipes during a spell of hot weather, whereas 30° is found in the sumps of cooling towers. The higher temperature of 40–60° represented the lower to higher temperatures that would be found within a hot water system. Temperatures of 5–20° required the use of a chiller unit (Grant LTD 6, UK) supplying a cooling coil wrapped around the chemostat body to maintain the required temperature. Typically, the temperature of the first or seed vessel was kept below 30° to provide a consistent inoculum of complex microbiota to the vessels connected downstream; these could

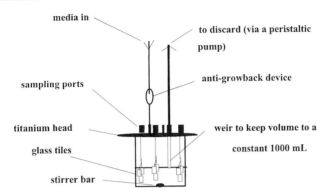

media in

to discard (via a peristaltic pump)

sampling ports

anti-growback device

titanium head

weir to keep volume to a

glass tiles

constant 1000 mL

stirrer bar

FIG. 4. Diagram of the model biofilm system with multiple assemblages of coupons suspended from rigid titanium wire inserted through silicone rubber bungs in the top ports. The weir system was used to maintain the volume at the required level. Temperature, oxygen, and pH probes are not shown [S. B. Surman, L. H. G. Morton, and C. W. Keevil, in "Biofilms in the Aquatic Environment" (C. W. Keevil, A. Godfree, D. Holt, and C. S. Dow, eds.), p. 160. Royal Society of Chemistry, Cambridge, UK, 1999].

be challenged with extremes of temperature, pH, or biocide concentration without destroying the seed supply, hence providing an open flow, microbiologically reproducible system.

The monitoring and control of all physicochemical parameters such as temperature, pH, dissolved oxygen concentration, and stirrer speed used the appropriate glass bodied electrodes linked to microprocessor control units (Anglicon Systems, Brighton, UK). The stainless steel–bodied temperature sensor was sheathed in Teflon before use. As before, the assemblages of 1 cm^2 coupons, consisting of plastic or metal materials that are in contact with water in the built environment, or photographic glass control surfaces, were suspended from titanium wire into the cultures (Fig. 4). The exceptions were fresh and aged copper coupons when copper wire (aged in distilled water for several weeks) replaced the titanium. The coupons were drilled with 1 mm holes, 1 mm from one end, to secure the wire.

Biofilm Enumeration and Identification

The coupons were periodically removed, aseptically, from the system via the sampling ports, in order to assess biofilm growth. Coupons were first rinsed gently in 10 ml of sterile water to remove any unattached microorganisms and the biofilms subsequently removed from both sides of the coupon by scraping with a dental probe into 2 ml of sterile water. The suspensions were then vortexed and serial dilutions made in sterile microfuge tubes. Duplicate spread plates were made on nonselective media and onto media selective for pathogens such as legionellae. Scraping and vortexing was chosen to remove the biofilm in preference to sonication, to prevent disruption of the more fragile aquatic bacteria.[27] The plank-tonic population was sampled by removing 10 ml from the aqueous phase of the chemostat and serially diluting and culturing as above.

The total heterotrophic population was assessed by growth on R2A medium; buffered charcoal yeast extract agar (BCYE) was used as an enriched nonselective medium to grow the more fastidious bacterial isolates including *L. pneumophila.* BCYE plates with added glycine, vancomycin, polymyxin, and cycloheximide (GVPC) were used as the selective medium for the isolation of legionellae. After inoculation, plates were sealed in polyethylene bags to prevent drying. The R2A, BCYE, and GVPC were incubated aerobically at 30° or 37° for up to 7 days before enumeration. Presumptive identification was determined by colony morphology and the use of the first stage diagnostic tables for gram-negative bacteria. Further identification was then carried out using either the API 20NE (API-Biomerieux, Basingstoke) or Biolog GN-Microplate system (Atlas Bioscan, Hayward, CA). Isolates that grew on GVPC and failed to grow on BCYE without cysteine were tentatively identified as *Legionella*. This result was confirmed by a latex aggluti-nation test using the "Prolex" rapid latex kit (Pro-Lab Diagnostics, UK).

Amebal identification utilized schemes described by Page.[28] Amebae were removed by washing the plate with amebal saline that was gently pipetted over the surface and collected in a sterile universal bottle. Several drops of the amebal suspension were streaked diagonally over lawns of *Klebsiella aerogenes* NCTC 7427 grown on amebal agar. The plates were examined frequently for trophozoites and cysts, and these were then tentatively identified by microscopic examination and with the aid of the identification keys. Fungal isolates were tentatively identified by morphological characterization.

Data Obtained

Using such systems, it was demonstrated that biofilms of high species diver-sity could be generated reproducibly for many months on a range of plastic and metal materials (Table III).[29] Moreover, *L. pneumophila* could survive and grow in the biofilms, even at temperatures up to 50° on plastic surfaces but not cop-per (Table IV). Survival and growth of legionellae occurred even in the apparent absence of eukaryotic grazing species, which had been suggested by many to be obligatory for the growth of this so-called "obligate intracellular parasite" in the environment.[29,30] Subsequently, Surman *et al.*[28] were able to confirm that it is a

[28] S. B. Surman, L. H. G. Morton, and C. W. Keevil, *in* "Biofilms in the Aquatic Environment" (C. W. Keevil, A. Godfree, D. Holt, and C. S. Dow, eds.), p. 160. Royal Society of Chemistry, Cambridge, 1999.

[29] J. Rogers, A. B. Dowsett, P. J. Dennis, J. V. Lee, and C. W. Keevil, *Appl. Environ. Microbiol.* **60**, 1842 (1994).

[30] J. Rogers and C. W. Keevil, *Appl. Environ. Microbiol.* **58**, 2326 (1992).

TABLE III

COLONIZATION OF PLUMBING MATERIALS AFTER 21 DAYS AT $30°$[a]

	Steel	SS	Latex	E-P	PP	PE	PVCu	PVCc
L. pneumophila	17	13	150	500	37	13	11	7.9
P. aeruginosa	30							
P. acidovorans						40		11
P. diminuta						2		
P. fluorescens				1000				3
P. maltophila	10	11	3000			10	10	
P. mendocina			13			40	0.01	
P. stutzeri	140	70	2000					0.1
P. testosteroni		180				20		8
P. vesicularis	250							3
P. xylosoxidans						7	40	
S. paucimobilis	30	36	5000	1600	790	170	140	362
Actinomyctes sp.	130	2	7000	8000		9	0.01	2.8
Aeromonas sp.		6000						
Alcaligenes sp.	10	10			320	80		30
Flavobacterium sp.	41		15,000	2400		0.2	90	50
Methylobacterium sp.	20	150			140	30	60	60
Klebsiella sp.			31,000					
Acinetobacter sp.	70	39	22,000	3100	400	180	40	60

[a] SS, stainless steel substratum; E-P, ethylene–propylene copolymer; PP, polypropylene; PE, polyethylene; PVCu, unpolymerized polyvinyl chloride; PVCc, chlorinated polyvinyl chloride [J. Rogers, A. B. Dowsett, P. J. Dennis, J. V. Lee, and C. W. Keevil, *Appl. Environ. Microbiol.* **60**, 1842 (1994)].

facultative intracellular parasite when they grew the aquatic microbiota in the biofilm model in the presence of cycloheximide, which eradicated all eukaryotic species. In these conditions, the numbers of legionellae actually increased >50-fold in the biofilms.

In the past 10 years the environmental version of the chemostat biofilm model has been applied to the study of a range of prokaryotic and eukaryotic species persisting in diverse environments. These have included potential coliform indicators of faecal pollution, *E. coli*, vero-cytotoxigenic *E. coli* O157 and *Klebsiella* spp., opportunistic pathogens such as *P. aeruginosa* and *Aeromonas hydrophila*, microaerophilic pathogens such as *L. pneumophila* and *Campylobacter jejuni*, and pathogenic protozoa such as *Cryptosporidium parvum* and *Acanthamoeba* spp. The model has provided great insight into how environmental biofilms provide a safe haven for these pathogens, protecting them from extremes of temperature, pH, redox, shear force, and nutrient depletion.

The model's ability to accurately define the growth parameters and apply chemostat theory has also provided valuable information into how coliforms such

TABLE IV
COLONIZATION OF PLUMBING MATERIALS BY AQUATIC
MICROBIOTA AND *Legionella pneumophila*[a]

Temp	Material	Microbiota	*L. pneumophila*
20°	Copper	2.2×10^5	0
	Polybutylene	5.7×10^5	665
	PVCc	1.8×10^5	2130
40°	Copper	8.1×10^4	2000
	Polybutylene	1.2×10^6	112,000
	PVCc	3.7×10^5	68,000
50°	Copper	2.3×10^4	0
	Polybutylene	3.2×10^6	890
	PVCc	1.2×10^5	60
60°	Copper	4.5×10^2	0
	Polybutylene	4.3×10^4	0
	PVCc	5.2×10^3	0

[a] Adapted from Ref. 33.

as *Klebsiella oxytoca* may be capable of regrowth in a mains distribution supply. It is normally assumed that their presence indicates a recent fecal ingress due to cross connections with waste pipes, breakthrough in old or cracked pipes, or back siphonage into the pipework downstream. However, using Eq. (9) described earlier, Keevil and colleagues[31,32] were able to show that *E. coli, A. hydrophila,* or *K. oxytoca* did not wash out of a potable water biofilm model system operating at a fast dilution rate of 0.2 hr^{-1} (temperature of 15°) within the predicted 72 hr (Fig. 5). The mathematical analysis indicated that it was growing, albeit more slowly than the rate allowed by the dilution rate. It may therefore persist for long periods in the biofilms of water supplies, and as such might not be a good indicator of a recent fecal pollution event.

Advantages

The chemostat models described have the benefit of a simple design, with few internal mechanical parts. Their design incorporates an external stirrer motor,

[31] C. W. Mackerness, J. S. Colbourne, P. J. Dennis, T. Rachwal, and C. W. Keevil, *in* "Society for Applied Bacteriology Technical Series" (S. P. G. M. S. S. Denyer, ed.), Vol. 30, p. 217. Blackwell Scientific Publications, Oxford, 1993.

[32] P. J. Packer, D. M. Holt, J. S. Colbourne, and C. W. Keevil, *in* "Coliforms and *E. coli:* Problem or Solution" (D. Kay and C. Fricker, eds.), p. 189. Royal Society of Chemistry, London, 1997.

[33] J. Rogers, A. B. Dowsett, P. J. Dennis, J. V. Lee, and C. W. Keevil, *Appl. Environ. Microbiol.* **60,** 1585 (1994).

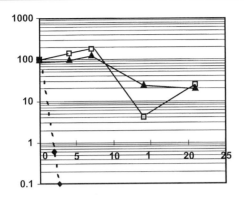

Time (days)

FIG. 5. Growth above theoretical washout (◆) for *E. coli* (▲) and *A. hydrophila* (□) in continuous culture at 15° ($D = 0.2 \text{ hr}^{-1}$). The vertical axis shows percent survivors.

heater pad, and cooling coils, together with an external pump to remove excess culture. The wide top plate houses additional ports for the aseptic insertion of biofilm coupons. The use of titanium materials to replace stainless steel offers the advantages of no Fe, Ni, Mn, or Cr leaching that would affect the medium chemistry and growth of sensitive organisms. Moreover, it is practicable to undertake limitation studies for iron, an important nutrient that may be restricted by transferrin and lactoferrin iron binding proteins *in vivo* and at pH above neutral in the environment. The continuous culture design permits efficient rapid mixing of the incoming limiting nutrient, enabling chemostat theory to be applied for growth kinetics and microbial physiology. Although biofilms can be removed to study their subsequent immersion in antibacterial solutions, such biocides or antibiotics can be fed in a pulse or continuous fashion into the cultures, permitting *in situ* assessment of their activity in a defined fashion. In particular, they can be assessed to determine if biofilm formation can be prevented by beginning the challenge before fresh coupons are added, or young or old biofilm can be generated before the challenge is initiated.

Although the shear rate of liquid across the biofilm surfaces immersed in the chemostat is not as accurately controlled as, say, in a rotatorque, this is offset by the better nutrient mixing and gas transfer parameters for kinetic analysis. Moreover, the biofilms of complex aerobic and anaerobic microbial communities can be generated over many months or years and are easily withdrawn for microscopy and microbiological analysis. Last, some laboratories have attempted a compromise by growing microorganisms in defined continuous culture before passing them through modified Robbins devices. These give some degree of shear control, modified by the rate of flow through the apparatus, but shear perturbations occur at the entry to the devices, and the studs can be difficult to remove and may damage the biofilms in the process. Biofilms on studs sampled near the entry of the device are frequently different in structure, composition, and function from those sampled farther along. Perhaps of greater importance, the cells have been removed from near instantaneous contact with their limiting nutrient in the chemostat (as defined by the equations earlier) and are therefore starving as they pass, sometimes very slowly, through the external devices in a physiologically uncontrolled fashion with little or no control over the gas supply and redox environment.

   This paper has concentrated on the use of continuous culture biofilm models for the reproducible generation of defined biofilms containing microbial pathogens. Nevertheless, these models have also shown great versatility for the study of general biofouling, biofilm structure and function at the molecular level, corrosion processes, and perturbation of water quality, some of which have been discussed elsewhere by this author and others.

Section III

Biofilm Growth in Special Environments

[9] Biofilms in Unsaturated Environments

By Patricia A. Holden

Introduction to Unsaturated Environments

Unsaturated environments contain three dynamic phases: solid, water, and air. In nature, the solid phase contains a combination of inert organomineral complexes and biotic materials. The dominant fluid phase is normally air but is transiently water. For example, the unsaturated terrestrial subsurface (including the rhizosphere) or "vadose zone"[1] consists of microbially colonized organomineral solids coated with thin fluid water films; the spaces between the particles are normally gas-filled (Fig. 1A). The air-filled pore spaces in the vadose zone become water-filled during precipitation and infiltration (Fig. 1B). Subsequent drainage and evaporative drying causes the system to become gas-filled again (Fig. 1C). Other examples of transiently wet, normally dry environments include the surfaces of plants, food, and rocks or monuments. In all of these unsaturated environments, the solid phase is the substratum for adherent microbes that are encapsulated in exopolymeric substances (EPS), which, by definition, is biofilm.[2] What differentiates these biofilms from biofilms in environments where the main fluid phase is always water (e.g., rivers and water distribution pipes) is that "unsaturated" biofilms directly exchange mass with the air phase and are routinely subjected to fluctuations in water content and water potential. Just as fluid flow characteristics are controlled when culturing saturated biofilms, water content and potential should at least be considered and at best be controlled when culturing unsaturated biofilms.

Fundamentals of Water Potential and Water Activity

The potential energy per unit volume of water is called total water potential (ψ) or, in soils, total *soil* water potential.[3,4] The concept of ψ can easily be understood by considering its two important components: (1) matrix water potential (ψ_m) due to capillarity and the attraction of thin water films to solids[5] and (2) solute water potential (ψ_s) due to the association of water with dissolved salts. In nonsaline

[1] R. A. Freeze and J. A. Cherry, "Groundwater." Prentice-Hall, Inc., Englewood Cliffs, NJ, 1979.

[2] W. G. Characklis and K. C. Marshall, *in* "Biofilms: A Basis for an Interdisciplinary Approach" (W. G. Characklis and K. C. Marshall, eds.), p. 3. John Wiley & Sons, Inc., New York, 1990.

[3] W. A. Jury, W. R. Gardner, and W. H. Gardner, "Soil Physics," 5th ed. John Wiley & Sons, New York, 1991.

[4] R. I. Papendick and G. S. Campbell, *in* "Theory and Measurement of Water Potential" (J. F. Parr, W. R. Gardner, and L. F. Elliot, eds.), p. 1. Soil Science Society of America, Madison,WI, 1981.

[5] L. J. Waldron, J. L. McMurdie, and J. A. Vomocil, *Soil Sci. Soc. Am. Proc.* **25**, 265 (1961).

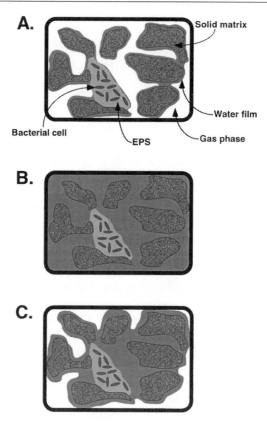

FIG. 1. A conceptual progression of soil wetting and drying at the pore scale. Initially dry soil contains thin water films and air-filled pore space (A). Pore spaces become filled with infiltrated rainwater (B) and become air-filled again with subsequent drainage and evaporative drying (C). Hysteresis in the water characteristic curve results from redistribution of water and retention of air in pore spaces during the wetting–drying cycle.

systems, ψ_m is the dominant component of ψ.[4] Both ψ_m and ψ_s are variables in the unsaturated environment that surrounds, yet is in continuous communication with, bacteria and biofilm matrices.

Water content and ψ_m or ψ are empirically related through the hysteretic "characteristic" or moisture release curve,[3,4] which is unique for a given soil or other water-holding material. The characteristic curve is a useful relationship: given ψ_m and the characteristic curve, water content (θ) can be estimated; with θ and specific surface area, the average water film thickness on particles can be estimated.[6] There are other components of ψ that should be considered for deep

[6] S. A. Taylor and G. L. Ashcroft, "Physical Edaphology: The Physics of Irrigated and Nonirrigated Soils." W. H. Freeman and Company, San Francisco, 1972.

soils; these are discussed in standard texts on soil physics[3,4,7] and are not commonly replicated in laboratory culture.

In unsaturated systems, water and gas are in local equilibrium and ψ is related to relative humidity as follows[4]:

$$\psi = \frac{RT}{V_w} \ln\left(\frac{\text{r.h.}}{100}\right) \qquad (1)$$

where

$$R = \text{gas constant} \left(\frac{P - L^3}{\text{mol} - T}\right)$$

$$T = \text{temperature } (T)$$

$$V_w = \text{molal volume of water} \left(\frac{\text{mol}}{\text{liter}^3}\right)$$

$$\text{r.h.} = \text{relative humidity } (\%)$$

The relationship of ψ to water activity, a_w, is straightforward: $a_w = (\text{r.h.}/100)$ and thus a_w can be substituted into Eq. (1) to calculate ψ. The restatement of (1) in terms of a_w may be a more meaningful way to express ψ to a wider range of scientists, including food scientists, biotechnologists,[8] plant scientists, and soil scientists. As per (1), ψ is negative when r.h. $< 100\%$ or when $a_w < 1$. Alternatively, ψ is 0 when the r.h. is equal to 100% (i.e., when $a_w = 1$), which applies to unsaturated environments over a range of water contents that qualitatively span from moist to fully saturated. In surface soils, the diurnal fluctuation of ψ can range widely, depending on season and soil texture. Although it depends on the environment that one wishes to reproduce in laboratory culture, operating over a range of ψ between 0 and -1.5 MPa (15 bar) is a reasonable span for culturing soil microorganisms because -1.5 MPa is the wilting point for certain cultivated plants,[4] and this is the low end of water potential tolerated by many bacteria.[9] However, many fungi are tolerant of water potential far below -1.5 MPa.[9]

In aqueous systems $\psi = \psi_s$, and the van't Hoff equation can be used to calculate ψ from the solute concentration[4]:

$$\psi_s = -\phi\gamma cRT \qquad (2)$$

where

$\phi = $ the osmotic coefficient for the salt at a molality c (dimensionless)

$\gamma = $ number of osmotically active particles per solute molecule (e.g., 2 for NaCl)

$c = $ molal concentration of solute (mol/kg)

[7] D. Hillel, "Fundamentals of Soil Physics." Academic Press, San Diego, 1980.

[8] B. Hahn-Hagerdal, *Enzyme Microb. Technol.* **8,** 322 (1986).

[9] R. F. Harris, *in* "Effect of Water Potential on Microbial Growth and Activity" (J. F. Parr, W. R. Gardner, and L. F. Elliott, eds.), p. 23. Soil Science Society of America, Madison, WI, 1981.

TABLE I
CHARACTERISTICS OF SOME AQUEOUS SATURATED SALT SOLUTIONS

Solid phase	r.h. / 100^a	$\psi\,(-MPa, 25°)^{b,c}$
$K_2Cr_2O_7$	0.9800	2.73
KNO_3	0.9248	10.6
$ZnSO_4 \cdot 7H_2O$	0.8710	18.7
KCl	0.8426	23.2
$(NH_4)_2SO_4$	0.7997	30.2
NH_4Cl	0.7710	35.2
$NaCl$	0.7528	38.4
$NaNO_3$	0.7379	41.1
NH_4NO_3	0.6183	65
$LiNO_3 \cdot 3H_2O$	0.4706	102
$K(C_2H_3O_2) \cdot 1.5H_2O$	0.2245	202
$LiCl \cdot H_2O$	0.1105	298
$NaOH \cdot H_2O$	0.0703	359

[a] R. A. Robinson and R. H. Stokes, "Electrolyte Solutions." Academic Press, New York, 1959.
[b] Calculated using Eq. (1) in this chapter.
[c] Although many of these water potential conditions are too low for cultivating bacterial biofilms, the data convey the reduction in liquid vapor pressure above many common salts.

Through equating (1) and (2) above, it follows that equilibrium r.h. and a_w (and thus ψ) in a closed system can be poised by adding a salt solution at the appropriate concentration. If the salt concentration exceeds the solubility (K_{sp}), the salt solution is "saturated" and the maximum effect of the salt on water vapor pressure is achieved. The use of salt solutions to control water vapor pressure and thus ψ is called "isopiestic control."[10,11] The r.h. in equilibrium with various saturated salt solutions are provided in Table I; tabulated values of osmotic coefficients (Eq. 2) for various salts at different aqueous concentrations can be found in either Papendick and Campbell[4] or Robinson and Stokes.[10]

In either liquid culture or in soil slurries, the ψ_s of the aqueous media can be varied by adding a nontoxic solute. By varying ψ_s, the effects of high osmolality (i.e., low ψ_s) on bacterial physiology can be studied independently of the matrix-associated thin film effects that occur in dry soil.[12–15] However, the full effects on

[10] R. A. Robinson and R. H. Stokes, "Electrolyte Solutions," 2nd ed. Butterworths, London, 1959.
[11] R. F. Harris, W. R. Gardner, A. A. Adebayo, and L. E. Sommers, *Appl. Microbiol.* **19,** 536 (1970).
[12] J. M. Stark and M. K. Firestone, *Appl. Environ. Microbiol.* **61,** 218 (1995).
[13] E. F. Soroker, Ph.D. Dissertation, University of California, Berkeley, 1990.
[14] P. A. Holden, L. J. Halverson, and M. K. Firestone, *Biodegradation* **8,** 143 (1997).
[15] L. J. Halverson, T. M. Jones, and M. K. Firestone, *Soil Sci. Soc. Amer. J.,* in press (2000).

biota of changing water potential and water content in unsaturated porous media are from at least three processes:[16] (1) water content displacing air and thus limiting gas-phase mass transfer, (2) thin water films at low ψ restricting solute mass transfer, and (3) low water potential impairing microbial physiology. Additionally, swelling and shrinking of the biofilm matrix, if it occurs during wetting and drying, could alter biofilm ultrastructure in ways that could affect nutrient mass transfer through the biofilm matrix.

The basic principles of water content and water potential need to be considered when culturing bacterial biofilms in unsaturated conditions. Further, if a culture system is intentionally designed as a simplified model of the natural environment, e.g., to eliminate the confounding effects of pore processes in soil, then the results have to be interpreted in the context of the experimental system and their extrapolation to describe more complex systems should be qualified. Table II summarizes ways in which ψ and its components are controlled experimentally. Note that the methods of controlling ψ vary in their applicability to studying either low ψ effects on cells, local effects on biofilms, or the three processes (delineated above) that change simultaneously in porous media with changing ψ.

General Principles for Cultivating Unsaturated Biofilms

We have used three methods for culturing unsaturated biofilms. In the first and simplest method, the ψ of solid media is controlled through the addition of an osmoticum such as polyethylene glycol MW 8000 (PEG-8000, Table III). Because PEG-8000 molecules are too large to pass the cytoplasmic membrane, PEG-8000 in either solid or liquid media effectively simulates a reduced ψ_m condition to bacteria without the toxic effects of specific ions.[17] Gelrite gellan gum (Sigma Chemical, St. Louis, MO) must be used as the solidifying agent because agar will not solidify in the presence of PEG.[18] Typically, we culture organisms directly on the ψ-amended solid medium; however, others have poured solid medium inside the petri dish lid to achieve isopiestic control of soil or other media in close proximity, i.e., in the base of the dish[11] (Table II).

In the second method, biofilms are cultured on a semipermeable (ultrafiltration or UF) membrane [13,19,20] that allows passage of water and small solutes between the biofilm environment on one side and an aqueous solution on the other (Fig. 2). The aqueous solution contains the osmoticum which is either a permeating solute (e.g.,

[16] P. A. Holden, J. R. Hunt, and M. K. Firestone, *Biotechnol. Bioeng.* **56,** 656 (1997).

[17] J. V. Lagerwerff, G. Ogata, and H. E. Eagle, *Science* **133,** 1486 (1961).

[18] J. Mexal and C. P. P. Reid, *Can. J. Bot.* **51,** 1579 (1975).

[19] B. Zur, *Soil Sci.* **102,** 394 (1966).

[20] L. J. Waldron and T. Manbeian, *Soil Sci.* **110,** 401 (1970).

TABLE II

METHODS FOR CONTROLLING ψ WHEN CULTURING BIOFILMS

Description	Depiction	Application and biofilm effect
Biofilm on ψ-controlled solid media[a,b]	ψ-Controlled solid / Biofilm / Petri dish	ψ_m of cells is controlled with PEG; ψ_s of biofilm is set by the solute concentration
Culture directly under ψ-controlled solid media (isopiestic control)[c]	Biofilm or porous media / ψ-Controlled solid / Petri dish	ψ is controlled by the effect on r.h.
Saturated salt solution in a closed chamber (isopiestic control)[c,d]	Desiccator / Porous media or biofilm culture / Base of desiccator containing saturated salt solution	ψ is controlled by the effect on r.h.
Controlled water content of porous media	Porous media	Either ψ or ψ_m is controlled, depending on the basis for the characteristic curve
Culture on semipermeable membrane overlaying solution of constant osmolality	Biofilm or porous media / Ultrafiltration membrane / ψ-Controlled solution	ψ_m is controlled with PEG; ψ_s is set by the salt concentration
Continuous flow of r.h.-controlled gas	Gas inflow / Exhaust / Gas canula / ψ-Controlled solution	ψ is controlled by the effect on r.h.

[a] P. A. Holden, Ph.D. Dissertation, University of California, Berkeley, 1995.

[b] L. J. Halverson and M. K. Firestone, *Appl. Environ. Microbiol.* **66,** in press (2000).

[c] R. F. Harris, W. R. Gardner, A. A. Adebayo, and L. E. Sommers, *Appl. Environ. Microbiol.* **19,** 536 (1970).

[d] R. I. Papendick and G. S. Campbell, *in* "Theory and Measurement of Water Potential" (J. F. Parr, W. R. Gardner, and L. F. Elliot, eds.), p. 1. Soil Science Society of America, Madison, WI, 1981.

TABLE III

WATER POTENTIAL OF AQUEOUS PEG-8000
SOLUTIONS[a,b]

PEG-8000 (g/liter)	ψ ($-$MPa, 25°)
100	0.25
150	0.5
262	1.0
330	1.5

[a] B. E. Michel. *Plant Physiol.* **72**, 66 (1983).
[b] P. A. Holden, J. R. Hunt, and M. K. Firestone, *Biotechnol. Bioeng.* **56**, 656 (1997).

NaCl) and/or a nonpermeating solute (e.g., PEG) for controlling either ψ_s and/or ψ_m, respectively. Whether or not the solute passes through the semipermeable membrane depends on the selected osmoticum and the molecular weight cutoff criterion for the particular membrane. For example, when PEG-8000 is added to the aqueous solution, the biofilm adhering to the opposite side of a MW 3000-cutoff membrane experiences a reduced ψ_m that is quantitatively related to the amount of PEG in solution (Table III). On the other hand, if NaCl is added to the aqueous solution to reduce ψ_s [Eq. (2)], NaCl molecules diffuse across the membrane and ψ_s is reduced on both sides of the membrane by the interaction of salts with water.

For the third method, biofilms are cultured in unsaturated porous medium (e.g., fine quartz sand). At least two ways can be used to set the water potential in porous media. First, ψ_m and ψ_s can be controlled across a semipermeable membrane where the biofilm-colonized porous medium is spread over the membrane and the membrane is in intimate and continuous contact with the aqueous nutrient/osmoticum reservoir below.[13,19] Alternatively, ψ can be poised by the gravimetric moisture content through adding aqueous mineral medium. In the latter case, it is necessary to first develop the characteristic curve based on total water potential (i.e., ψ vs θ). This is done by using thermocouple psychrometry[21] to measure r.h. at a range of moisture contents for the material of interest. Depending on the thermocouple psychrometer in use, these r.h. values are then converted to ψ using a calibration curve that has been generated from a set of standard salt solutions; i.e., ψ_s is known from Eq. (2) and r.h. is measured for each salt standard. The value for water content that will adjust the initially dry sample to a desired ψ can be extracted from the characteristic curve.

[21] S. L. Rawlins and G. S. Campbell, *in* "Water Potential: Thermocouple Psychrometry" (A. Klute, ed.), 2nd ed., p. 597. Soil Science Society of America, Madison, WI, 1986.

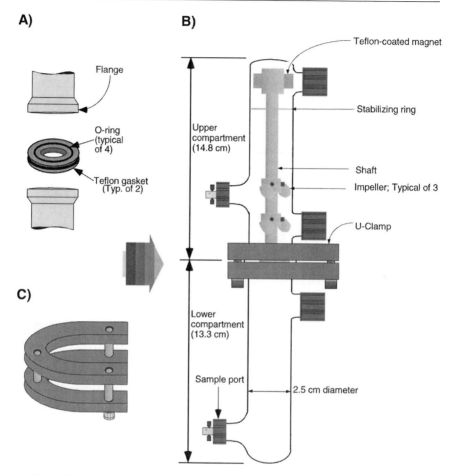

FIG. 2. Membrane reactor for cultivating unsaturated biofilms (not to scale). (A) gasket design for the flanged center joint, (B) reactor, (C) U-clamp. The flanges are connected end-to-end by the U-clamp.

With these three techniques for culturing biofilms as a function of ψ, when should one be chosen over another? Cultivating unsaturated biofilms either on solid medium or on semipermeable membranes is similarly useful when the study goal is to examine the development and characteristics of either (1) biofilms that grow on exposed surfaces (plant surfaces, food, rocks, etc.), or (2) biofilms that grow on solid surfaces within porous media. In the latter case, the surrounding porous medium matrix is intentionally absent; thus, biofilms are intentionally studied without the confounding processes of either restricted solute or restricted gas transport that can occur in pores. It should be noted that solid medium culture may still result in restricted nutrient supply, which is observed in studies of antibiotic diffusion zones,[22] but membrane culture solves the problem of restricted liquid mass transfer of nutrients when the aqueous nutrient reservoir is stirred. Still, before extrapolating results from the simplified system of membrane culture to the complex matrix of porous media, there should be a quantitative analysis of matrix-associated water potential effects [Eqs. (1) and (2), last section] to determine if they potentially outweigh local effects of reduced water availability on biofilms. Finally, if the goal of studying biofilms in unsaturated systems is to understand biofilm development in the context of all three interacting processes (see previous section), then the culture system should mimic the natural substrata as much as possible; e.g., instead of a semipermeable membrane, porous media should be used as the substrata for biofilm growth if the actual growth environment is soil or sand. The following provides guidelines for cultivating unsaturated bacterial biofilms using each of the three described methods.

Cultivation on Solid Media

Bacterial colonies on solid media surfaces are the simplest examples of unsaturated biofilms: there is a continuous air biofilm interface and cells are attached to each other and to the surface by an EPS matrix. The ψ of the biofilm environment is established by the ψ_s of the solid media formulation. The solidifying agent does not affect ψ, and for most standard solid medium formulations ψ is approximately 0 in the absence of high concentrations of osmoticum.[11,23]

For controlling ψ_m of the biofilm grown on solid media, a nonpermeating solute is used.[24,25] Two solutions are prepared and sterilized separately, then each warmed to between $65°$ and $75°$ before mixing. Solution A consists of PEG-8000 (Table III) plus 500 ml purified (18 MΩ-cm) water. Fully dissolve the PEG-8000 by stirring at room temperature. After the PEG-8000 has fully dissolved, measure the total volume of the solution and record this volume. The PEG-8000 increases solution volume, and therefore additional nutrients above the typical solid medium formulation are needed so that nutrient concentrations are equivalent at each ψ. Filter sterilize (0.2 μm) solution A, then bring to temperature ($65–75°$) while stirring. Solution B contains desired mineral salts, a carbon source, purified (18 MΩ-cm) water, plus 9.3 g gellan gum. For solution B, use 500 ml water for any ψ less than 0; for solid medium without PEG, use 1000 ml water. Calculate the mass of mineral salts to be added by multiplying the desired mass per liter by the fractional increase in medium volume due to the PEG addition in solution A. Again, additional mineral

[22] K. E. Cooper, *in* "The Theory of Antibiotic Diffusion Zones" (F. Kavanagh, ed.), Vol. II, p. 13. Academic Press, New York, 1972.

[23] W. J. Scott, *Adv. Food Res.* **7**, 83 (1957).

[24] P. A. Holden, Ph.D. Dissertation, University of California, Berkeley, 1995.

[25] L. J. Halverson and M. K. Firestone, *Appl. Environ. Microbiol.* **66**, in press (2000).

salts will be necessary in PEG-amended media to provide the same concentration (mass/volume) of mineral salts in each solid media formulation. Mix solution B to dissolve, autoclave to sterilize, and then bring to temperature. While continuously stirring solution A, slowly add warmed solution B. Because the media solidifies quickly, the plates should be poured immediately after mixing the two solutions. Also, because of the rapid solidification rate, only one ψ_m-amended solid medium formulation should be prepared at a time. Store like-ψ_m plates together in a sealed plastic bag in the refrigerator until needed.

Cultivation in Membrane Reactors

Reactor System Description. Figure 2 is a schematic of a "membrane reactor" we developed for cultivating unsaturated biofilms with ψ_m control.[16] As shown in Fig. 2, the reactor consists of two compartments: a stirred upper compartment that contains the liquid phase, and a lower compartment that contains gas phase plus biofilm. The biofilm is cultivated in the gas compartment on a UF membrane. By having the liquid compartment resting above and on top of the UF membrane, the contact between the overlying liquid medium and the UF membrane is continuous and without gas pockets. The stirrer consists of a fabricated 316 stainless steel (SST) shaft with three purchased 316 SST impellers (Cole-Parmer, Vernon Hills, IL) held by set screws. A magnetic stir bar is situated crosswise through one end of the stirrer shaft and a stabilizing ring is welded to the upper shaft. The magnetic stir bar is attracted through the glass to an external magnet at the outside top of the reactor (Fig. 3). The external magnet is part of the drive assembly for the stirrer: a constant speed motor with a toggle (ON/OFF) switch is directly coupled to the external magnet, which in turn drives the stirrer magnet.

An experienced glass blower is required for fabricating the reactor. The body of the reactor is flanged Pyrex glass tubing (Corning, Corning, NY) that is fused on each end to create the desired geometry (e.g., flat at the magnet end and round on the gas compartment end). The design includes sampling ports made of 13 mm threaded Pyrex glass test tubes (Corning) that are fused to the reactor body in various places. At a minimum, there should be two liquid compartment sampling ports (one at the base of the liquid column and one in the headspace) and two gas compartment sampling ports (one near the membrane and another near the rounded end of the gas compartment). The caps on the test tubes are Mininert (Valco Instruments Co., Inc., Houston, TX), and the threads are wrapped with Teflon tape (E. I. Dupont de Nemours, Wilmington, DE) to ensure gas tightness. The center seal of the reactor consists of two Teflon gaskets, one seated inside each of the two glass flanges. The Teflon gaskets each have two Viton O-rings: (1) one around the perimeter of the gasket to seal the gasket against the inside of the flange; and (2) one on the top (membrane) face of the gasket to seal the two compartments together. A fabricated two-part plastic U-clamp with three points of compression (Fig. 2) is used to connect the two reactor flanges. When the U-clamp is secured

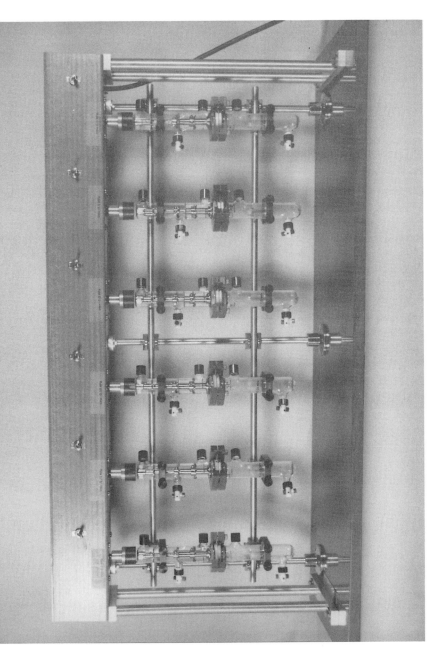

FIG. 3. Rack for six membrane reactors. Single-speed (150 rpm) motors are concealed behind the top front plate. Toggle (ON/OFF) switches for motor operation are located above and to the right of each reactor. Each motor is directly coupled to an external magnet that turns with the stirrer inside the reactor.

and the O-rings on the gaskets are fully compressed, a completely gas-tight seal is created at the center where the flanges abut.

The complete membrane reactor system consists of the reactors (Fig. 2), a support rack with drive motors and magnets (Fig. 3), and an assembly stand (Fig. 4) that enables one person to align, assemble, and inoculate the reactor in a sterile hood. Various small tools are also necessary, including metal spring clamps (Fisher Scientific, Tustin, CA) for holding the reactors together during autoclaving, electron microscopy forceps for handling smaller membranes, membrane forceps, and

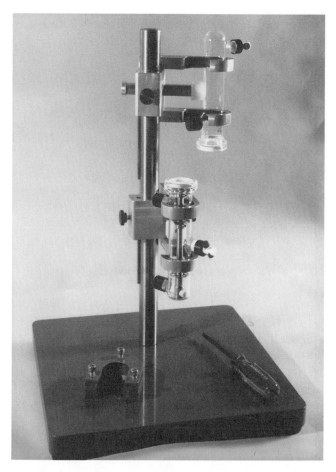

FIG. 4. Membrane reactor assembly stand holding a separated reactor. The liquid compartment is at the bottom so that culture medium can be added. Foam fabric (for cushion) lines the interior of the two moveable grips. The ball-point hex driver and a U-clamp are lying on the base of the stand.

a ball-point hex driver (McMaster Carr, Los Angeles, CA; e.g., Fig. 4) for tightening the screws on the three-point U-clamp.

Procedure for Cultivation. The reactor is autoclaved with the stirrer, Mininert caps, and Teflon gaskets in place, but the O-rings and three-point U-clamps are not autoclavable. The reactor is assembled for autoclaving by holding the two flanged ends together, wrapping the flanged joint with aluminum foil, then clamping with a spring clamp and tightening the spring clamp screw. To assemble the autoclaved reactor, the assembly stand is first swabbed with 70% ethanol and moved with other assembly tools into a sterile hood. The inoculum is prepared from either liquid culture or solid media culture and consists of a suspension of cells of desired type and population density. Once the reactors, tools, and inoculum are in the hood, assembly proceeds as follows:

1. Tighten the Mininert caps by hand, being careful not to overtighten. Overtightening will crack the delicate glass threads.

2. Place the reactor on the assembly stand such that the liquid compartment (stirrer side) is on the bottom. Align the reactor vertically and secure both compartments to the stand using the adjustable arms and set screws.

3. Remove the center foil wrap that was added for autoclaving. Lift the top arm (holding the top half of the reactor) slightly and swing the reactor half (i.e., the gas compartment oriented at the top for setup purposes) away from the lower half (i.e., the liquid compartment for setup purposes).

4. Pipette 30 ml of ψ_m-adjusted liquid medium (Table III) into the lower reactor half (i.e., the liquid compartment).

5. Sterilize the O-rings in 70% ethanol for 2 minutes. Using sterile forceps, shake off the residual ethanol and place the O-rings on the gaskets. Seat the gaskets in the flanges using sterile forceps.

6. Cut a UF membrane (Amicon YM, Millipore, Bedford, MA) to size (approximately 3 mm diameter). Sterilize the precut UF membrane in 70% ethanol for 2 min, rinse in sterile purified (18 MΩ-cm) water, shake off residual water, then place the membrane with sterile forceps on top of the O-ring such that the UF-skin side of the membrane rests on top of the liquid-compartment side of the reactor. If desired, an additional sterile, moist membrane, e.g., nylon, can be overlaid on the UF membrane. This other membrane may be necessary for bacteria that do not readily colonize the UF membrane backing. If two membranes are used as a composite, it is important that the two membranes stay in intimate contact so the biofilm receives continuous nutrition from the liquid compartment.

7. Pipette 10–20 μl of cell suspension to inoculate the center of the membrane. Ensure that the inoculum drop stays in the center of the membrane during the final steps of assembling the reactor.

8. Swing the top reactor half (gas compartment) such that it is vertically aligned with the lower half (liquid compartment). Lower the gas compartment down gently while ensuring that the O-rings remain seated.

9. Slide one half of a three-point U-clamp over the liquid compartment flange. Slide the other half over the gas compartment flange. Hold the clamp halves together and tighten the hex screws first by hand then by the ball-point hex driver. The screws are tightened sequentially, a turn at a time, so that the final compression is equivalent around the flanges.

10. The assembled reactor is removed from the assembly stand and turned vertically so that the liquid compartment is on top. The small drop of inoculum will absorb into the membrane. The reactor is then placed on the rack. When the stir motor is switched ON, the stirrer typically seats itself with the internal magnet pulling up tightly against the inside top of the glass reactor. This is the proper orientation of the internal magnet and results in a gap between the lower impeller and the membrane.

Measuring Diffusion of a Volatile Compound. During cultivation of the biofilms, the diffusion of a volatile compound across the biofilm and membrane can be measured. The volatile compound is added, in excess, as a liquid in the bottom (gas compartment); this sets the lower boundary condition in the gas phase at a constant concentration corresponding to the saturation vapor pressure of the compound. After the compound is added to the gas compartment, the diffusion of the compound across the biofilm can be measured by analyzing, over time, for the volatilized compound in the headspace above the liquid compartment. In our studies with *P. putida,* we used benzene as the conserved diffusing compound because toluene was a carbon source for the bacteria.[16] We compared the non-steady-state accumulation of benzene above the liquid compartment of an inoculated reactor with the non-steady-state accumulation in an uninoculated reactor to enable the calculation of mass transfer resistance due exclusively to the biofilm. The data is interpreted using mathematical models based on Fick's law and the conservation of mass in the reactor. The mathematical description of diffusion and reaction processes occurring in membrane-cultivated unsaturated biofilms are described previously.[16]

Procedure for Harvesting Biofilms. The membrane-cultivated biofilms are harvested from the reactor in an order that essentially parallels the assembly process. The reactor is transferred to the sterile hood and turned upside down, with the liquid compartment now oriented downward as in the assembly procedure. The reactor is aligned vertically and is secured to the assembly stand using the assembly stand arms and set screws. The three-point U-clamp is removed and the reactor halves are separated without damaging the biofilm. If the biofilm has been cultured on a nylon membrane, this membrane is lifted away from the ultrafiltration membrane with sterile forceps. The biofilm, still attached to the membrane, can be imaged with light microscopy. The biofilm can also be washed from the membrane using a 1% SDS solution. The protein content of the biofilm can then be quantified by the method of Bradford[26] using detergent-compatible reagents (Bio-Rad, Hercules, CA).

Cultivation in Unsaturated Sand

It may be desirable to cultivate unsaturated biofilms on either quartz sand or other porous media, particularly if the objective is to study the formation of biofilms on solid surfaces in pore spaces. If so, the quartz sand must be washed thoroughly beforehand to remove clay minerals and organic matter. Various washing methods have been recommended. One multistep method involves (1) soaking in 50% HNO_3 for 24 hr, then complete rinsing in distilled water; (2) sonication in 1.0% sodium polyphosphate, then distilled water rinsing; (3) 20 min sonication in 1.0 mM $NaNO_3$ (pH 6.6), then complete rinsing with distilled water; and (4) oven drying (110°).[27] At a minimum, sand should be washed with 0.1 M HCl[28] to solubilize carbonates, then rinsed thoroughly and treated with 1% H_2O_2 to remove organic material. Following the cleaning process, the porous medium should not leach acidity or alkalinity; this can be confirmed by soaking 1 volume of sand in 2 volumes of distilled water and noting the stability of the pH over the course of several days. Clean quartz sand should be autoclave-sterilized. The characteristic curve of the clean, sterile sand can then be determined. If the ψ is desired to be less than 0, it can be set either by adjusting the water content using aqueous mineral solution or by isopiestic equilibration of the moistened sand. The end result will be slightly different with these two methods: in the first, the total mass of nutrients added will vary with the volume of aqueous mineral solution added and, in the second, the total mass of nutrients is the same, but the concentration of nutrients in the water films on the sand grains changes during isopiestic equilibration. Very fine sand (e.g., 200 μm) is recommended, as it will be easier to control the ψ of fine sand (high specific surface) than that of coarse sand. The inoculum (suspension of bacterial cells) is added in the aqueous mineral medium, which also may contain the carbon source. The moistened sand should be well mixed using a roller table. We cultivate biofilms in batch sand cultures in two different vessels and use 24 mm Mininert caps on each: either 40 ml glass vials (Supelco, Inc., Bellefonte, PA) or 250 ml boston rounds (Fisher, Tustin, CA). The ψ of the system will be maintained as long as it stays tightly closed. However, if the system is opened for removing sand samples during a time course experiment, the moisture content and ψ can be maintained by flushing the system continuously with sterile (0.2 μm), humidity-controlled air through a sterile gas cannula (Popper & Sons, New Hyde Park, NY).

[26] M. M. Bradford, *Anal. Biochem.* **72**, 248 (1976).
[27] J. Wan, J. L. Wilson, and T. L. Kieft, *Appl. Environ. Microbiol.* **60**, 509 (1994).
[28] C. Chenu, *Geoderma* **56**, 143 (1993).

High Resolution Microscopy of Unsaturated Biofilms

We have used three methods for visualizing bacteria at high resolution in unsaturated biofilms. The first two, transmission electron microscopy (TEM) and atomic force microscopy (AFM), are used with membrane-cultivated biofilms. The last method, environmental scanning electron microscopy (ESEM), is used for imaging biofilms cultivated in porous media.

TEM of Membrane-Cultivated Biofilms

Described here are the TEM methods we have used for membrane reactor-grown biofilms.[16,24] To grow unsaturated biofilms for TEM, we precut cigarette paper disks (1/8-inch diameter, 23 μm thick) with a metal punch and autoclave-sterilize the disks in a closed glass petri dish. Several sterile disks are overlain on the UF membrane in the gas compartment (i.e., biofilm side) of a reactor. The inoculum is pipetted onto the paper disks and the reactor assembly is completed as usual. During the course of the incubation, the biofilm grows over and covers the disks. The disks are removed with electron microscopy forceps and kept under isopiestic conditions to maintain ψ of the biofilm culture environment. The biofilms, still on the paper disks, are cryopreserved using a propane jet freezer (model MF 7200, RMC Inc., Tucson, AZ). The frozen samples are maintained in liquid nitrogen until freeze substitution with 1% osmium and 0.1% uranyl acetate in anhydrous acetone. Embedment, sectioning, poststaining, and TEM are then performed according to standard methods.[16] The biofilms shown in Fig. 5 were cultivated using the membrane reactor system described herein, and TEM was performed by Doug Davis and Kent McDonald (U.C. Berkeley).

AFM of Membrane-Cultivated Biofilms

We have used atomic force microscopy (AFM) for high resolution imaging of unsaturated *Pseudomonas putida* biofilms.[29,30] We culture the biofilms on Nuclepore (Corning Separations Division, Corning, NY) track etch polyester membranes (13 mm diameter, 0.1 μm pore size) overlaying solid (agar) media. We have found that these membranes provide a suitably smooth substratum for subsequent AFM imaging. Generally, the image stability is improved if the biofilms are dried slightly.[30] Biofilms can be dried on the membranes by removing the membrane from the agar and placing the membrane inside a plastic bell jar (Nalgene, Rochester, NY) that contains a saturated salt solution (Table I) for isopiestic control of equilibrium r.h. (Table II). The membrane is then adhered to a mica substrate using double-stick tape.[30] No other preparation of the biofilm is necessary.

[29] H. G. Hansma, L. I. Pietrasanta, I. D. Auerbach, C. Sorenson, R. Golan, and P. A. Holden, *J. Biomaterials Sci., Polymer Edition,* in press (2000).

[30] I. D. Auerbach, C. Sorenson, H. G. Hansma, and P. A. Holden, *J. Bacteriol.* **182,** in press (2000).

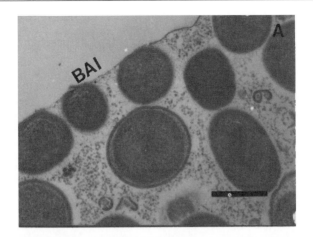

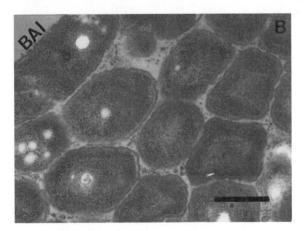

FIG. 5. Transmission electron micrographs of *Pseudomonas putida* mt-2 unsaturated biofilms cultivated in membrane reactors with glucose as the carbon source. The water potential of aqueous media in the upper reactor chamber is controlled by PEG-8000: (A) −0.25 MPa, (B) −0.5 MPa. Scale bar is 1 μm; BAI = biofilm air interface.

Operating parameters for the Nanoscope AFM (Digital Instruments, Santa Barbara, CA) are described elsewhere.[29,30]

ESEM of Biofilms in Porous Media

ESEM is a high resolution imaging technique that is applicable to biofilms in porous media because there is no need for the excessive sample preparation (e.g., drying or coating with a conductant) that is typically required in conventional EM.[31] ESEM is a particularly intriguing EM technique for biofilms cultivated in

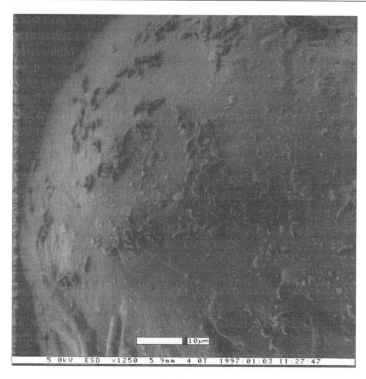

FIG. 6. Environmental scanning electron micrograph (ESEM) of *Rhodococcus erythropolis* cultivated in unsaturated sand with hexadecane as the sole carbon source. The moisture content is 3.4% (w/w). The sample was imaged without pretreatment.

unsaturated conditions because there is the possibility to wet and dry the sample, thus mimicking the natural biofilm environment, while observing the biofilm morphological changes in real time. We have used ESEM to search for bacteria in native vadose zone material and to examine microbial colonization of sand in laboratory preparations (Fig. 6). There is no sample preparation for ESEM; the sample can be wet, oily, and uncoated. The instrument we have used is similar or identical to the XL30 ESEM FEG (FEI Company, The Netherlands).

Assaying Water Potential Effects on Bacterial Physiology

In unsaturated biofilm science, culturing organisms in well-mixed liquid is necessary for obtaining Michaelis–Menten kinetic parameters related to substrate transformation[16] or Monod growth parameters.[14] With media alterations, the water potential conditions experienced by unsaturated biofilm microbes can be simulated in liquid culture so that the purely physiological responses to water stress can be

---

[31] B. Little, P. Wagner, R. Ray, R. Pope, and R. Scheetz, *J. Indust. Microbiol.* 213 (1991).

determined. The information gained from liquid culturing under water stress conditions is without the confounding effects of mass transfer limitations in porous media, etc. This short section is to make the reader aware of the various approaches that have been used to control ψ in liquid culture. Solute stress, i.e., low ψ_s, can be imposed by adding nontoxic salts to the growth media. As before, the total water potential of the medium is confirmed using a thermocouple psychrometer. A reduction in ψ_m can be imposed in liquid culture by amending the aqueous mineral medium with high molecular weight PEG[14,32,33] (Table III). However, PEG is a cosolvent[34] and its effects on mass transfer of various chemicals in the culture flask should be accounted for.[14,24] When we culture bacteria using ψ_s- or ψ_m-controlled medium, we use a nephelo flask that can be made gas-tight, such as the 500 ml triple-baffled side arm flask made by Bellco (Vineland, NJ). We make the flask gas-tight by wrapping the top and side arm threads with Teflon tape. The side arm port is capped with an 18 mm Mininert cap so that headspace and liquid can be easily sampled.

Conclusions

Unsaturated biofilms can be cultivated in many ways that effectively simulate the dry or transiently moist environments in which they naturally grow. These natural environments include the surfaces of plants, food, rocks, monuments, and the vadose zone (including surface soil). What is required for planning unsaturated biofilm cultivation is first an understanding of water relations for the natural system being simulated, and second a solid understanding of how the cultivation scheme will enable one to address the questions of interest. At the same time, the limitations, if any, of the selected cultivation scheme for replicating the full complexity of the natural environment should be acknowledged so that the results of experimentation can be interpreted appropriately. Although unsaturated biofilms are understudied relative to their saturated counterparts, the variety of useful methods available for cultivating and studying them will hopefully stimulate more researchers to examine their various forms and functions in the many systems they occupy.

Acknowledgments

The preparation of this manuscript was supported by funding from the U.S. EPA (R827133- 01) and the U.C. Water Resources Center (W-904). The membrane reactor system was developed as part of P. A. Holden's doctoral dissertation in the lab of Mary K. Firestone at the University of California, Berkeley. The membrane reactor rack and assembly stand was machined by Jim Thornbury, U.C. Berkeley. The design of the membrane reactor system was assisted by Michael D. Guthrie.

[32] K. J. McAneney, R. F. Harris, and W. R. Gardner, *Soil Sci. Soc. Am. J.* **46,** 542 (1982).
[33] M. D. Busse and P. J. Bottomley, *Appl. Environ. Microbiol.* **55,** 2431 (1989).
[34] S. Banerjee and S. H. Yalkowsky, *Anal. Chem.* **60,** 2153 (1988).

[10] Biodegradable Organic Matter Measurement and Bacterial Regrowth in Potable Water

By CHRISTIAN J. VOLK

Introduction

The issues of drinking water biostability, coliform regrowth control, and compliance with water quality regulations are major concerns to water producers. After water treatment, alteration of water quality can occur during distribution to the customer's tap. The presence of low concentrations of organic nutrients in water is sufficient to support biofilm growth within the pipe network. The consequence of this bacterial regrowth can include the following: degradation of bacterial water quality (i.e., high heterotrophic bacteria counts, occurrence of coliform bacteria, noncompliance with existing water regulations), amplification of corrosion, generation of bad taste and odors, customer complaints, and development of a food chain leading to the presence of invertebrates.[1-3] Therefore, the control of biodegradable organic matter level is becoming recognized as an important part in the operation of drinking water plants and distribution systems. It is critical to dispose of simple methodologies in order to routinely determine the levels of biodegradable organic matter in finished waters. This chapter presents methods available to evaluate biodegradable organic matter concentrations in water, and describes and compares several bioassays commonly used in the drinking water field to assess the bacterial regrowth potential of waters.

Biodegradability Estimation

Natural organic matter (NOM) in water represents a complex mixture of various compounds (such as humic substances, hydrophilic acids, carbohydrates, amino acids, and carboxylic acids) generally present at low concentrations, making qualitative and quantitative analysis of specific molecules difficult.[4,5] The total amount of organic materials can be determined by measuring concentrations of total and dissolved organic carbon (TOC, DOC). NOM can also be divided into two fractions. The first, called biodegradable organic matter (BOM), can be utilized by bacteria as a source of energy and carbon. The second fraction is refractory to

[1] J. Coallier, P. Lafrance, D. Duschesne, and J. Lavoie, *Sci. Tech. Eau* **22,** 63 (1989).
[2] E. E. Geildreich, *in* "Microbial Quality of Water Supply in Distribution Systems." CRC Press, Boca Raton, FL, 1996.
[3] R. W. Levy, F. L. Hart, and R. D. Cheetham, *J. Amer. Water Works Assoc.* **78,** 105 (1986).
[4] G. L. Amy, *in* "Proc. Workshop on NOM in Drinking Water," Chamonix, France, Sept. 1993.
[5] E. M. Thurman and R. L. Malcolm, *Environ. Sci. Technol.* **45,** 463 (1981).

0076-6879/00 $35.00

biodegradation (e.g., nonbiodegradable) and has little effect on bacterial regrowth. The BOM that is not removed during water treatment results in growth of bacteria in distribution systems. Several biological tests have been developed to assess the level of biodegradable organic matter in water.[6,7] Experimental setup varies from one assay to another; however, they include the following steps[8]:

Pretreatment of glassware: Carbon contamination must be avoided when performing BOM measurements because water samples contain generally low carbon levels. The glassware and materials in contact with water samples are rendered carbon free after heat treatment or washing in a strong oxidant solution.

Pretreatment of water samples: Depending on the methodology and sample type, pretreatment of samples is required to inactivate indigenous bacteria present in the water sample, remove suspended matters, or neutralize any toxic chemicals such as disinfectant residuals. Inactivation of microorganisms can be performed using a pseudo pasteurization. Water samples are heated at 60° for 30 min to inactivate autochthonous bacteria without drastically altering the organic matrix. Sample bacteria can also be retained after filtration using 0.2 μm membranes. Filtration is also performed for raw water samples containing suspended materials. Filtration membranes used to filter water samples are made of materials that do not release organic compounds, such as fiberglass or polycarbonate membranes. Treated waters often contain disinfectants that are neutralized using sodium thiosulfate.

Preparation of a bacterial inoculum: Since there is no standardized inoculum, results can be affected by the inoculum type. Inoculum can be composed of pure cultures of one or several bacterial strains used at certain densities or mixed natural microbiota. Pure culture inocula involve bacterial strains well adapted to low nutrient concentration conditions found in waters. Natural inocula are generally composed of several unknown species; it is difficult to assess the number of active cells, species diversity, and bacterial affinity toward organic molecules. The origin of the inoculum does not affect the result when using mixed natural inocula.[9]

Inoculation and incubation of water samples: Inoculum density varies from a few hundreds to millions of bacteria, depending on the method. Incubation periods that allow test bacteria to grow or to degrade organic molecules vary from a few hours to 1 month.

Measurement of the test parameter for biodegradation: Bacterial growth or carbon utilization is monitored during the incubation of inoculated water samples.

[6] P. M. Huck, *J. Amer. Water Works Assoc.* **82,** 78 (1990).

[7] J. C. Joret and M. Prevost, *in* "Biodegradable Organic Matter in Drinking Water" (J. C. Joret and M. Prevost, eds.). In preparation, 2001.

[8] L. Mathieu, Ph.D. thesis in water chemistry and microbiology, University of Nancy, France, 1992.

[9] J. C. Block, L. Mathieu, P. Servais, D. Fontvielle, and P. Werner, *Wat. Res.* **26,** 481 (1992).

Increase in bacterial counts can be evaluated using bacterial counts on agar, microscopic numeration in epifluorescence, measurement of cell energy such as ATP, or turbidity monitoring. Carbon consumption by bacteria is assessed by performing dissolved organic carbon measurements.

Calculation of the final biostability value: Depending on the parameter used to evaluate the biodegradable organic matter, results can be expressed using an index of biodegradability (μg acetate carbon equivalents/liter), a cell production (growth yield or maximum growth), or a carbon consumption (expressed in mg C/liter).

The biological assays developed are based on two concepts (Fig. 1): (i) the assimilable organic carbon (AOC) when bacterial growth is monitored, and (ii) the biodegradable dissolved organic carbon (BDOC) when organic carbon consumption is measured. The AOC refers to a fraction of TOC that can be utilized by bacteria for growth. The inoculum is a mixture of pure bacterial strains cultivated under laboratory conditions (*Pseudomonas fluorescens* P17, *Spirillum*

AOC

Inoculation:- Pure strains
- Natural microbiota

BACTERIAL GROWTH Monitoring

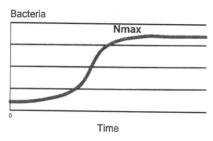

Monitored parameter:
-Plate counts
-ATP

Results: μg Acetate C eq. /l

BDOC

Inoculation: suspended or
fixed natural microbiota

DOC REDUCTION Monitoring

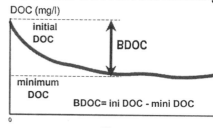

Monitored parameter:
- DOC concentration

Results: mg C /l

FIG. 1. AOC and BDOC concepts for biodegradable organic matter determination.

NOX).[10,11] Bacterial growth is monitored in the water samples by performing plate counts of bacteria and the maximum growth (N_{max}) observed during the incubation is converted into assimilable organic carbon using the growth yield of the bacteria from calibration curves performed from known concentrations of standard organic compounds (acetate, oxalate) (Fig. 1). Since the pioneering work of Van der Kooij,[10,11] various AOC methodologies have been developed (Table I). The AOC method was simplified[12] to be included in the "Standard Methods."[13] LeChevallier et al.[14] substituted fastidious plate counts with ATP measurements to evaluate bacterial levels. The time needed to perform the AOC test was reduced by increasing the incubation temperature and inoculum size. Other variations were developed to determine AOC concentrations. Kemmy et al.[15] used a mixture of four bacterial strains with a growth yield based on a standard mixture of organic compounds. Water samples can also be inoculated with environmental bacteria.[16] Bacterial growth is recorded using ATP measurements. Finally, other tests were developed to determine the regrowth potential of waters.[17-19] One method is based on recording the growth kinetics of a natural mixed inoculum using a sensitive turbidimeter.[17] When turbidity is measured for monitoring bacterial growth, two parameters are determined: the growth yield (μ) recorded during the exponential growth phase (indicating the level of biodegradability of the organic matter) and the growth factor (log Y/Yo) corresponding to the increase in bacterial population. Another method evaluates the ability of waters to support the growth of coliforms.[18] The coliform growth response assay (CGR) uses Enterobacter cloacae as the seed organism. Changes in bacterial numbers are monitored over a 5-day period using plate counts. The index of available nutrient to support coliform regrowth is calculated by log transformation of the ratio between colony density at the end and the beginning of the incubation (CGR = log N5/N0). On the other hand, the BDOC content represents the fraction of DOC that can be assimilated and mineralized by heterotrophic microbes (Fig. 1). The inoculum consists of environmental bacteria suspended in water or fixed on a surface. BDOC is the difference between initial

[10] D. Van der Kooij, A. Visser, and W. A. M. Hijnen, *J. Amer. Water Works Assoc.* **74**, 540 (1982).

[11] D. Van der Kooij and W. A. M. Hijnen, *Appl. Environ. Microbiol.* **47**, 551 (1984).

[12] L. A. Kaplan, T. L. Bott, and D. J. Reasoner, *Appl. Environ. Microbiol.* **59**, 1532 (1993).

[13] American Public Health Association, *in* "Standard Methods for the Examination of Water and Wastewater," 20th ed. American Public Health Association, Washington, D.C., 1999.

[14] M. W. LeChevallier, N. E. Shaw, L. A. Kaplan, and T. L. Bott, *Appl. Environ. Microbiol.* **59**, 1526 (1993).

[15] F. A. Kemmy, J. C. Fry, and R. A. Breach, *Water Sci. Technol.* **21**, 155 (1989).

[16] G. Stanfield and P. H. Jago, *WRC Environment Report,* PRU 1628-M, Manheim, UK (1987).

[17] P. Werner and B. Hambsch, *Water Supply* **4**, 227 (1986).

[18] D. J. Reasoner and E. W. Rice, *in* "Proc. Workshop on Measurement of AOC in the Field of Drinking Water Treatment," Karlsruhe, RFA, Nov. 1989.

[19] S. M. Bradford, C. J. Palmer, and B. H. Olson, *Wat. Res.* **28**, 427 (1994).

TABLE I

AOC METHODOLOGIES[a]

Authors (reference)	Sample preparation	Inoculum (concentration)	Incubation conditions	Parameter measured	Result expression
Van der Kooij et al.[10,11]	Pasteurization	Pure strains Ps. fluorescens P17 and Spirillum NOX (500 CFU/ml)	20 days 15°	CFU/ml	AOC calibration in known solutions of sodium acetate (µg C eq. acetate/liter)
Kaplan et al.[12]	Pasteurization	P. fluorescens P17 + Spirillum NOX (500–1000 CFU/ml)	9 days 20°	CFU/ml	AOC
LeChevallier et al.[14]	Pasteurization	P. fluorescens P17 + Spirillum NOX (10^4 CFU/ml)	5 days 22°	ATP	AOC
Bradford et al.[19]	Filtration	P. fluorescens P17	12 hr 20°	Cell elongation Microscopic counts	AOC
Kemmy et al.[15]	Filtration	Mixture of 4 strains: Pseudomonas fl. + Curtobacterium + Corynebacterium + 1 species of coryneform type	6 days 20°	CFU/ml	AOC calibration in solution of mixed organic compounds (µg C/liter)
Stanfield and Jago[16]	Filtration	Bacteria from: Raw water Sand filtered water	Until max. growth 20°	ATP	AOC—standard conversion factor (µg C/liter)
Werner and Hambsch[17]	Filtration	Water sample bacteria retained on the filter (5×10^4 cells/ml)	60 hr 20°	Turbidity	Growth yield (µ) Growth factor (logY/Yo)
Reasoner and Rice[18]	Filtration	Enterobacter cloacae, Escherichia coli, Klebsiella oxytoca	5 days 20°	CFU/ml	log N5/NO

[a] Adapted from Huck.[6]

TABLE II
BDOC METHODOLOGIES[a,b]

Author (references)	Sample preparation	Inoculation	Incubation conditions	Parameter measured	Result expression
Servais et al.[21]	Filtration	Suspended bacteria from river water	28 days 20°	DOC	$BDOC = DOC_i - DOC_f$
Joret and Levi,[20] Volk et al.[26]	—	Bacteria fixed on sand	1 week 20°	DOC	$BDOC = DOC_i - DOC_{mini}$
Mogren et al.[22]	—	Bacteria fixed on sand	5 days 20°	DOC	$BDOC = DOC_i - DOC_{5 \text{ days}}$
Frias et al.[23] Kaplan et al.[24]	Filtration	Column with bacteria fixed on porous glass particles	2.5 hr 20°	DOC	$BDOC = DOC_{inflow} - DOC_{outflow}$

[a] Adapted from Huck.[6]
[b] DOC_i, initial DOC concentration; DOC_f, final DOC concentration; DOC_{mini}, minimum DOC concentration.

DOC of the water sample and minimum DOC observed during the incubation period (Table II).[20-24]

Methodology

Four common methodologies will be described below, one for measuring AOC and three for evaluating BDOC levels in waters.

Rapid AOC Methodology using ATP

Test Principle. A pasteurized water sample is inoculated with approximately 10^4 CFU/ml of either *Pseudomonas fluorescens* P17 or *Spirillum* NOX. After incubation, ATP levels are measured and converted into assimilable organic carbon.[14]

Inoculum Preparation. Cultures of *Pseudomonas fluorescens* strain P17 and *Spirillum* strain NOX can be obtained from D. van der Kooij, KIWA, The

---

[20] J. C. Joret and Y. Levi, *Trib. Cebedeau.* **39**, 3 (1986).

[21] P. Servais, G. Billen, and M. C. Hascoet, *Wat. Res.* **21**, 445 (1987).

[22] E. M. Mogren, P. Scarpino, and R. S. Summers, *in* "Proc. Annual AWWA Conf.," Cincinnati, OH, June 1990.

[23] J. Frias, F. Ribas, and F. Lucena, *Wat. Res.* **26**, 255 (1992).

[24] L. A. Kaplan, F. Ribas, J. C. Joret, C. Volk, J. Frias, and F. Lucena, *in* "An Immobilized Biofilm Reactor for the Measurement of Biodegradable Organic Matter in Drinking Water: Validation and Test Field." AWWA Research Foundation, Denver, 1996.

Netherlands. Cultures are stored in a solution of 20% glycerol and 2% peptone at $-70°$. Prior to use, the cultures are retrieved from the freezer and streaked for purity on R2A agar (Difco Laboratories, Detroit, MI) incubated at room temperature (20–22°) for 3–5 days. A portion of the pure culture is inoculated into 100 ml of a sodium acetate solution. The sodium acetate solution contains 11.34 mg sodium acetate (2000 µg acetate carbon) in the following buffer: 7.0 mg K_2HPO_4, 3 mg KH_2PO_4, 0.1 mg $MgSO_4 \cdot 7H_2O$, 1.0 mg $(NH_4) SO_4$, 0.1 mg NaCl, 1.0 µg $FeSO_4$ (per liter). The inoculated acetate solution is incubated at room temperature for 7 days. Bacterial counts are generally in the range of 8.2×10^6 and 2.4×10^7 CFU/ml for stains P17 and NOX, respectively.

Sample Collection. To determine AOC concentrations, 15 precleaned 40 ml borosilicate glass vials with polytetrafluoroethylene (PTFE) baked silicone septa are used as sample containers. All vials and septa are rendered carbon free by either combustion (muffled at 525° for 6 hr) or oxidant soaking (for 1 hr at 70° in 100 g/liter sodium persulfate solution), respectively. The vials are filled to the shoulder with approximately 40 ml of water and heat-treated at 70° in a water bath for 30 min.

Test Procedure. Twelve vials are inoculated with approximately 10^4 CFU/ml, final density, of either strain P17 or NOX (6 vials each strain). Inoculated vials are incubated at 22° for 5 days. One vial is set aside as an uninoculated control. The controls are stored at 4° to prevent growth of indigenous bacteria surviving pasteurization. Analysis of ATP in the uninoculated control is used to adjust background levels. For each bacterial strain, ATP measurements are performed on 2 vials after 3, 4, and 5 days of incubation. To measure the ATP level of the test strains, the content of the vial is filtered through a 25 mm diameter, 0.2 µm pore size, cellulose acetate membrane filter (Sartorius, Hayward, CA) under a vacuum of 15.2 mm (6 in) Hg. Without being allowed to dry, the filter is placed into a solution of 0.5 ml HEPES buffer (pH 7.75) and 0.5 ml ATP releasing agent with phosphatase inhibitor (Turner Designs, Sunnyvale, CA) and vortexed for 10 sec. Care is taken to ensure that the filter is completely immersed in the buffered releasing reagent. After 20 min contact time, 150 µl is placed into a 8×50 mm polypropylene test tube, and the tube is placed into the luminometer (Turner Designs, Model TD-20e). Luciferin–luciferase (100 µl) is injected into the sample, and after a 5 sec delay, the full integral of light is recorded (10 sec measurement). Luminometer measurements are converted into AOC units by determining the level of ATP per cell (either P17 or NOX) and multiplying the value by the standard yield values of 4.1×10^6 CFU/µg acetate-C (P17) and 1.2×10^7 CFU/µg acetate-C for NOX.[10] The total concentration of AOC in µg/liter of equivalent acetate carbon is the sum of P17 AOC and NOX AOC.

Quality Control. The remaining two sample vials are used as a growth control and spiked with 100 µg/l acetate carbon (final concentration) in a mineral salt solution (17.1 mg K_2HPO_4, 76.4 mg NH_4Cl, and 144 mg KNO_3 per liter of

HPLC-grade water). The growth vials are used to detect growth-inhibitory substances in the water. In addition, a blank vial containing just the mineral salt buffer without added carbon is used to detect carbon contamination of the glassware. A yield vial containing 100 μg/liter acetate carbon and mineral salt buffer is used to check the growth yield of the cultures.

Method Variation to Monitor Bacterial Growth. ATP measurements can be accompanied or replaced by plate counts for evaluating bacterial levels. After dilution, samples are plated (spread plate technique on standard plate count agar). Plates must be incubated for 3–5 days at 22° before colonies are counted. Cell counts are converted into AOC using the P17 and NOX growth yields, and the sum of P17 AOC and NOX AOC defines the total AOC concentration.

BDOC Determination Using Suspended Bacteria

Principle. A water sample is inoculated with suspended bacteria and incubated at 20° ± 2° in the dark for 28 days. Bacterial inoculum consists of indigenous bacteria contained in a small volume of surface water. Dissolved organic carbon is measured at the beginning of the test and at the end of the incubation period by using a total organic carbon (TOC) analyzer. The BDOC concentration represents the difference between initial and final DOC readings.[21]

Preparation of Glassware. Since the water samples incubated may range in volume from tens to hundreds of milliliters, incubation vessels can vary in size and shape, but must be free of organic carbon. For example, suitable vessels may include 40 ml borosilicate glass vials or 500 ml capped bottles.[25] The glassware is cleaned by using a detergent wash, two acid rinses, and three organic carbon-free water rinses (precleaned commercially available vessels can also be used.) Glassware is dried, capped with aluminum foil, and muffled at 525° for 6 hr to remove residual organic contamination. Glassware that cannot be muffled (volumetric flasks or Teflon apparatus) is treated by soaking either with 100 g/liter sodium persulfate solution (1 hr at 70°) or sulfochromic acid (30 min in a concentrated solution of H_2SO_4 saturated with $K_2Cr_2O_7$), and rinsed with carbon-free water. Various types of filtration devices may be used to filter water samples and surface water inoculum, including glass vacuum filtration systems, peristaltic pumps, or syringes with 22 or 47 mm diameter filter holders. To limit organic contamination of the sample by the filters, glass fiber filtration membranes are heat-treated, and filtration devices are prewashed before use.

Water Collection and Preparation. Water samples are collected in carbon-free bottles and stored at low temperature (4°) before analysis. Untreated surface waters or partially treated waters that contain high concentrations of suspended solids should be prefiltered through carbon-free glass fiber filters or other carbon-free

[25] L. A. Kaplan, D. J. Reasoner, and E. W. Rice, *J. Amer. Water Works Assoc.* **86**, 121 (1994).

filters. Disinfectant residuals contained in treated potable waters must be neutralized with sodium thiosulfate (10%, w/V, 100 g of $Na_2S_2O_3$ in 1 liter of ultrapure water). It is not recommended to add more than 0.20 ml of a 10% sodium thiosulfate solution per 1000 ml of water sample. Water samples that have high DOC concentrations (raw or treated waste waters, certain surface waters) are diluted with blank water (spring water with a DOC level <0.2 mg/liter) to obtain DOC concentrations of diluted waters below 10 mg C/liter.

Sample Inoculation and Incubation. The inoculum consists of surface water with a DOC concentration that is preferably equal to or lower than that of the test water. The use of surface waters with higher carbon contents is acceptable but may increase the risk of carbon contamination of water samples. Inoculum water is filtered through 2.0 μm porosity polycarbonate membrane to remove particles and any protozoa that could graze bacteria. Inoculation ratio is 1 ml of surface water per 100 ml of water sample [ratio of 1% (v/v)]. The DOC of the water sample is checked prior (labeled DOC_0) and after inoculum addition (labeled DOC_1) to detect potential contamination of the water samples by the inoculum. The difference between DOC_1 and DOC_0 must be less than 0.05 mg C/liter. Incubation of samples is performed at $20° \pm 2°$, for 28 days in the dark. The DOC concentration of water samples is measured at the end of the incubation period (labeled DOC_f). The sample is filtered. The concentration of BDOC is expressed in mg C/liter according to the following formula: $BDOC = DOC_i - DOC_f$. DOC_i is the DOC concentration at the beginning of the test (average DOC_0 and DOC_1).

Quality Control. Quality control specific to the BDOC assay using suspended bacteria includes testing the incubation vessel and procedure for filtration contamination, checking the activity of the inoculum, and testing for inhibition of the assay bacteria.

Determination of BDOC Concentration with Bacteria Attached to Sand

Principle. A water sample is incubated with sand colonized by bacteria. Incubation of water sample is performed at $20° \pm 2°$ in the dark, in an environment free of organic carbon contamination. Water samples can be aerated with organic carbon free air during the incubation (Fig. 2). Dissolved organic carbon concentration is measured daily from the beginning of the test using a TOC analyzer. The assay is stopped when DOC concentration has reached a minimal value. The minimum DOC value, which corresponds to the nonbiodegradable DOC, is commonly reached after a 5- to 7-day incubation period. BDOC concentration is defined as the difference between initial and minimum DOC values.[20,26]

---

[26] C. Volk, C. Renner, C. Robert, and J. C. Joret, *Env. Technol.* **15**, 545 (1994).

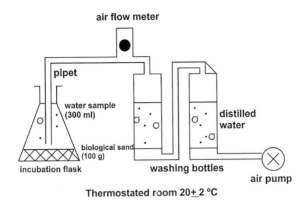

FIG. 2. Experimental setup to measure BDOC levels with bacteria attached to sand.

Test Preparation and Apparatus. Cleaning procedure, preparation of glassware, sample collection, and preparation are identical to the steps used in the method using suspended bacteria. The incubation vessels consist of organic-carbon-free 500 ml Erlenmeyer flasks (aerated test) or borosilicate glass bottles with cap (ground glass or Teflon seal) (nonaerated test). The aeration system includes an aquarium air pump, Pyrex glass washing bottle (250 or 500 ml), air flow meters (or aquarium multivalve system) to regulate air flow rate, tubing, and a sterile plugged borosilicate glass disposable pipette (5 ml) (Fig. 2). A filtration device is necessary for collecting DOC samples during water incubation. A 30 ml glass syringe, single use syringes, filter holders, and glass fiber filters (25 mm diameter), or single use filters (polycarbonate or polyvinylidene fluoride filter units; 25 mm diameter, pore size = 0.45 μm) are suitable.

Inoculum Preparation and Storage. The inoculum is a quartz sand (with a granulometry of 1–2 mm) colonized with bacterial biomass sampled from a sand filter of a water treatment plant without a prechlorination stage (no oxidant residual present), or colonized sand from a freshwater fish tank. When a pool of biological sand is freshly sampled from a treatment plant or an aquarium, the sand is prewashed with dechlorinated drinking water, then with organic-carbon-free water until the wash water is colorless and clear. The biological sand is ready for use as inoculum when no detectable DOC is released by the sand. The pool of washed biological sand is stored in dechlorinated tap water or under a stream of nonchlorinated water at room temperature (a nonturbid surface water or partially treated water containing no residual oxidant can be used to store the sand).

Test Starting and Sample Inoculation. The initial DOC (labeled DOC_0) of the water sample to be analyzed is checked. Before use, the pool of biological sand is rinsed with dechlorinated tap water. The water is drained from the sand. For sample inoculation, a sand/water sample ratio of $1/3$(w/v) is used (e.g., 100 g of sand per 300 ml test water). Drained sand (100 ± 10 g) is weighed into each incubation flask. Then, approximately 100 ml of water sample is poured into each incubation flask. The flasks are swirled gently to suspend the sand. After 10 min, the water is drained. This rinsing step avoids carbon exchanges (adsorption–release) between the biological sand and the water sample. Then, 300 ml of water is gently poured into the flask containing the sand. The DOC of the water sample in the presence of sand is confirmed after a 3–5 min sand/sample contact time (labeled DOC_1). The difference between DOC_1 and DOC_0 should not be higher than 0.2 mg C/liter. The incubation is started by capping the incubation flask (test without any aeration) or aerating the water samples as described in Fig. 2 (air flow rate 4.0 ± 0.5 liter/hr). Each washing bottle is filled with "ultrapure" water and every incubation flask is covered with clean aluminum foil.

BDOC Determination. The DOC concentration of each water sample is measured on a daily basis. The incubation is continued until a minimum value is reached (labeled DOC_m). The incubation is then prolonged 2 days after the minimum DOC concentration is obtained. The concentration of biodegradable dissolved organic carbon is expressed in mg C/liter according to the following formula: $BDOC = DOC_i - DOC_m$ (DOC_i is the DOC concentration at the beginning of the test: average of DOC_0 and DOC_1).

BDOC Determination using Bioreactors

Test Principle. The plug flow biofilm reactor was developed to measure biodegradable organic matter levels within a few hours.[23,24] The reactor column media is heavily colonized with bacteria from the environment. After the bioreactor is loaded with a water sample, the BDOC concentration is defined as the difference between the DOC concentration in the influent and effluent water of the reactor.

Reactor Description. A BDOC reactor consists of two glass chromatography columns (2.5×60 cm, Kontes, Vineland, NJ) containing porous glass particles (Siran; 1–2 μm diameter, 60–300 μm in pore size, Schott America, Yonkers, NY) colonized by bacteria (Fig. 3). Columns are protected from light with foam pipe insulator to prevent reactor colonization by algae. Bioreactors are continuously fed with water at a flow rate of 4 ml/min using a variable speed peristaltic pump (Masterflex pump, Cole Parmer, Vernon Hills, IL). The hydraulic retention time of the BDOC reactor is 2 hr. Tubing consists of PTFE and PharMed tubing (Cole Parmer). Two 3-port injection valves (Hamilton, Reno, NE) allow sampling water at the inlet and outlet of the reactor (Fig. 3).

Colonization. Columns are set in a convenient place to work close to a sink (for water waste) and at room temperature (approximately $20°$). At first, columns are colonized using water containing high levels of nutrients and becteria. Organic and mineral nutrients contained in the water will be used by bacteria for their growth

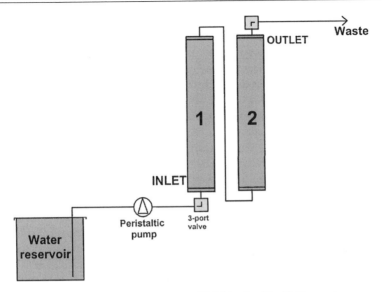

FIG. 3. Experimental setup to measure BDOC levels with a biofilm reactor.

and development. River water is the feed water of choice to allow development of a biofilm within the reactor. River water has to be filtered to limit reactor plugging using a multistage filtration system (Balston filter cartridge, Whatman, Haverhill, MA) or an on-line filtering device (Collins Products, Livingston, TX). Filtered river water can be stored in a polyethylene water reservoir and continuously pumped through the reactor at 4 ml/min. Instead of filtered river water, GAC filtered water containing no disinfectant residual can be used to colonize reactors at a water utility. Using GAC filter effluent as feed water is simple, as no water preparation such as filtration or dechlorination is required. As colonization progresses, a yellow coloration (biofilm growth) develops on the surface of the glass beads at the bottom of the first column. DOC consumption within the reactor (corresponding to the difference in DOC levels between the inlet and outlet of the reactor) can be regularly checked and compared to the BDOC concentration of the feed water determined using the BDOC–sand method. The colonization phase is completed when BDOC levels determined by the bioreactors are similar to those obtained using the sand method.

Maintenance. After colonization, bioreactors can operate for several years, but they require continuous water flow (feed water: filtered river water, GAC filtered water, or dechlorinated tap water). Routine checks include pump functioning, absence of leaks, and flow through the reactor. If water flow is interrupted, flow must be reestablished promptly. Columns might handle flow interruption for a few days. A gray/black color will develop, indicating anaerobic conditions.

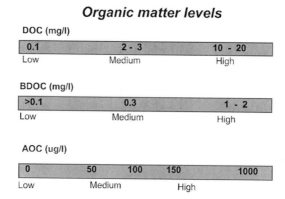

FIG. 4. Typical organic matter concentrations recorded in waters.

Sampling. Sampling, glassware, and sample preparation are identical to the BDOC methods previously described. Approximately 5 liters of water are collected in a clean glass container (5 liter Pyrex bottle). Turbid samples are filtered and disinfectant residuals are neutralized with sodium thiosulfate. Water flow rate is checked and adjusted to 4 ml/min if necessary. The test is started by loading the bioreactor with the water sample. The column inflow and outflow are sampled at least 2 hr following the beginning of the test. Precleaned 40 ml vials are used to collect DOC samples. The filtering device to prepare DOC samples is the same as for the BDOC–sand method. BDOC concentrations are determined from changes in DOC concentrations from the reactor inflow and outflow.

Nutrient Levels in Waters

The different methods have shown that nutrient concentrations vary dramatically between water sources over the seasons and from one year to another. Figure 4 depicts typical organic matter concentrations found in waters. In general, DOC levels are low (<1 mg/liter) in ground water. Surface waters contain generally from 2 to 10 mg/liter of DOC[5]; however, higher values (20–30 mg/liter) can be recorded in some water supplies (reservoir waters or high-humic content waters from the southeastern of the United States). BDOC concentrations range generally from <0.1 mg/liter to 1–2 mg/liter. For example, a study conducted to assess nutrient levels in various water types showed that river waters had DOC values of 3.2 to 4.0 mg/liter and BDOC concentrations above 1.2 mg/liter.[27] Partially treated and finished water BDOC levels ranged between 0.2 and 1.0 mg/liter (DOC between 0.9 and 2.9 mg/liter). BDOC levels were lower than 0.1 mg/liter for waters with low organic matter levels (mineral, spring, and ultrapure waters with DOC

[27] C. Volk, C. Renner, and J. C. Joret, *Sci. Eau* **5**, 189 (1992).

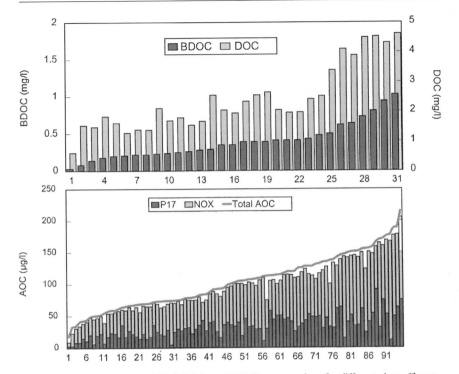

FIG. 5. Geometric means of AOC, DOC, and BDOC concentrations for different plant effluents. (The *x* axis represents the site number from 1 to 31 for BDOC, and from 1 to 95 for AOC, respectively, after ranking.)

levels <0.5 mg/liter).[27] Figure 5 summarizes data collected during a large survey that evaluated BDOC and AOC concentrations in 31 and 95 treatment plant effluents, respectively.[28] The geometric mean of BDOC concentrations was 0.32 mg/liter for all the sites. When BDOC levels were tabulated site by site, mean concentration of plant effluents ranged between 0.03 and 1.03 mg/liter while DOC levels ranged from 0.61 to 4.53 mg/liter. On the average, BDOC represented from 5 to 21% of the DOC. Forty-five percent of the sites, on average, had low BDOC concentrations (<0.30 mg/liter). Thirty-five percent of the sites had moderate BDOC levels (0.3–0.6 mg/liter), and 10% had high BDOC concentrations (>0.60 mg/liter).[28] In general, AOC concentrations vary from a few to several hundred micrograms per liter (Fig. 4). The AOC survey conducted at 95 sites (Fig. 5) showed that AOC concentrations entering distribution systems were generally high. Overall, average plant effluent AOC levels were close to 100 μg/liter (geometric mean of 94 μg/liter for all the sites). When data were averaged for each site, mean AOC

[28] C. J. Volk, N. Welch, and M. W. LeChevallier, unpublished data, 1998.

levels ranged from 18 to 214 μg/liter, depending on the site. Nine percent of the sites had average plant effluent AOC levels lower than 50 μg/liter, whereas 75% of the plants had moderate AOC concentrations (50–150 μg/liter), and 16% had high AOC levels (>150 μg/liter).[28] These finished water results are comparable to other AOC data collected from 53 source waters including surface or ground waters in North America.[29] Raw water AOC concentrations varied from 18 to 322 μg C-acetate/liter, or from 2.4 to 44% of the DOC content. Sixty-two percent of the raw water AOC values were above 100 μg/liter, and the mean and median values were 145 and 143 μg/liter, respectively. Bradford $et~al.$[19] reported AOC variations from 94 to 275 μg/liter (mean of 168 μg/liter) in a surface water reservoir, and from 75 to 731 (mean of 317 μg/liter) river water (Santa Ana River, CA). The mean AOC concentration for three ground waters was 54 μg/liter. Much lower AOC concentrations (<10 μg/liter) were recorded after bank or slow sand filtration.[10,11]

Factors Affecting BOM Results

BOM levels can be estimated using various bioassays, and each test has particular strengths and weaknesses.[6,7] Most of the methods have not been standardized, and many factors such as the inoculum (pure strain vs mixed inocula from the environment), glassware used, the monitored parameter (bacterial growth vs DOC), the use of suspended or fixed bacteria, and the experience of the laboratory performing the test can influence the test results.[7,9]

Parameters Affecting BDOC Results

BDOC concentrations vary depending on the method applied and the operating conditions (inoculum size, incubation time, sample aeration). For the BDOC determination using bacteria fixed on sand, both inoculum size and aeration have been shown to affect the result of BDOC measurement.[26,30] A sand : water ratio of 100 g : 300 ml (w : v) allows a rapid decrease in DOC concentration and optimal biodegradation of the organic matter. However, results of BDOC measurements increase with the sand : water ratio, showing biosorption of a low fraction of DOC (0.1–0.2 mg of DOC per 100 g of sand). Aeration mixes the water and provides a more uniform exposure to the sand, promotes faster kinetics of DOC consumption, increases dissolved oxygen concentrations, and sometimes leads to a greater consumption of DOC.[26] Concerning the method using suspended bacteria, inoculum size (from 0.04 to 5%, v/v) and aeration (4 liter/hr) have not been shown to have any effects on BDOC determination.[26,30] On the other hand, incubation time affects the result of

[29] L. A. Kaplan, D. J. Reasoner, E. W. Rice, and T. L. Bott, $Sci.~Eau$ **5**, 207 (1992).

[30] C. Volk, Ph.D. thesis in "Sciences et Techniques de l'Environnement," Ecole Nationale des Ponts et Chaussees, Paris, 1994.

BDOC measurement. Biodegradation could be incomplete after 28 days of incubation for certain water types; BDOC levels determined after 28 days increased by 0–125% when the incubation was extended to 85–120 days.[26,30] Precautions have to be taken regarding the dechlorination agent. MacLean et al.[31] reported that sodium thiosulfate levels higher than 20 mg $Na_2S_2O_3$/liter interfered with the 30-day BDOC test determined by the Servais[21] method. In the extended incubation associated with the suspended bacteria test, the presence of $Na_2S_2O_3$ promoted the growth of sulfur oxidizing bacteria, resulting in the production of sulfuric acid and a depressed pH. The low pH inhibited the activity of heterotrophic bacteria to utilize organic carbon.[31]

The BDOC column method is attractive for use by water utilities because by simply measuring the DOC at the inlet and outlet of the columns, the BDOC could rapidly be determined on a continuous basis. The problem, however, with the bioreactor method is that a long period is required for the columns to be colonized with the variety and density of bacteria necessary to rapidly utilize trace levels of organic carbon. When using Mississippi River water, a colonization period of 9 months was necessary for BDOC removals to be observed within the bioreactors that were similar to those measured by the sand method.[28] Similarly, Kaplan et al.[24] reported that steady-state colonization required 6–7 months using low organic level water (DOC of 1 to 2 mg/liter) from White Clay Creek in Pennsylvania. For other bioreactors,[32] a delay of 8 months was necessary for stable biodegradation to be observed within reactors fed with Marne River water (France). Prevost et al.[33] also reported that it took 7 months to complete colonization. Alternatively, colonization was possible within a few weeks when using Llobregat River water (Spain) at temperatures >22°.[34] Several factors such as temperature, organic carbon composition and level, and bacterial densities in the feed water may affect bacterial colonization rates of reactors. The colonization rate might be slowed down when using a reservoir to store the feed water. Suspended bacteria in the reservoir may consume some of the BOM prior to loading the column. Prevost et al.[33] reported that a large amount of DOC (0.5 mg/liter) was biodegraded after continuous passage of the source water through a prefilter and storage in a reservoir. Bioreactor are also highly site-specific; changing the feed water source may require an acclimation period of several months. Columns colonized with Mississippi River water required up to 7 months to equilibrate once shipped to new locations with different types of source waters.[28] Natural organic matter is composed of a mixture of hundreds of simple molecules and complex polymers

[31] R. G. MacLean, M. Prevost, J. Coallier, D. Duchesne, and J. Mailly, Wat. Res. 30, 1858 (1996).

[32] J. C. Joret, unpublished data, 1994.

[33] M. Prevost, G. Dubreuil, R. Dejardin, and R. G. MacLean, in "Proc. AWWA Water Quality Technol. Conf.," Denver, CO, Nov. 1997.

[34] F. Ribas, J. Frias, and F. Lucena, J. Appl. Bacteriol. 71, 371 (1991).

such as humic substances that are highly specific to a particular watershed. When colonization occurs, organic matter leads to the development of unique and specific populations of microbial species. A succession of species may also develop from the inlet to the outlet of the reactor with the capability of biodegrading the most to the least assimilable compounds. This high specificity of bacteria to the site organic material was reported to be a limitation of the BDOC-column methodology.[24,35] Bioreactors do not perform well in a stand-alone analyzer for samples from various origins. During interlaboratory experiments of bioreactors colonized with different water sources, maximum biodegradation was observed at the site where the microbiota were specific to tested water samples. At the other sites, when bioreactors were exposed for a few hours to the new water conditions, the reactor biofilm could not metabolize any of the organic matter.[24] Several phenomena that require time could occur to deal with a dramatic change in the characteristics of organic matter. Bacterial populations may be able to adapt to new nutrient conditions after modifying enzymatic pathways (enzyme induction) and developing appropriate catabolic properties. In addition, bacterial populations could shift and species with the highest biodegradation capabilities would become predominant while other strains would become dormant or decrease. Other bacterial species (autochthonous bacteria originating from the source water) may also colonize the bioreactors. On the other hand, the bioreactor columns require constant attention and maintenance. Particular attention must be focused on the pumping equipment to ensure the bioreactors are constantly supplied with low turbidity water that does not contain disinfectants. The BDOC columns are very sensitive to the presence of oxidants. Accidental exposure of the biofilm reactors to potassium permanganate required an additional 9 months for the columns to be recolonized.[28] When reactors are fed continuously with finished water, dechlorinated tap water can be stored in a reservoir rather than connecting the reactor directly to the plant effluent line. The "reservoir solution" could be preferred because of its simplicity, low maintenance, and ease of sampling, and because no additional equipment is required for feeding a dechlorination reagent. If a feed water reservoir is used, the water within the reservoir must be periodically replenished. Problems were also uncounted with precipitation and plugging due to the high levels of sodium thiosulfate normally added to water samples to neutralize disinfectant residuals. Reducing the levels of sodium thiosulfate to 20 mg/liter resolved the problem.[28]

Variation between Different BDOC Methods

The estimation of BDOC concentration depends on the applied method. BDOC concentrations obtained with a suspended and a fixed inoculum are well correlated (Fig. 6). However, BDOC values are generally higher (twice as high) when using

[35] L. A. Kaplan and J. D. Newbold, *Wat. Res.* **29**, 2696 (1995).

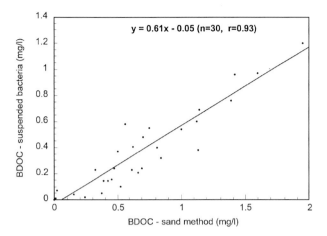

FIG. 6. Relationship between BDOC values evaluated with the suspended bacteria and sand methods.

attached bacteria.[30] Mathieu[8] also observed that BDOC values measured using bacteria attached to sand were 1.2–1.5 times higher than those obtained with suspended bacteria. Several hypotheses could explain discrepancies between the suspended and fixed inocula. The bacterial density in freshly inoculated water samples is 100–1000 times higher when using colonized sand than when using river water as inoculum (10^7–10^8 bacteria/gram of sand vs 10^6 bacteria/ml of river water). This difference in the inoculum size could affect the ratio (So/Xo) between the initial substance concentration (So) and the initial substrate concentration (Xo). So/Xo ratio was found to affect batch biodegradation in experiments with mixed bacterial populations.[36,37] An attached inoculum might be able to degrade a larger range of organic compounds. Thus molecules are refractory to biodegradation with a suspended inoculum could appear to be biodegradable by a fixed inoculum. This observation could be due to a greater bacterial diversity in the sand biofilm than in a river sample, a higher adaptation of sand bacteria (fixed species are selected according to their ability to degrade a large spectrum of organic compound), and an advantage due to the fixation because cometabolism, synergy mechanisms can occur on sand particles.

Both the bioreactor and the sand methods yield to similar BDOC results. Figure 7 shows a significant correlation between BDOC concentrations evaluated by the bioreactors and by the sand assay. Sand method data ranged between 0 and 1.13 mg/liter (mean of 0.30 mg/liter), and the bioreactor BDOC levels varied from 0 to 0.94 mg/liter (mean of 0.22 mg/liter). The differences between the

[36] P. Chudoba, B. Capdeville, and J. Chudoba, *Water Sci. Technol.* **26,** 743 (1992).
[37] S. Simkins and M. Alexander, *Appl. Environ. Microbiol.* **50,** 816 (1985).

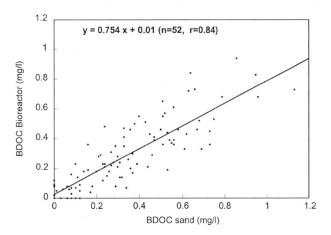

FIG. 7. Relationship between BDOC values evaluated with the bioreactor and sand methods.

bioreactors and sand method BDOC levels were less than 0.1 mg/liter for 50% of the samples. The difference was between 0.1 and 0.2 mg/liter for 17% of the samples. By comparison, Joret[32] reported differences <0.1 mg/liter and between 0.1 and 0.2 mg/liter in 49% and 30%, respectively. Kaplan et al.[24] and Prevost et al.[33] found that the bioreactor results were higher than the BDOC levels determined using suspended bacteria. This later observation is not surprising, since the BDOC–sand method gives higher results than the BDOC–suspended bacterial method.[26]

Variation between AOC and BDOC Data

AOC measures a biological response to assimilable carbon, but unlike BDOC, it is not a direct measure of the carbon level itself. Correlations observed between AOC and BDOC can be strong in some cases and weak in the others.[7] Figure 8 shows the comparison of AOC and BDOC values, based on a monthly analysis of 31 plant effluent samples.[28] Although AOC and BDOC values were significantly correlated, the variation between the test results was high (correlation coefficient of only 0.36). When the test results were grouped, a more clear relationship was observed between the two tests (Fig. 9). Volk et al.[27] observed an excellent correlation ($r = 0.996$) between AOC and BDOC levels in a series of dilutions of a single river water. However, the correlation between the tests was not as strong ($r = 0.77$) when performed on 31 different water types. Kaplan et al.[29] reported a significant but weak correlation ($r = 0.59$) between AOC and BDOC data in a survey involving 109 samples from 79 drinking water supplies. AOC is an index of regrowth potential that is strongly related to specific groups of organic molecules. Moreover, the origin and changes in the characteristics of organic matter might

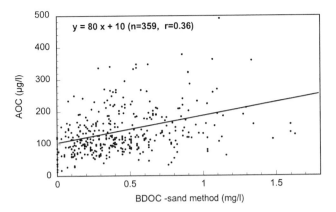

FIG. 8. Relationship between AOC and BDOC concentrations.

affect the relationship between the two methods of BOM measurement. Because NOM characteristics vary from one water type to another and seasonally for a given water, it is not surprising that waters with similar BDOC concentrations might have low or high levels of AOC depending on the types of organic molecules present in the water and constituting BDOC. BDOC concentrations are generally much higher than AOC levels (Figs. 8 and 9). This observation is due to a difference in the range of metabolic capabilities contained in the AOC test using two bacterial stains (P17 and NOX), versus the BDOC–sand bioassay using a wide variety, and a greater density of microorganisms able to degrade a large spectrum of organic compounds. It is admitted that the AOC measure the "easily" assimilable carbon,

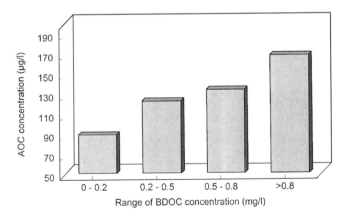

FIG. 9. Mean AOC concentrations for different ranges of BDOC levels.

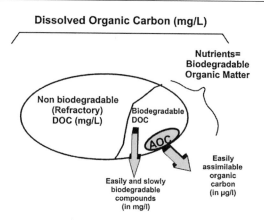

FIG. 10. Characterization of organic matter.

while the BDOC fraction includes easily and slowly biodegradable compounds (Fig. 10). Because of the complexity of the organic matter, it has not been possible to determine precisely the chemical characterization of its biodegradable part. Biodegradable organic materials have generally been considered to be carbohydrates and small molecular weight compounds, whereas complex molecules such as humic substances in water were assumed nonbiodegradable. However, some studies have shown that this classification should be used with caution.[38–40] It has been found that biodegradable compounds include both low and high molecular weight molecules. Some low molecular weight molecules can be directly utilized for metabolism. Other compounds such as humic substances that constitute most of the DOC can be partially degraded by bacteria after enzymatic action. A study[40] showed that 27% of humic substances in stream water were utilized by biofilm bacteria. On average, BDOC was composed of 75% humic substances, 30% carbohydrates, 4% amino acids, and 39% of molecules >100,000 Dalton.[40] Finally, BDOC can also be fractionated into three pools (named S, H1, and H2) depending on degradation kinetics.[41] The S fraction includes monomeric substrates directly assimilable by bacteria. The H1 fraction represents the polymeric organic matter that can be rapidly hydrolyzed by bacterial exoenzymes, whereas H2 includes organic molecules slowly hydrolyzed. Although BDOC can include larger molecules, humic substances, and some humic bound compounds, it appears that for most water samples, the AOC is mainly composed of small (<1,000 Dalton) and nonhumic molecules.[28]

[38] L. A. Kaplan and T. L. Bott, *Freshwater Biol.* **13**, 363 (1983).
[39] R. M. W. Amon and R. Benner, *Limnol. Oceanogr.* **41**, 41 (1996).
[40] C. J. Volk, C. B. Volk, and L. A. Kaplan, *Limnol. Oceanogr.* **42**, 39 (1997).
[41] P. Laurent, P. Servais, M. Prevost, and D. Gatel, *J. Amer. Water Works Assoc.* **89**, 92 (1997).

BOM Removal during Water Treatment to Control Bacterial Regrowth

The control of bacterial regrowth is achieved when the amount of BOM entering the distribution system is limited. Van der Kooij[42] showed that heterotrophic bacterial levels in non-chlorinated systems did not increase when AOC levels were lower than 10 μg/liter. LeChevallier et al.[43] suggested that regrowth of coliform bacteria in chlorinated systems may be limited by AOC levels less than 50–100 μg/liter. Block et al.[44] recommended an absence of biodegradable organic material after water treatment to limit bacterial regrowth. Servais et al.[45] associated biological stability (corresponding to no BDOC consumption within the distribution system) with a BDOC concentration of 0.16 mg/liter in the finished water. Volk and Joret[46] indicated that BDOC levels should be less than 0.15 mg/liter at 20° (warm temperature) and <0.30 mg/liter at 15° for achieving biological stability in the Paris suburb distribution systems. Coliform occurrences were related to consumption of BDOC within the distribution system of more than 0.10–0.15 mg/liter.[46] All the above objectives are difficult to achieve consistently. A variety of treatment processes can be used to control BOM levels in drinking water. The degree of organic carbon removal during drinking water production depends on several parameters, such as the treatment train design, operational conditions, and the seasonal fluctuations of the source water quality. In general, coagulation and sedimentation, biological filtration, or membrane filtration reduce BOM levels, whereas oxidation with ozone or chlorine form biodegradable compounds.

Coagulation and sedimentation can be described as physicochemical processes that follow chemical addition and rapid mixing. The amount of organic removal depends on the characteristics of the water matrix and the coagulation conditions (coagulant type and dose, pH, temperature, overflow rate, etc.).[47] Coagulation removes mostly the hydrophobic organic fraction (humic substances) and large molecules, whereas hydrophilic compounds are little affected by coagulation.[47–49] BDOC removals from 30 to 76% have been achieved through coagulation and settling.[7,28] However, AOC is typically not affected by coagulation, probably because AOC

---

[42] D. Van der Kooij, J. Amer. Water Works Assoc. **84,** 57 (1992).

[43] M. W. LeChevallier, W. Shulz, and R. G. Lee, Appl. Environ. Microbiol. **57,** 857 (1991).

[44] J. C. Block, P. Servais, and P. Werner, in "Proc. Technology Conf. on Bacterial Regrowth—Bugs, Molecules and Surfaces." Big Sky, MT, Aug. 1993.

[45] P. Servais, G. Billen, P. Laurent, Y. Levi, and G. Randon, in "Proc. AWWA Water Quality Technol. Conf., " Miami, FL, Nov. 1993.

[46] C. Volk and J. C. Joret, Sci. Eau **7,** 131 (1994).

[47] J. G. Jacangelo, J. DeMarco, D. M. Owen, and S. J. Randtke, J. Amer. Water Works Assoc. **87,** 64 (1995).

[48] J. K. Edzwald, in "Proc. AWWA Water Quality Technol. Conf.," San Francisco, Nov. 1994.

[49] S. J. Randtke, J. Amer. Water Works Assoc. **80,** 40 (1988).

targets small molecular weight, nonhumic compounds that are not amenable to coagulation.[28]

Organic removal can be maximized by biological filtration. When there is no prechlorination stage prior to filtration, a large amount of biomass is allowed to attach to the filter medium in the form of a biofilm, leading to the assimilation and removal of biodegradable materials contained in the water. Organic removal during filtration is affected by the organic matter quality and quantity and will vary by season, the media type, the contact time, and the backwashing strategy. Water temperature can limit bacterial activity within the GAC filters. Moreover, season affects the composition and concentration of the organic matrix and its subsequent biodegradation kinetics. The bacterial density and biomass activity showed differences between winter and summer periods. Better elimination of the most easily biodegradable compounds could be observed at higher temperatures.[50-52] Filtration steps may include mono, dual, or multimedia filters (sand, GAC, anthracite–sand, GAC–sand, anthracite–sand–garnet, etc.). The type of bacterial medium support is of primary importance for bacterial colonization and activity.[7,53,54] Filtration with biologically active carbon is more efficient than biological sand filters for the removal of biodegradable organic carbon.[53] Compared to microporous carbon, macroporous activated carbons provide bacteria with a large number of attachment sites protected from the abrasive action of backwash.[50] The water-filter medium contact time affects the amount of removed organics. The higher the contact time, the better the removal.[51,55,56] The effect of backwashing on biofilter bacteria varies depending on the type of support media and backwash strategy. Backwashing with chlorinated water may remove a significant portion of the fixed biomass and decrease the performance of the filter for removal of organic carbon.[57,58]

Different membrane technologies are available for production of potable water, including reverse osmosis (removal of ions and organic matter), nanofiltration membrane (removal of polyvalent ions and organic compounds larger than

[50] M. Prevost, R. Desjardins, N. Arcouette, D. Duchesne, and D. Coallier, *Sci. Tech. Eau* **23,** 25 (1990).

[51] N. Merlet, M. Prevost, Y. Merlet, and J. Coallier, *Sci. Eau* **5,** 143 (1992).

[52] P. Servais, G. Billen, C. Ventresque, and G. Bablon, *J. Amer. Water Works Assoc.* **83,** 62 (1991).

[53] G. Bablon, C. Ventresque, and R. Benaim, *J. Amer. Water Works Assoc.* **80,** 47 (1988).

[54] D. A. Reckhow, J. E. Tobiason, M. S. Switzenbaum, R. McEnroe, Y. Xie, Q. W. Zhu, X. Zhou, P. McLaughlin, and H. J. Dunn, *in* "Proc. Annual AWWA Conf.," Vancouver, BC, Canada, 1992.

[55] M. W. LeChevallier, W. C. Becker, P. Schorr, and R. G. Lee, *J. Amer. Water Works Assoc.* **84,** 136 (1992).

[56] P. Servais, G. Billen, P. Bouillot, and M. Benezet, *Aqua* **41,** 163 (1992).

[57] W. G. Characklis, *in* "Bacterial Regrowth in Distribution Systems." AWWA Research Foundation, Denver, 1988.

[58] R. J. Miltner, R. S. Summers, J. Wang, J. Swertfeger, and E. W. Rice, *in* "Proc. AWWA Water Quality Technol. Conf.," Toronto, Canada, Nov. 1992.

400 Dalton), ultrafiltration (UF) (removal of colloids and molecules larger than 10,000 Dalton), and microfiltration (MF) (removal of particles >0.2 μm, including turbidity, parasites, bacteria, and some viruses). Microfiltration and ultrafiltration may not be suitable for organic matter removal unless pretreatments such as ozone and powdered activated carbon are applied.[59–61] Nanofiltration membranes produce high removal efficiencies of DOC (from 50 to 99%), BDOC, precursors of chlorination by-products, pesticides/herbicides, color, and UV absorbance.[62–64] However, not all AOC molecules could be retained during membrane filtration treatment.[65]

Oxidation using ozone may be applied at different stages of the treatment chain for the removal of water pollutants, odors, color, and tastes, for improvement of the flocculation, and for disinfection.[66] Ozone can be applied in conjunction with hydrogen peroxide, UV light, or heterogeneous catalysts to oxidize a larger spectrum of organic compounds or to increase the oxidation level of molecules.[67–69] Ozonation increases acidic functional groups and polarity and decreases the size of organic molecules. Using the different biological tests for estimating BOM levels, several researchers[7,66] have reported an increase in biodegradability of dissolved natural organic matter following ozonation. For example, the increase in BDOC concentrations ranged from 19 to 94% following the ozonation of river or sand filtered waters.[70] Ozonation is often combined with biological filtration to remove the biodegradable molecules produced after oxidation. The effects of chlorination on organic levels have been less investigated. However, it was reported that biodegradability seemed to increase after chlorination.[55,71,72]

[59] J. M. Laine, M. M. Clark, and J. Mallevialle, *J. Amer. Water Works Assoc.* **82,** 82 (1990).

[60] J. Mallevialle, V. Mandra, I. Baudin, and C. Anselme, *in* "Proc. Filtech Conf.," Karlsruhe, Germany, Oct. 1993.

[61] Y. Richard, *in* "Proc. IWSA Regional Conf.," Zurich, May 1994.

[62] J. S. Taylor, D. M. Thompson, and J. K. Carswell, *J. Amer. Water Works Assoc.* **79,** 72 (1987).

[63] L. Tan and G. L. Amy, *J. Amer. Water Works Assoc.* **83,** 74 (1991).

[64] K. M. Agbekodo, B. Legube, and P. Cote, *J. Amer. Water Works Assoc.* **88,** 67 (1996).

[65] I. Escobar and A. Randall, *J. Amer. Water Works Assoc.* **91,** 76 (1999).

[66] B. Langlais, D. A. Reckhow, and D. R. Brink, *in* "Ozone in Water Treatment: Application and Engineering" (B. Langlais, D. A. Reckhow, and D. R. Brink, eds.). Lewis Publishers, Boca Raton, FL, 1991.

[67] J. Hoigne and H. Bader, *Water Res.* **10,** 377 (1976).

[68] H. Paillard, M. Dore, and M. M. Bourbigot, *in* "Proc. 10th Ozone World Congress, IOA," Monaco, Mar. 1991.

[69] C. Volk, P. Roche, J. C. Joret, and H. Paillard, *Water Res.* **31,** 650 (1997).

[70] C. Volk, C. Renner, P. Roche, H. Paillard, and J. C. Joret, *Ozone Sci. Eng.* **15,** 389 (1993).

[71] D. Van der Kooij, *in* "Proc. 2nd National Conf. on Drinking Water," Edmonton, Canada (P. M. Huck and P. Toft, eds.). Pergamon Press, New York, 1987.

[72] B. Hambsch, U. Schmiedel, P. Werner, and F. H. Frimmel, *Acta Hydrochem. Hydrobiol.* **21,** 167 (1993).

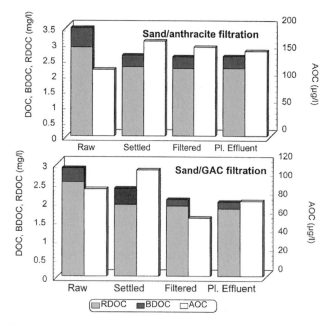

FIG. 11. Changes in DOC, BDOC, and AOC concentrations during water treatment using sand/anthracite or GAC filtration (DOC = RDOC + BDOC).

BOM levels can be altered (increased or decreased) depending on the treatment practices. Figure 11 shows the fate of organic compounds at different points within a conventional treatment prior to and following granular activated carbon (GAC) implementation. Overall after prechlorination and anthracite/sand filtration the site recorded a DOC removal of 25% and an increase of AOC concentration of 50% through the treatment process. Removal of DOC occurred during ferric coagulation and settling, and primarily affected the refractory part (nonbiodegradable) of the DOC. The geometric mean of the raw water AOC level increased from 122 to 174 μg/liter after preoxidation and settling processes, suggesting that chlorine pretreatment resulted in a large production of AOC. DOC, BDOC, and AOC were not removed by mixed media anthracite/sand filtration, because a continuous chlorine residual was maintained through the filters. GAC implementation modified organic fate during treatment (Fig. 11). Overall, the plant recorded a reduction in AOC and BDOC levels after implementing GAC filtration. AOC was increased by preoxidation; however, AOC levels were reduced by 58% by GAC filtration. BDOC concentrations also decreased after GAC filtration. Biological activity can take place during conventional treatment when chlorine levels are low on top of the GAC filters.

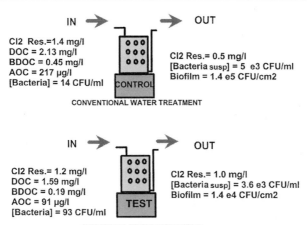

FIG. 12. Average water quality data at the inlet and outlet of the pipe reactors after conventional (control) or biological treatment (test).

Biofilm density and bacterial water quality are related to the amount of biodegradable material entering the system. Servais et al.[45] observed a relationship between biofilm bacteria and the concentrations of BDOC at the point of entry of several full-scale distribution systems. Another study[73] compared bacterial water quality in two pilot distribution systems supplied with different levels of organic matter. The first system was fed with ozonated water, whereas the second was supplied with biologically filtered water. Lowering the nutrient with biological filtration led to lower biofilm densities. Biofilm counts were 10^6 to 10^7 CFU/cm^2 for the ozonated system, compared to 10^5 to 10^6 CFU/cm^2 for the system fed with biologically filtered water. A pilot study using annular reactors evaluated the effect of reducing nutrient levels on bacterial water quality in drinking water.[74] A system fed initially by conventionally treated water received biologically treated water. Biodegradable organic matter levels were reduced approximately by half after biological treatment (Fig. 12). On the average, biofilm densities were reduced by 1 log unit by biological treatment. Interestingly, the effect of the treatment changes was not immediate. It required several months (approximately 6 months) of biological treatment before there was an observable impact on bacterial water quality, suggesting that the impact of the treatment changes was influenced by other uncontrolled factors.[74] Distribution systems represent complex reactors. In addition to nutrient levels, distribution system bacterial water quality is

[73] A. Rompre, M. Prevost, J. Coallier, C. Gauthier, R. Serkedjieva, P. Lafrance, and P. Laurent, in "Proc. AWWA Water Quality Technol. Conf.," New Orleans, Nov. 1995.

[74] C. Volk and M. W. Lechevallier, Appl. Environ. Microbiol. 65, 4957 (1999).

affected by many other parameters (pipe condition and corrosion levels, disinfection, etc.).

Conclusions and Recommendations

Because a high BOM level is one factor related to bacterial regrowth, BOM removal should be emphasized during water treatment to improve water quality. There is a weak correlation between AOC and BDOC concentrations. The BDOC test attempts to identify the entire pool of BOM, including very labile constituents to slowly biodegradable molecules (requiring the successive steps of cell adsorption, exoenzymatic hydrolysis, and cell consumption), whereas the AOC test detects easily assimilable organic compounds. Since both tests emphasize different fractions of biodegradable organic matter, both parameters should be monitored when studying nutrient changes during water treatment and distribution. Because of a long incubation period, using the BDOC suspended bacteria method is not practical at a water utility. A water utility can use either the sand or bioreactor methodology, depending on its testing objectives. Both methods yield to a similar BDOC result. The BDOC sand measurement is recommended for a short study or a limited number of samples. If the monitoring involves frequent measurements over a long period of time (>1 year), it is worth investing time and effort in setting up the bioreactor assay. To obtain frequent and rapid information on the level of BOM, the water utility should also purchase and become proficient with the use of a TOC analyzer. Research should continue to help transfer biostability monitoring technologies to operating companies.

Acknowledgments

The author thanks Mark LeChevallier (American Water Works Service Co.) and Louis Kaplan (Stroud Water Research Center) for their help.

[11] Development of *Thiobacillus* Biofilms for Metal Recovery

By GUSTAVO CURUTCHET, EDGARDO DONATI, CRISTIAN OLIVER, CRISTINA POGLIANI, and MARISA R. VIERA

Introduction

The bacteria most commonly isolated from inorganic mining environments are *Thiobacillus ferrooxidans, Thiobacillus thiooxidans,* and *Leptospirillum ferrooxidans.* These microorganisms are considered to be the most important in industrial bioleaching processes because they are adapted to growth under conditions such as low pH and iron- or/and sulfur-rich environments in which bioleaching reactions occur.[1–6]

These bacteria are all gram-negative, mesophilic, autotrophic, and obligately acidophilic. *T. ferrooxidans* is able to use ferrous iron or reduced sulfur compounds as electron donors, whereas *T. thiooxidans* is only capable of using reduced sulfur compounds and *L. ferrooxidans,* only ferrous iron. Ferric iron and sulfuric acid are the final products of these oxidation processes. The ability of acidophilic bacteria to assist in the recovery of metals by the dissolution of sulfide minerals is well known, but the mechanism is not fully understood. Two different mechanisms have been proposed to explain bacterial action: a direct attack on the sulfide by the bacteria (both *Thiobacillus*) and the indirect mechanism that entails the oxidation of the sulfides by the ferric iron produced by bacterial oxidation of ferrous iron, almost always contained in ores (*T. ferrooxidans* and *L. ferrooxidans*).[3]

Two novel indirect leaching mechanisms have been proposed to explain degradation of sulfides. Both mechanisms combine characteristics of the previously differentiated direct and indirect mechanisms. One is based on the oxidative attack of ferric iron on acid-insoluble metal sulfides involving thiosulfate as the main intermediate.[7] The other mechanism is started by proton and/or ferric iron attack on acid-soluble metal sulfides with polysulfides and sulfur as intermediates.[8]

Thiobacillus cells have shown a great tendency to attach to different surfaces. Although the mechanism of adhesion and the formation of biofilms of these

[1] A. D. Agate, *World J. Microbiol. Biotechnol.* **12,** 487 (1996).

[2] H. L. Ehrlich and C. L. Brierley, "Microbial Mineral Recovery." McGraw-Hill, New York, 1990.

[3] G. Rossi, "Biohydrometallurgy." McGraw-Hill, Hamburg, 1990.

[4] J. Barrett, M. N. Hughes, G. I. Karavaiko, and P. A. Spencer, "Metal Extraction by Bacterial Oxidation of Minerals." Ellis Horwood, Chichester, 1993.

[5] D. E. Rawlings and S. Silver, *Biotechnology* **13,** 773 (1995).

[6] D. E. Rawlings, *J. Ind. Microbiol. Biotechnol.* **20,** 268 (1998).

[7] A. Schippers, P. G. Jozsa, and W. Sand, *Appl. Environ. Microbiol.* **62,** 3424 (1996).

[8] A. Schippers and W. Sand, *Appl. Environ. Microbiol.* **65,** 319 (1999).

bacteria have not received much attention, the use of immobilized *T. ferrooxidans* cells seems a promising method, especially to improve the rate of ferrous iron oxidation.[9,10]

With this aim, various immobilization methods, reactors, ion exchange resins, glass beads, activated carbon particles, polystyrene, polyurethane, etc., have been employed.[11-19] However, there are a few reports about *Thiobacillus* biofilms producing sulfuric acid, and they were almost exclusively done by our research group.[20,21]

In this chapter, we describe the preparation of two types of biofilms: (1) *T. ferrooxidans*, immobilized by the two more widely used techniques—attachment and entrapment—for the production of sulfuric acid, and (2) attached *T. ferrooxidans* cells for the ferrous iron oxidation. In both cases, we used whole and viable cells. In addition, we relate the application of these biofilms to recover metals from the leaching of a sulfide ore.

Thiobacillus Biofilms to Oxidize Ferrous Iron

Biofilm Preparation using Glass Beads as Support Matrix

A strain of *T. ferrooxidans* (DSM 11477) growing in 9K medium was used throughout this study.[22] Inocula are obtained by harvesting the cells during the late exponential growth phase and filtering to remove jarosite [basic iron(III) sulfate]. Culture is filtered with a 0.22 μm filter and the cells are washed several times with acidified water (pH 1.5) and suspended in an iron-free medium. These suspensions, with a bacterial population of 1×10^8 cells ml^{-1}, are used as inocula at 10% (v/v).

[9] A. B. Jensen and C. Webb, *Process Biochem.* **30**, 225 (1994).

[10] D. E. Rawlings, "Biomining: Theory, Microbes and Industrial Processes." Springer-Verlag, Berlin, 1997.

[11] K. Nakamura, T. Noike, and J. Matsumoto, *Water Res.* **20**, 73 (1986).

[12] S. Grishin and O. Tuovinen, *Appl. Environ. Microbiol.* **54**, 3092 (1988).

[13] M. J. Garcia, I. Palencia, and F. Carranza, *Process Biochem.* **24**, 84 (1989).

[14] H. Armentia and C. Webb, *Appl. Microbiol. Biotechnol.* **36**, 697 (1992).

[15] S. Porro, C. Pogliani, E. Donati, and P. Tedesco, *Biotechnol. Lett.* **15**, 207 (1993).

[16] N. Wakao, K. Endo, K. Mino, Y. Sakurai, and H. Shiota, *J. Gen. Microbiol.* **40**, 349 (1994).

[17] C. Webb and G. A. Dervakos, "Studies on Viable Cell Immobilization." Academic Press, Austin, 1996.

[18] F. K. Crundwell, *Minerals Eng.* **9**, 1081 (1996).

[19] M. Nemati and C. Webb, *Biotechnol. Lett.* **19**, 39 (1997).

[20] E. Donati, L. Lavalle, V. de la Fuente, P. Chiacchiarini, A. Giaveno, and P. Tedesco, *in* "Biohydrometallurgical Processing" (T. Vargas, C. A. Jerez, J. V. Wiertz, and H. Toledo, eds.), p. 293. University of Chile, Santiago, 1995.

[21] C. Cerruti, G. Curutchet, and E. Donati, *J. Biotechnol.* **62**, 209 (1998).

[22] M. P. Silverman and D. G. Lundgren, *J. Bacteriol.* **77**, 642 (1959).

Cultures of *T. ferrooxidans* in 9 K medium (sterilized by filtration with a 0.22 μm filter) with 20% w/v (pulp density) of glass beads are placed in 250 ml Erlenmeyer flasks maintained at 30° in an orbital shaker at 180 rpm. Before use, glass beads are treated with 5 *M* HCl for 2 hr, washed until neutral, and finally autoclaved at 120° for 20 min. When iron(II) is exhausted, cultures are filtered to retain solids (colonized by cells) and supernatants are discarded. Iron(II) is determined by titration with 0.002 *M* KMnO$_4$ in acid medium, whereas total soluble iron is analyzed by atomic absorption spectrophotometry. Cultures without glass beads and sterile culture medium with glass beads are used as controls.

The solids obtained at this stage are characterized as potassium jarosite [KFe$_3$(SO$_4$)$_2$(OH)$_6$] or ammoniojarosite [(NH$_4$Fe$_3$(SO$_4$)$_2$(OH)$_6$] by X-ray diffraction. Residue analysis through Mossbauer spectroscopy has proved the existence of amorphous ferric hydroxysulfate sediments, although jarosite is the main solid phase, in agreement with previous reports.[9,23] The deposit formation is more copious when the cultures are kept in contact with the glass beads, even after the total oxidation of iron(II).

It was observed that iron(II) oxidation is similar in cultures with or without glass beads, whereas there is no significant iron(II) oxidation in the sterile control (independent from the presence of glass beads). For this reason, we assume there is no inhibition or any important changes in bacterial physiology influencing iron(II) oxidation due to the presence of glass beads. In contrast, although there is similar total bacterial population, there are significant differences in free bacterial population (determined by direct counting) in cultures with or without glass beads. In the presence of glass beads, there is a decrease in bacterial population shortly after inoculation, probably due to the initial bacterial attachment to the support. When 65–70% of iron(II) is oxidized, abundant precipitation of jarosite occurs in culture with glass beads and the free bacterial population begins to be lower than in the control culture (without glass beads). When iron(II) is exhausted, the nonattached bacterial population is only 40% of that in the control culture.

An increase in the pulp density results in a higher amount of deposits (evaluated through the deposit weight and the total soluble iron concentration) and a decrease in the number of suspended bacteria. Thus, glass beads contribute to jarosite precipitation, acting as seed crystals around which crystallization may occur. However, only a small amount of the iron(III) deposits is attached to the glass beads or to the flask walls. At the end of the experiment, the jarosite adherent to the glass beads contains most of the viable bacterial population. A microphotograph of the glass beads taken when the disappearance of soluble iron and the formation of deposits were starting can be seen in Fig. 1a. It is possible to observe that bacteria were located on the deposits, but bacteria were absent in those areas where there

[23] S. Grishin, J. M. Bigham, and O. Tuovinen, *Appl. Environ. Microbiol.* **54**, 3101 (1988).

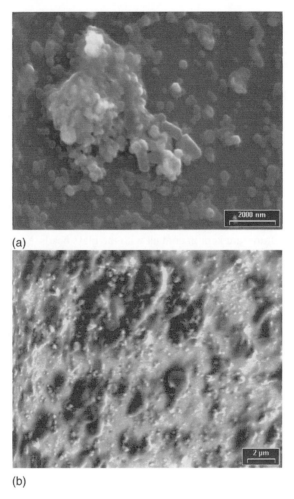

FIG. 1. SEM microphotographs of surface of glass beads in contact with a *T. ferrooxidans* culture when iron deposits began to form over the glass (a) and at the end of the growth (b). Note the presence of cells associated with the deposit (a) and coated by the deposit (b).

was no deposit. At the end of the growth phase, the beads were completely covered with deposits, underneath which cells could be detected (Fig. 1b).

The EDAX diagram of the surface (Fig. 2) shows that the deposit is mainly jarosite. Such a large number of cells attached to jarosite could be explained by the high porosity of this substance.[24] It has been reported that this kind of precipitate

[24] D. G. Karamanev, *J. Biotechnol.* **20**, 51 (1991).

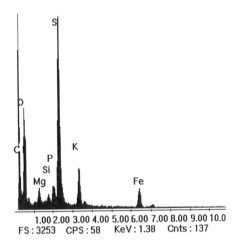

FIG. 2. EDAX scan of the deposit showing that the surface is coated with potassium jarosite $[KFe_3(SO_4)_2(OH)_6]$.

has a natural tendency toward accepting protons, modifying the surface character-istics, thus increasing the tendency of bacteria to attach to it.[25] In agreement with previous reports,[24,26] a linear correlation between the attached bacterial population and the amount of precipitated iron was obtained.

In another set of experiments, jarosite or glass beads from these cultures were placed in flasks with fresh medium without further inoculation. A fast iron(II) oxidation was observed.

The possibility that jarosite by itself alters iron(II) oxidation rate was discarded by adding jarosite from a previous culture to a new one. Jarosite was previously washed with sulfuric acid solution of pH 1.8 and dried at room temperature to kill the attached bacteria. Iron(II) oxidation rate was not altered by the presence of jarosite compared with a control.

The formation of iron(III) precipitates, especially jarosites, is highly dependent on pH. Because low pH represses iron(III) hydrolysis and subsequent jarosite precipitation, at pH 1.4 negligible jarosite precipitation and iron disappearance were observed. Thus, at this pH value, free bacterial population is similar to a control culture without glass beads. Moreover, microphotographs taken at the end of the growth showed small amounts of deposits over the glass beads. When the glass beads from a previous culture are put into fresh medium (without new inoculation), iron(II) oxidation shows a long lag phase. This lag phase increases at

[25] M. Fletcher, *in* "Bacterial Adhesion. Molecular and Ecological Diversity" (M. Fletcher, ed.), p. 30. John Wiley & Sons, New York, 1996.

[26] D. Emerson, J. V. Weiss, and J. P. Megonigal, *Appl. Environ. Microbiol.* **65,** 2758 (1999).

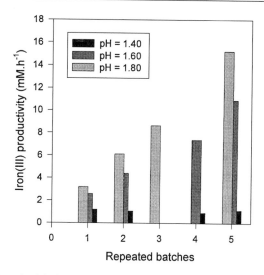

FIG. 3. Iron(III) productivity in repeated batches of *T. ferrooxidans* biofilm on glass beads (shaken flasks) at different pH values. [Reprinted from C. Pogliani and E. Donati, *Proc. Biochem.* **35**, 997 (2000) with permission from Elsevier Science.]

every successive batch. This probably means that there is only a small immobilized bacterial population that is even further diminished by successive washing. In the case of initial pH values higher than 1.4, an increase in iron(III) productivity until a constant value—within the range of experimental error—was observed (Fig. 3). Thus, the immobilized bacterial population is not easily washed out, but it increases step by step. As most bacteria are immobilized on the iron(III) deposits and they, in turn, increase step by step, it may be expected that the number of attached bacteria, the iron(III) oxidation rate, and iron(III) productivity increase constantly. Nevertheless, constant values are reached, among other reasons, because: new iron(III) deposits cover bacteria previously immobilized, thus preventing their access to nutrients, and gradual increase in bacterial number seriously affects nutrient availability, especially of oxygen and carbon dioxide.

Extracellular polymeric substances could play an important role in biofilm formation, but,[10,27,28] in agreement with a model proposed by Karamanev,[24] our results suggest that in this biofilm bacteria are basically adsorbed to the surface of jarosite. Crundwell[18] suggested that this type of biofilm could also be produced on substrates such as pyrite. In the layer of jarosite formed on pyrite, iron would be cycled between the cells and the mineral surface, contributing to the mineral leaching.

[27] J. W. Costerton, *J. Ind. Microbiol.* **15**, 137 (1995).
[28] C. Pogliani and E. Donati, *J. Ind. Microbiol. Biotechnol.* **22**, 88 (1999).

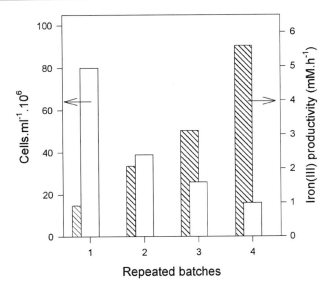

FIG. 4. Iron(III) productivity in repeated batches of *T. ferrooxidans* biofilm on glass beads (packed bed reactor).

The glass beads colonized by cells (from cultures at initial pH value of 1.8) are separated from jarosite deposits and washed four times with 5 ml of H_2SO_4 to eliminate nonattached bacteria. A 160 g amount of these colonized glass beads is placed in a packed bed reactor (made of glass, length 250 mm, inner diameter 40 mm) to allow biofilm formation. A 200 ml volume of fresh medium is added without further inoculation. An air current of 120 liter hr^{-1} is used to facilitate the medium recirculation in the column and to supply the gaseous nutrients. The reactor is maintained at 30°. Culture medium is replaced by fresh medium when the complete oxidation of iron(II) is detected. Figure 4 shows the iron(III) mean productivities and the bacterial number in the first four repeated batches. Iron(III) productivities are lower than those obtained in the repeated batches in shaken flasks. However, it can be seen that cells continue to accumulate within the biofilm, from batch to batch, so that iron (III) productivities are increased beyond the levels achieved in a single operation.

After the fourth cycle (data not shown), iron(III) productivity reaches a nearly constant value, whereas the suspended bacterial population is so low that iron(II) oxidation can be ascribed exclusively to the biofilm bacteria. From this cycle, the amount of precipitated iron reaches a maximum value of ca. 2500 mg $liter^{-1}$, and more than 60% of the total iron remains soluble.

To study biofilms in conditions resembling those of an industrial application, i.e., with continuous flow of fresh medium, a reactor similar to the one described

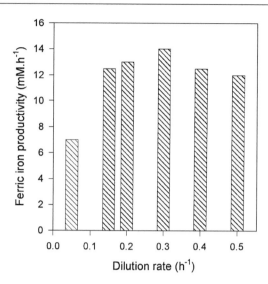

FIG. 5. Iron(III) productivity of *T. ferrooxidans* biofilm on glass beads (packed bed reactor) at different dilution rates.

above is used. When the reactor, operating in the "repeated batches" mode, reaches its highest iron(III) productivity, the culture medium is replaced by fresh medium and the continuous flow of medium through a peristaltic pump is started. Figure 5 shows iron(III) productivity as a function of the dilution rate. The productivities are higher than those obtained in the batch mode. Dilution rates higher than those in the figure cause a constant decrease in iron(III) productivity. It was also observed that the higher the dilution rate, the lower the amount of iron(III) precipitate. This decrease in the amount of iron deposits produces a progressive "washout," since cells cannot be retained effectively.

Although glass beads do not play a role in industrial practice, results with glass beads suggest that it is very useful to determine the conditions needed for important iron(III) productivity by biofilms. In mineral bioleaching processes, the most suitable immobilization support is the sulfide ore, which is oxidized by the bacteria attached to it. Our experiments using a low-grade copper ore are described below.

Biofilm Preparation using Low-Grade Copper Ore as Support Matrix

A reactor similar to the one described above is used. The glass beads are replaced by 80 g of sulfide ore from Bajo la Alumbrera (Belén-Catamarca, Argentina) with particle size between 2.0 and 3.4 mm. This ore contains magnetite, hematite,

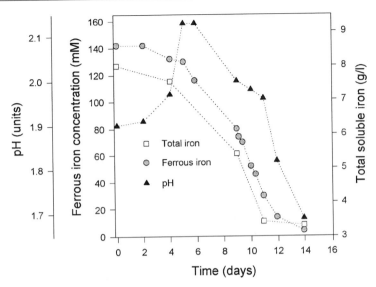

FIG. 6. Iron(II), total iron concentration, and pH during the growth of *T. ferrooxidans* culture in the presence of particles of low-grade sulfide ore.

and pyrite as the main sulfide. Covellite is the major copper constituent of the samples, although the ore also contains chalcopyrite and chalcocite. The chemical composition of the ore is 78.30% Si (as SiO_2); 10.0% Fe (as Fe_2O_3); 2.85% S; 1.25% Cu; 0.035% Zn; and 0.08% Mn (as MnO_2). This sulfide ore has been previously bioleached by *T. ferrooxidans,* and after that colonized similarly to the glass beads. Copper content decreases from 1.25% (original ore) to 0.41% (bioleaching residue).

Figure 6 represents the evolution of iron(II), total iron, and pH during the colonization stage, before the mineral is placed in the column. As can be seen, there is high iron precipitation and very low iron(III) productivity (of about 0.4 mM hr^{-1}). This low value of iron(III) productivity is probably due to the presence of a significant amount of sulfide in the mineral. The sulfides can act as an energy source for bacteria, decreasing iron(II) oxidation rate, and/or they can react with the iron(III) produced by bacteria. These assumptions have been confirmed by the decrease in the copper quantity (less than 0.2%) in the mineral just after the end of the colonization stage and before it is placed in the packed bed reactor. Figure 7 shows the iron(III) mean productivities obtained in repeated batches during cultivation in the packed bed reactor. As can be seen, productivity increases with every batch up to a maximum value of 5.1 mM hr^{-1}; afterward it remains constant.

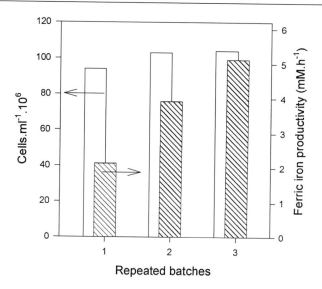

FIG. 7. Iron(III) productivity in repeated batches of *T. ferrooxidans* biofilm on particles of low-grade sulfide ore (packed bed reactor).

Biofilms of *Thiobacillus* on Sulfur

The generation of *T. ferrooxidans* biofilms on sulfur to produce sulfuric acid is described in this section. Two different techniques of immobilization have been used: attachment and entrapment of cells.[16,17,29,30]

Attachment of Cells

The first air-lift percolator (sulfuric acid producing bioreactor) was prepared as follows: 160 g of elemental sulfur (particle size: about 2–4 mm) was added to a percolation column. A flow of 120 liters hr^{-1} air was continually fed to the solution. A 180 ml volume of an iron-free 9 K medium (initial pH 2.0) was added to the column and inoculated with 20 ml of *T. ferrooxidans* culture in the exponential stage of growth. The system was maintained at 30°. When the pH value was between 0.7 and 1.0, the same volume of fresh medium (without new inoculation) replaced the entire exhausted medium. The procedure was repeated until a constant rate of sulfuric acid production was reached (this situation indicates the maximum attached bacterial population). Thereafter, the bioreactor produced sulfuric acid continually and at a constant rate.

[29] P. S. J. Cheetham, K. W. Blunt, and C. Bucke, *Biotechnol. Bioeng.* **21**, 2155 (1979).
[30] A. López, N. Lázaro, and A. M. Marqués, *J. Microbiol. Methods* **30**, 231 (1997).

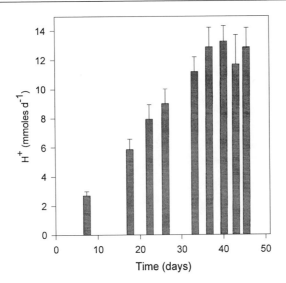

FIG. 8. Sulfuric acid productivity in repeated batches of *T. ferrooxidans* biofilm on sulfur (packed bed reactor).

Figure 8 shows the mean productivity of sulfuric acid obtained in the *T. ferrooxidans* bioreactor. As can be seen, the productivity increased to a maximum value after 35 days (65 mmol H^+ liter^{-1} d^{-1}). This sulfuric acid productivity (probably correlated with a maximum attached bacterial population) was approximately constant (data not shown) until the end of the experiment. Scanning electron microscopy (SEM) (Fig. 9) shows elemental sulfur taken from the reactor. It is possible to see the surface colonized by *T. ferrooxidans* cells.

In the bioreactor, the maximum productivity was reached after a long time because the support surface was increased by bacterial action. This is because the sulfur crystals are first converted to colloidal sulfur, which is able to enter the periplasmic space of cells.[3] When the bioreactor reached a constant production rate, it could produce about 406 mmol H^+ liter^{-1} kg^{-1} d^{-1}, which was calculated assuming a linear dependence between the amount of support and production. Although the productivity was calculated by direct extrapolation of the experimental rate of acid production, it obviously depends on the bioreactor design, the exposed sulfur surface (related to particle size), the attached cell population, the uniform contact between liquid and solid, and the availability of oxygen, among others.

Entrapment of Cells

Because of the very low solubility of elemental sulfur, during sulfur oxidation small particles and colloidal sulfur are produced through the phospholipid

FIG. 9. SEM microphotograph of sulfur colonized by *T. ferrooxidans* cells.

action.[3,31] In a continuous bioreactor with cells immobilized on sulfur to produce sulfuric acid, a continuous washout of colloidal sulfur and bacteria is observed.[20,21]

In order to prevent washout, we have used calcium alginate as gel matrix to entrap a high number of *T. ferrooxidans* cells and elemental sulfur under strongly acidic conditions. The efficiency of the gel beads with immobilized cells for stable sulfuric acid production has also been analyzed.

A solution containing sodium alginate is added to an equal volume of *T. ferrooxidans* cell suspension (bacterial population of about $1.6-4.5 \times 10^9$ cells ml^{-1}), including residual elemental sulfur. This mixture is pumped into 100 mM calcium chloride solution through a syringe tip needle in order to form 2–4 mm diameter beads that gel on entering the calcium chloride solution.

With alginate concentration between 1 and 3%, the gelation technique has efficiency higher than 85–90%. Alginate concentrations lower than 1% do not allow bead formation. After 10–15 days, beads shrink to 50% of their original volume. Beads containing total sulfur are more efficient in acid production than those containing only small particles (size <3 μm) of sulfur (indicated as "colloidal" sulfur) at 1.5% alginate concentration. To prepare beads with only "colloidal" sulfur, the cell suspension is previously filtered through blue ribbon filter paper to remove particles of elemental sulfur larger than 3 μm.

Beads (1500) entrapping cells and sulfur are added to 250-ml flasks containing 100 ml of medium (pH 2.0) without sulfur and incubated in an orbital shaker operating at 100 rpm at 30°. Beads varying in alginate concentration presented an average wet weight of 0.022 g (1%), 0.033 g (1.5%), 0.039 g (2%), and 0.061 g

[31] J. Rojas, M. Giersig, and H. Tributsch, *Arch. Microbiol.* **163**, 352 (1995).

(3%) containing between 92–95% (w/w) water. Each bead contains approximately 9.5×10^7 cells.

It was observed that the lower the alginate concentration, the faster the acid production rate (acid production rates were 5.3, 3.3, 2.5 and 1.2 mM d^{-1} for 1%, 1.5%, 2%, and 3% alginate concentrations, respectively). That was probably because higher bead densities (less permeable gel) limit the diffusion of nutrients or increase the adsorption of protons by the beads.

In repeated batch cultures transferring the same gel beads (1% alginate, approximately 2.4×10^8 cells per bead) to fresh medium, a similar rate of acid production was found in the first two steps. Initially, an increment of pH was observed, probably due to the proton diffusion into the beads. In a third batch culture, acid production decreased significantly. After the end of this batch, residual sulfur in the beads was not detected by chemical analysis. This fact was confirmed by photographs of the cross section of calcium alginate beads compared with others corresponding to freshly prepared beads. In the last two steps, free bacterial population increased progressively.

Beads (2200) and 80 ml of medium (pH = 2.0) without sulfur were placed in a percolation column. An air flow was continually fed into the solution. When the pH was low enough, the same volume of a fresh medium (without new beads or new inoculation) replaced the exhausted medium. Figure 10 illustrates the acid

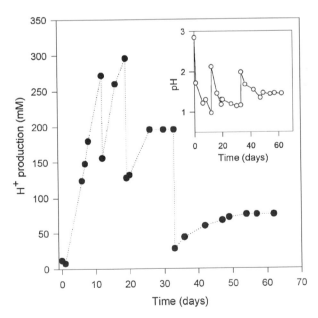

FIG. 10. Sulfuric acid production in repeated batches using *T. ferrooxidans* cells and sulfur entrapped in calcium alginate beads (packed bed reactor).

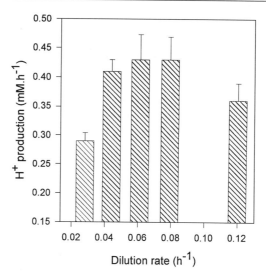

FIG. 11. Sulfuric acid productivity using *T. ferrooxidans* cells and sulfur entrapped in calcium alginate beads at different dilution rates (packed bed reactor).

production (outer graph) and pH (inner graph) obtained in successive batch cultures in the percolation column (1% alginate concentration, approximately 2.4×10^8 cells per bead). The number of free cells increased significantly in the last two batch cultures, and there was a total loss of activity after the fifth batch culture (residual sulfur was not detected in the beads). Acid production rate in the percolation column was slower than that in shaken flasks (3.4 and 9.8 μmol H^+ d^{-1} bead^{-1}, respectively), probably due to a high nutrient diffusion in the second case.

A percolation column prepared in a similar way as in batch studies was used at 30° in a continuous experiment. Fresh medium (pH 2.0) was pumped through the column at different dilution rates. Figure 11 shows the acid production rate in the pseudo steady state at different dilution rates. Acid production increased until a rate of 10.3 m*M* d^{-1} was reached (at a dilution rate of approximately 0.06 hr^{-1}). Thereafter, the rate decreased, probably because there was not enough sulfur in the beads.

Applications of *Thiobacillus* Biofilms in Metal Recovery

Capillitas (Catamarca, Argentina) is a low-grade sulfide ore containing 9.87% zinc, 1.89% copper, 11.6% iron, and 11.4% manganese. The main minerals are galena, sphalerite, pyrite, chalcocite, covellite, rhodocrosite, and pyrolusite. Because of the high content of carbonate minerals, bacterial leaching is only feasible after neutralization of the carbonate. For this reason, a combined attack using two packed bed reactors was employed: one of the reactors contained the biofilm formed on sulfur to produce sulfuric acid, whereas the other contained the biofilm formed on glass beads intended for iron(III) production. The effluents generated by these reactors were used for mineral leaching.

The biofilm formed on glass beads was preferred to the one formed on low grade copper ore for iron(III) productivity, not only because of its high iron(III) production, but also because the latter produced a high amount of iron(III) precipitates that reduce the leaching capacity. Besides, the biofilm formed on sulfur was preferred to the one formed by entrapment in calcium alginate because although the former can be washed out, it has a higher productivity.

The experimental setup consisted of two packed bed reactors and an air-lift percolator containing 162 g of ore (particle size of about 1–2 mm). Initially, sulfuric acid produced by the *T. ferrooxidans* biofilm on sulfur was circulated. The medium was frequently replaced by fresh medium to avoid an increase of the pH above 3.5. This procedure was repeated for 29 days until a constant pH value was reached. Afterward, the medium was replaced by fresh medium with iron(III) (produced by *T. ferrooxidans* biofilm on glass beads); periodically, biogenerated sulfuric acid was added to allow a pH of about 1.0 (to prevent precipitation of jarosites). In the percolating column, the liquid phase was recirculated by means of the air flow.

Figure 12 shows the concentration of the leached metals as a function of time. Metal ion concentrations were assayed by atomic absorption spectrometry. After 75 days, 69.7% of copper, 39.5% of zinc, and 99.5% of manganese was recovered.

As shown in Fig. 12, manganese extraction was due almost exclusively to acid action; thus, manganese could not be present mainly as pyrolusite. This was confirmed by X-ray because rhodochrosite (manganese carbonate) was the main manganese mineral. After bioleaching, rhodochrosite disappeared in the X-ray spectrum. The addition of iron(III) increased the recovery of the other metals, but at a low rate.

During this experiment, more than 30 g of sulfuric acid was consumed either to neutralize the carbonate present in the mineral or to leach the acid-soluble components of the mineral. This suggests that the acid consumption is the most expensive aspect of hydrometallurgical processing of this mineral. Sulfuric acid can be obtained at lower costs by bacterial oxidation of sulfur in *T. ferrooxidans* biofilms. This also avoids the dangerous transport of sulfuric acid. *T. ferrooxidans* biofilms also produce iron(III) capable of oxidizing other minerals not dissolved by acid attack. Strictly, in the case of this mineral, the addition of iron(II) would not have been necessary, as the acid leaching released part of the iron present in the mineral. This iron(II)-containing liquid could be oxidized by passing through *T. ferrooxidans* biofilm.

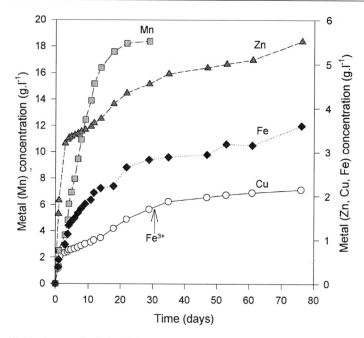

FIG. 12. Metal recoveries during bioleaching of Capillitas ore by combined attack using *T. ferrooxidans* biofilms.

Percentages of extraction as high as those found in this experiment cannot be achieved in direct bioleaching using "*in situ*" inoculation because (1) the pH rises rapidly (reflecting the high content of carbonate minerals) and inhibits the bacterial action; (2) the iron(II) oxidation occurs on the ore surface. Because precipitates cover mineral sites on ore particles, microbial leaching is adversely affected, impeding further copper dissolution; (3) *T. ferrooxidans* is inhibited by toxic metals in the leaching solution. This inhibition of iron(II) oxidation by *T. ferrooxidans* was found for the three first media that were in contact with the ore, but not for the others.

Summarizing, this bioleaching methodology using *Thiobacillus* biofilms for iron(III) production as well as sulfuric acid production is more efficient and economic than direct bioleaching and, obviously, than chemical leaching, especially when the ore contains large amounts of carbonate minerals.

[12] Methods Used to Assess Biofouling of Material Used in Distribution and Domestic Water Systems

By STEVEN PERCIVAL and JAMES T. WALKER

Introduction

The inner surfaces of water distribution pipes and domestic plumbing pipe systems are potential sites for biofilm development.[1–6] The term biofilm is used to describe a complex mixture of microbial cells enclosed within a matrix of extracellular polymers. Microbial cells found as part of a biofilm contribute to the contamination of the water bulk phase possibly due to sloughing as a result of water shear.[7] These sloughed sections of biofilm may harbor opportunistic pathogens that can multiply in poorly chlorinated waters, particularly in domestic plumbing systems.[4,8,9] Whereas drinking water is one of the most extensively monitored systems in terms of microorganisms relevant to public health,[10] the dominant nonpathogenic bacterial populations present within potable water cannot be ignored, since these play a major role in biofilm formation.[6,11] These microorganisms are generally regarded as nonpathogenic and considered harmless to the normal individual. However, they present a potential significant public health risk with respect to the young, the elderly, and immunocompromised individuals.

Problems are caused by the presence of biofilms in potable water with respect to the increased resistance of sessile bacteria to disinfectants when compared to their planktonic counterparts.[9] The resultant biofilms can generate taste and odor problems[11,12] and also corrosion of the pipe material substrata.[13]

[1] M. Allen, R. Taylor, and E. Gledriech, *J. Amer. Water Works Assoc.* **72,** 614 (1980).

[2] H. Ridgway and B. Olson, *Appl. Environ. Microbiol.* **41,** 274 (1981).

[3] R. Donlan, W. Pipes, and T. Yohe, *Water Res.* **28,** 1497 (1994).

[4] S. Percival, I. Beech, R. Edyvean, J. Knapp, and D. Wales, *J. Chart Inst. Water Environ. Manage.* **11,** 289 (1997).

[5] S. Percival, J. Knapp, D. Wales, and R. Edyvean, *Water Res.* **32,** 243 (1998).

[6] A. L. Percival and J. T. Walker, *Biofouling* **14,** 99 (1999).

[7] M. W. LeChevallier, C. D. Cawthon, and R. G. Lee, *Appl. Environ. Microbiol.* **54,** 2492 (1988).

[8] A. Parent, S. Fass, M. Dincher, D. Reasoner, D. Gatel, and J. Block, *J. Chart Inst. Water Environ. Manage.* **10,** 442 (1996).

[9] M. W. LeChevallier, C. D. Cawthon, and R. G. Lee, *Appl. Environ. Microbiol.* **54,** 649 (1988).

[10] W. Pipes, in "Drinking Water Microbiology" (G. McFeters, ed.), p. 428. Plenum Press, New York, 1995.

[11] M. W. LeChevallier, T. M. Babcock, and R. G. Lee, *Appl. Environ. Microbiol.* **53,** 2714 (1987).

[12] B. H. Olson and L. A. Nagy, *Adv. Appl. Microbiol.* **30,** 73 (1984).

[13] M. Lechevallier, C. Lowry, R. Lee, and D. Gibbon, *J. Amer. Water Works Assoc.* **84,** 111 (1993).

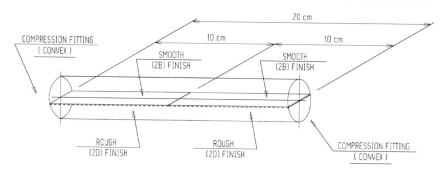

FIG. 1. Schematic of the mains water simulation system designed to investigate the effects of flow rate on the microbiology of biofilms, using stainless steel grades 304 and 316 as the metal substratum.

This study was established to evaluate how different grades and different surface finishes of stainless steel could affect the microbiology of biofilms at different water flow rates within potable water systems.

Materials and Methods

Experimental Rig to Study Water Flow Rate on Biofilm on Stainless Steel

A mains water simulation system was designed to investigate the effects of flow rate on the microbiology of biofilms, using stainless steel grades 304 and 316 as the metal substratum (Fig. 1). Each stainless steel pipeline rig is 2 m long, with an internal diameter of 20 mm and wall thickness of 2 mm. The 2-m pipelines are sectioned every 20 cm and joined together with 22-mm brass compression joints. Within each 20-cm pipe section are two 10-cm stainless steel slides (Fig. 2). Each slide has a wall thickness of 2 mm and a diameter of 1.9 cm with a rough (Fig. 2, 2D) underside and smooth (Fig. 2, 2B) finish topside.[5,14]

Flow rates are maintained at water velocities of 0.32, 0.96, and 1.75 m sec^{-1} using a three-speed central heating pump (Grundfos, UK) together with valve systems. Water is continually supplied to the system and wastewater is removed at known flow rates. The planktonic bacterial populations in potable water are measured before each new velocity is commissioned (the system is cleaned between each velocity study). Key parameters, including chlorine levels, pH, viable cell counts, and total cell counts, are measured monthly.

Preparation of Materials Used to Study General Development in Biofilms

All stainless steel slide surfaces used in the mains water simulation system are cleaned in 70% alcohol for a period of 1 hr prior to use. The whole simulation

---

[14] S. Percival, J. Knapp, D. Wales, and R. Edyvean, *Brit. Corros. J.* **33,** 121 (1998).

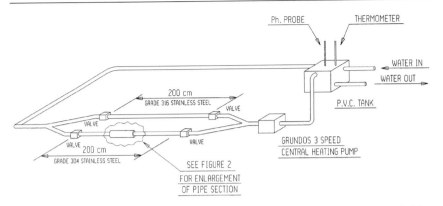

Fig. 2. Representation of each 20-cm pipe section that contained two 10-cm stainless steel slides for biofilm development.

system is also sterilized with peracetic acid according to recommended procedures supplied by the manufacturers (Solvay Interox, Widnes, UK).

Biofilm Development on Stainless Steel Grades 304 and 316

Stainless steel slide sections are removed monthly from the test site over a 5 month period. Four slide sections of rough (Fig. 2, 2D) and smooth (Fig. 2, 2B) stainless steel (2 cm long by 1.9 cm wide) are used to compare viable cell counts and bacterial genera development (heterotrophic). An additional four sections (2 cm by 1.9 cm) are used for total counts and another four (4 cm by 1.9 cm) to monitor dry weight and extracellular slime polysaccharide.

Biofilm Removal from Surfaces

Stainless steel slide sections are rinsed in sterile distilled water to remove any loosely attached bacteria. The surfaces, after exposure to potable water, are scraped using a sterile scalpel blade and swabbed with a sterile cotton wool swab. The biofilm samples are then suspended in 10 ml sterile distilled water and vortexed for 30 sec.

Viable Bacterial Counts

Serial dilutions of the suspended biofilm are prepared 10-fold in sterile distilled water and aliquots of 0.1 ml of each dilution are plated, in triplicate, on R2A agar (Difco, UK) 15 and the plates incubated at 28° for 7 days.

Detection of Biofilm on Stainless Steel using Epifluorescence Microscopy

Slides of rough (Fig. 2, 2D) and smooth (Fig. 2, 2B) stainless steel are washed gently in distilled water to remove any unattached or loosely bound microorganisms.

Samples are air dried and stained with filter sterilized (0.22 μm pore filter) 0.001% acridine orange (Difco, Detroit, MI) for 2 min. Samples are then rinsed with sterile distilled water and air dried before examination using epifluorescence microscopy (Olympus BH-2, Olympus Optical Co, London, UK). The numbers of adhered cells are estimated by randomly counting fluorescing cells in 120 fields of view of the total surface area; this is subsequently converted to cells cm^{-2} of metal surface.

Identification of Viable Sessile Bacteria

Representative isolates from the viable counts are selected and maintained at 4° on R3A medium.[15] Further details of the methods used to identify sessile bacteria can be located elsewhere.[14] Further identification of isolates is carried out using the API 20NE (Biomerieux, Basingstoke, UK) identification system.

Dry Weight, Extracellular Polysaccharide, and Total Carbohydrate Levels

Biofilms are removed from the slides, with the use of a sterile scalpel, and suspended in 10 ml double distilled water before freeze drying and being weighed. Doubly distilled water (2 ml) is added to the freeze-dried biofilm and the suspension vortexed for 2 min and then centrifuged at 30,000 rpm for 30 min to remove any cellular debris. This is repeated, following cleaning, to increase the purity of the biofilm extracellular material. The supernatant is then dialyzed, to remove all nonpolymeric material, using visking tubing (boiled in EDTA) for 24 hr in distilled water at 4°. The sample is then analyzed for extracellular slime polysaccharide levels using the phenol sulfuric acid method,[16] employing D-glucose as standard.

Statistical Evaluation of Results

All experiments involving statistical analysis are analyzed using the Students t-test and analysis of variance on SPSS (Statistical Package version 6.1).

Characteristics of Supply and Tank Water

The pH of the tank and tap water remained relatively constant throughout the study, the mean pH of the tank and tap water being 7 and 7.2, respectively. The temperature of the tap and tank water also remained relatively constant at 23°. Planktonic viable cell count was found to be 3.3×10^3 cfu ml^{-1}.

Biomass Development on Stainless Steel Grades 304 and 316

Stainless steel grade 304 was found to be colonized by microorganisms at a significantly higher ($p < 0.05$) level than grade 316 (Fig. 3) at each water velocity.

[15] D. J. Reasoner and E. E. Geldreich, *Appl. Environ. Microbiol.* **49**, 1 (1985).

[16] M. Dubois, K. Gilles, J. Hamilton, P. Rebers, and F. Smith, *Anal. Chem.* **28**, 350 (1956).

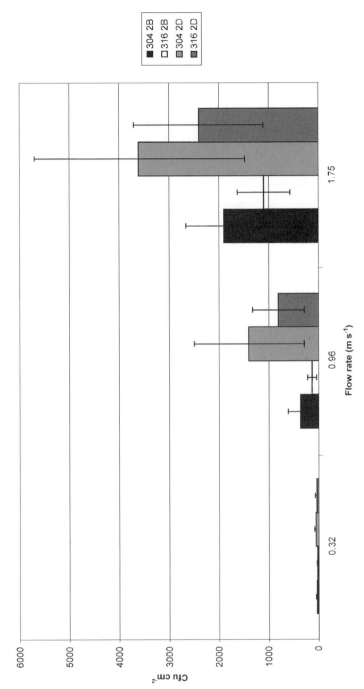

FIG. 3. Sessile mean viable counts (cfu cm^{-2}) on stainless steel after exposure to potable water at different flow rates over 5 months.

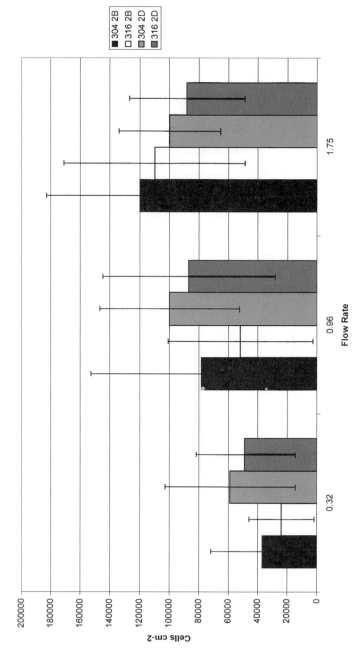

FIG. 4. Sessile total counts (cfu cm^{-2}) on stainless steel after exposure to potable water at different flow rates over 5 months.

Total viable cell counts of biofilms on all grades and surface finishes of stainless steel were significantly higher ($p < 0.05$) at water flow rates of 0.96 and 1.75 m sec^{-1} when compared to water flow at 0.32 m sec^{-1}. When a comparison between 0.96 and 1.75 m sec^{-1} was made, significant differences ($p < 0.05$) were evident, with an overall higher biofilm total viable cell count at the higher velocity of 1.75 m sec^{-1}, on both grades and finishes of stainless steel.

When a comparison of biofilm total cell counts on stainless steel grades 304 and 316 2B slides was made, between flow rates of 0.32 m sec^{-1} and 0.96 m sec^{-1}, it was found that cell counts were overall significantly higher ($p < 0.05$) at the flow rate of 0.96 m sec^{-1} (Fig. 4). Differences in total cell counts on the 2D slide sections, when a comparison was made between 0.32 m sec^{-1} and 0.96 m sec^{-1}, were also found. Cell counts were significantly higher ($p < 0.05$) on stainless steel grade 304 at the higher velocity of 0.96 m sec^{-1}. The 316 2D slides were only colonized at a significantly higher ($p < 0.05$) level at the higher flow rate of 0.96 m sec^{-1}. At a water flow rate of 1.75 m sec^{-1}, when compared to flow at 0.32 m sec^{-1}, it was found that total cell counts were significantly higher ($p < 0.05$) on 304 2D slides at the higher velocity of 1.75 m sec^{-1}. A similar result was obtained with the 2B slide sections. When a comparison of total cell counts on stainless steel between water velocities of 0.96 and 1.75 m sec^{-1} was made, it was found that most 2B and 2D surfaces were colonized at a significantly higher ($p < 0.05$) level at the higher flow rate of 1.75 m sec^{-1}.

Dry weight and exopolysaccharide levels, however, were not found to be significantly different ($p < 0.05$) between grades or finishes of stainless steel (Figs. 5 and 6). Overall, of the three different water velocities studied, no significant differences ($p < 0.05$) between flow rates and dry weight/extracellular slime polysaccharide levels was evident.

Effect of Flow Rates on Diversity of Colony Types

As numbers of bacterial genera isolated from biofilms were very similar for both grades of stainless steel, all data were combined in order to increase the statistical power for analysis. Colonization of *Acinetobacter* sp. on stainless steel at a flow rate of 0.32 m sec^{-1} was evident after 1-month exposure to potable water, indicating it is a primary colonizing bacterium. However, at month 2, numbers stabilized, suggesting the formation of a "quasi steady state." While *Acinetobacter* spp. were shown to be the initial colonizer of stainless steel at low flow rates, its growth or establishment within the biofilm becomes less evident over time.

Overall, *Acinetobacter* sp. has been shown to be the primary colonizer together with *Arthrobacter/Corynebacterium* spp. on stainless steel at low flow rates (Tables I and II). At higher flow rates the pioneering bacteria *Acinetobacter* sp. was also present with *Pseudomonas* spp. (which were not present initially at

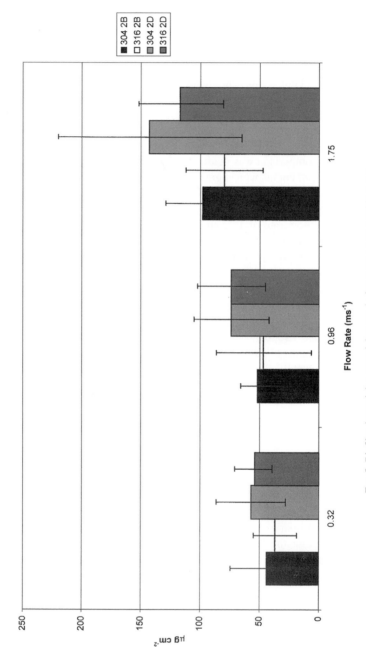

FIG. 5. Biofilm dry weights on stainless steel after exposure to potable water.

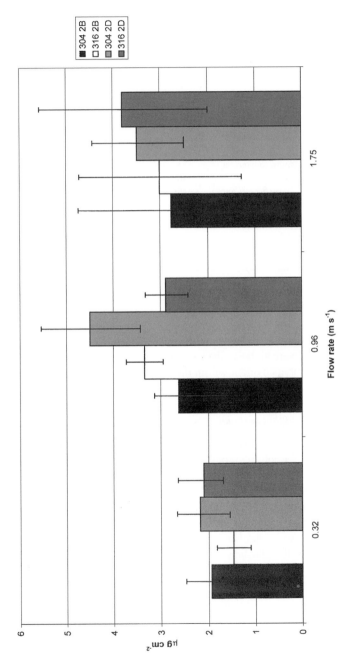

FIG. 6. Biofilm exopolymeric polysaccharide accumulation on stainless steel in potable water at different flow rates over 5 months.

TABLE I
MEAN FREQUENCY OF OCCURRENCE (%) OF SESSILE GENERA IN BIOFILMS DEVELOPED
ON STAINLESS STEEL IN POTABLE WATER[a]

Stainless steel	Genus	% isolated bacteria of each genus at a flow rate of		
		0.32 m sec^{-1}	0.96 m sec^{-1}	1.75 m sec^{-1}
2B (smooth)	Acinetobacter	43	11	20
	Corynebacterium/ Arthrobacter	43	25	8
	Pseudomonas	11	54	68
	Methylobacterium	3	10	4
2D (Rough)	Acinetobacter	48	11	26
	Corynebacterium/ Arthrobacter	34	18	9
	Pseudomonas	14	49	60
	Methylobacterium	3	22	3
	Flavobacterium	1		1
	Alcaligenes			1

[a] Mean % were calculated from all stainless steel grades 304 and 316 data combined. All values determined over a 5 month sampling period.

low flow rates despite quite high counts in the water supply). The numbers of *Pseudomonas* spp. were higher at the increased flow rates of 0.96 and 1.75 m sec^{-1}, despite similar numbers evident within the planktonic phase.

Effect of Surface Roughness on Diversity of Colony Types

No significant differences ($p < 0.05$) were evident with respect to grade or finish of stainless steel on the types of bacteria colonizing the surfaces (Table I). However, the following differences were observed. With respect to flow and surface finish it was found that the "washout" of *Acinetobacter* sp. was greater on the 2D (rough) surfaces. This could suggest that biofilm accumulation is greater on rougher surfaces considering greater evidence of sloughing. The number of sloughing *Pseudomonas* spp. were calculated to be very similar to that of *Acinetobacter* spp., between the 2B and the 2D slide surface finishes.

Influence of Time/Flow on Biofilm Development in Potable Water Rig

Throughout the 5-month studies other bacteria were present as part of the biofilms in addition to *Acinetobacter* and *Pseudomonas* spp. These included

TABLE II
BIOFILM BACTERIA IDENTIFIED USING API 20NE AND CONVENTIONAL METHODS[a,b]

Organism (conventional method)	Gram	O	M	P	°C growth 37	41	45	O\F	API 20NE
Acinetobacter	−ve, rods	−	−	White	+	+	−	Fermentative	Acinetobacter sp.
Alcaligenes	−ve, rods	+	+	White	+	−	−	No change	Alcaligenes xylosoxidans
Agrobacterium sp.	−ve, rods	−	−	Light yellow	−	−	−	Oxidative	Agrobacterium radiobacter
Arthrobacter spp.	+ve, rods	−	−	Orange	−	−	−	Fermentative	N/A
Bacillus	+ve, rods	+	+	Yellow	+	−	−	No change	N/A
Flavobacterium spp.	−ve, rods	+	−	Yellow	+	−	−	Oxidative	Flavobacterium breve
Methylobacterium	−ve, rods	−	−	Pink	+	−	−	Alkaline	Methylobacterium sp.
Pseudomonas spp.	−ve, rods	+	−	Yellow	+	−	−	Oxidative	Pseudomonas putida
	−ve, rods	+	+	Dark yellow	−	−	−	Oxidative	Pseudomonas cepacia
	−ve, rods	−	−	Yellow	+	−	−	Oxidative	Pseudomonas paucimoblis
Sphingomonas spp.	−ve, rods	−	−	Orange	+	−	−	No change	Sphingomonas paucimoblis

[a] P, Pigment; M, motility; O, oxidase, O\F, Hugh and Leifson.
[b] The remaining bacteria could not be identified using the API 20NE but identification was possible using the Vitek system.

Methylobacterium spp and *Bacillus* spp. Despite this, *Pseudomonas* spp. dominated the biofilms on stainless steel during the latter stages of biofilm growth at all three water velocities.

Overall, results suggest that *Acinetobacter* spp. and *Arthrobacter/Corynebacterium* spp., while dominating the biofilm at low velocities, are not able to compete with the robust characteristics evident with the *Pseudomonas* spp. that dominate the biofilms at the higher flow rates.

With all flow rates studied the decline of *Acinetobacter* spp. numbers is steady throughout the 5-month research period. Despite slightly "oscillating" numbers, as could be expected in potable water, the percentage colonization of *Acinetobacter* spp. declines over time at each flow rate studied.

The levels of *Corynebacterium/Arthrobacter* spp. fluctuated over the 5-month study periods and showed a decrease with respect to time. However, the passage of time indicated, in some cases, recovery of growth. Bacterial numbers were generally found to be low at higher flow rates, with numbers also low at month 5 on both surface finishes at a water velocity of 1.75 m sec^{-1}. This suggests, as with that of *Acinetobacter* spp., that *Corynebacterium/Arthrobacter* spp. are early colonizing bacteria at low flow rates but are "washed out" of the biofilm during maturation/development.

Pseudomonas spp. were found to increase over time, and this was evident at all three flow rates and also on the different surfaces and finishes studied.

Methylobacterium sp. was found to be a very early colonizing bacterium on all stainless steel surfaces and finishes studied. The sloughing of this organism was evident over time, with poor growth shown within biofilms during the latter stages of the studies.

Discussion

The results suggest surface finish and chemical composition have no major effects on the diversity of heterotrophic bacteria growing in a potable water biofilm. However, as the two steels under study are similar in chemical composition this is not surprising. Although a variety of microorganisms were evident as part of a biofilm, it was found that the pattern of microbial colonization was considerably different at the higher velocities when compared with the lower water velocities. The large diversity of microorganisms observed at low velocities was no longer evident toward the latter stages of the 5-month study period.

Budding hyphal bacteria and filamentous bacteria, only seen in very low levels on the surface at low velocities, were the dominant microbial colonizers at high velocities when observed under scanning electron microscopy. The mode of colonization also appeared to be different at high velocities and low velocities. Whereas numerous single cells were apparently able to attach to surfaces at low velocities and to develop into colonies in the biofilm, at high velocities the attachment of large clumps of cells and few single cells occurred.

At high velocities a slight microbial succession on stainless steel after only 5 months' exposure to mains water was apparent. Generally, the duration of microbial succession is inversely proportional to the range and concentration of available nutrients in the particular ecosystem under study. In drinking water ecosystems nutrient levels should be relatively low to prevent microbial growth.

In common with other studies,[17] the majority of bacteria that were isolated from this test rig were gram negative bacteria. After exposure to mains water at 0.32, 0.96, and 1.75 m sec^{-1} the pioneering bacteria were generally *Acinetobacter* spp. at the lower velocities and *Pseudomonas* spp. at the higher velocities on both stainless steel grades 304 and 316, closely followed by *Arthrobacter/Corynebacterium* spp. and then the pink colonies of *Methylobacterium* spp., *Corynebacterium/Arthrobacter* spp. and *Pseudomonas* spp. have also been isolated elsewhere in past research within distribution mains water.[6, 12, 18] *Methylobacterium* spp. have also been identified on plumbing material in other biofilm systems.[6, 19, 20] A likely explanation for the similarity between microorganisms on pipe surfaces and in drinking waters is that the detachment of microorganisms from the pipe

[17] D. Jain, *Water Res.* **29**, 1869 (1995).

[18] L. A. Nagy and B. H. Olson, *Zentralbl. Bakteriol. Mikrobiol. Hygiene* **182**, 478 (1986).

[19] M. W. LeChevallier, R. J. Seidler, and T. M. Evans, *Appl. Environ. Microbiol.* **40**, 922 (1980).

[20] J. Rogers, A. B. Dowsett, P. J. Dennis, J. V. Lee, and C. W. Keevil, *Appl. Environ. Microbiol.* **60**, 1585 (1994).

surface and their reentrapment into the passing water may account for the majority of bacteria in mains water.

Surface roughness is known to affect the transport and microbial cell attachment to a surface for many reasons. It is known to increase mass transport, provide shelter from shear force, and increase the surface area for the attachment of bacteria.[23] Substratum roughness seems to some extent to affect bacterial colonisation with an increase of adhesion with increasing roughness of the substratum.[6,21–23] Although substratum roughness probably plays an important role in the rate and extent of bacterial colonization, little conclusive evidence is available to quantitatively evaluate the effects within this study if we compare bacterial numbers with surface type. However, it is generally regarded that cells located inside pores on surfaces are sheltered from shear forces; therefore their removal rate is decreased and retention of a larger amount of cells is ensured.

One important aspect obtained from this work is that bacterial numbers and bacterial genera clearly change with time and with flow rate. Therefore, with respect to the control of biofilms within potable water there is a need for effective strategies to eradicate developing or established biofilms. Knowledge of the organisms to be killed within the biofilm is required to enable efficient use of biocides.[24] There are many conventional methods available to reduce biofilms, in particular chemical agents, physical agents, or both combined to reduce the extent of biofouling in potable water systems. A more efficient method of controlling the extent of biofouling could be by specifically selecting biocides to specific species of bacteria. This is because bacteria react differently to biocides, particularly because of differing cell wall properties. Therefore any form of control with the use of biocides would involve evaluation against specific sessile microbes at specific times and at the specific flow rate. This is due to the changes in dominant organism present within the biofilm. This would require continuous monitoring of the genera present in the biofilm so cocktails of biocides could be used at specific times for when the biofilm genera changes. Therefore, in order to increase the so-called "kill" potential of a biocide, more specific biocides need to be utilized to specific bacteria.

By modeling bacterial species, biofilm control systems could be set up to provide a more efficient and more economical system and so help reduce biofouling and therefore decrease any public health effects of biofilm sloughing on a long term basis. However, a cautionary note should be employed in that difficulties are encountered during identification of bacteria in potable water due to a number of bacteria that become unculturable after subculturing. This was evident in this study together with other problems being encountered with the use of the API 20NE strips

[21] G. Geesey and J. Costerton, *Can. J. Microbiol.* **25**, 1058 (1979).

[22] W. G. Characklis, *Biotechnol. Bioeng.* **23**, 1923 (1981).

[23] W. Characklis and K. Cooksey, *Adv. Appl. Microbiol.* **29**, 93 (1983).

[24] D. Allsop and K. Seal, "Introduction to Biodeterioration." Edward Arnold, London, 1986.

as identification tools. These were found not to be as reliable as the conventional nonrapid identification methods when comparisons were made between bacterial strains. This therefore suggests a need for better and more reliable identification systems for environmental microbes. Presently molecular biology techniques are developing that may be helpful in the future.[25,26]

We are presently studying how so-called "normal" biofilms, evident in potable water that contain nonpathogenic microbes, help in the entrapment and proliferation of potentially invading pathogens. This will have important implications for the control of "public health significant" biofilms now and in the future because of the development of emerging chlorine resistant pathogens that can enter the drinking water system.

[25] U. Szewzyk, W. Manz, R. Amann, K. Schleifer, and T. Stenstrom, *FEMS Microbiol. Ecol.* **13,** 169 (1994).

[26] J. Jess, N. Hight, H. Eptoin, D. Sigee, J. O'Neill, H. Meier, and P. Handley, *in* "Biofilms: Community Interactions and Control" (J. Wimpenny, P. Handley, P. Gilbert, H. Lappin-Scott, and M. Jones, eds.), p. 149. Bioline, Cardiff, 1997.

[13] Microbial Interactions to Intestinal Mucosal Models

By ARTHUR C. OUWEHAND, ELINA M. TUOMOLA, YUAN KUN LEE, and SEPPO SALMINEN

Introduction

The interaction of microbes with intestinal mucosa is important for many reasons. Binding to the intestinal mucosa may allow colonization of sites with a high flow rate, it is thought to be involved in the modulation of the intestinal immune system, it may play a role in the competitive exclusion by the normal microbiota, and binding is considered to be the first step in pathogenesis.

The first contact a microbe has with host tissue in the intestine is the mucus layer. This is a protective layer of varying thickness, 50–450 μm,[1] which covers the underlying epithelium. The mucus layer has many functions. Among these is the inhibition of the binding of certain bacteria while at the same time providing a habitat for others.[2] The main component of mucus is water, whereas its main organic component is mucin. The latter is responsible for the characteristic viscous

[1] A. Paerregaard, *Ugeskr. Laeger* **158,** 5423 (1996).

[2] J.-W. van Klinken, J. Dekker, H. A. Büller, and A. W. C. Einerhand, *Am. J. Physiol.* **269,** G613 (1995).

properties of mucus. Mucins are large glycoproteins (0.25–2 megadaltons). They consist of polypeptide backbones of which the central parts are heavily glycosylated tandem repeats, while the N and C termini are only slightly glycosylated.[1] At least nine different human mucin types have so far been isolated, all of which show a certain degree of tissue specificity.[2] Of these, the two main types found in the small and large intestine are MUC2 and MUC3 produced by goblet cells[3] interspersed between the enterocytes.[4] Also, enterocytes have been observed to produce mucins; small amounts of MUC3 are excreted and on the surface MUC1 is expressed as part of the glycocalyx.[3]

The mature enterocytes that line the intestine exhibit fingerlike protrusions (about 1 μm long) on the luminal side, the microvilli, and exhibit enzyme activity.[4] The main function of enterocytes is the uptake of nutrients and water. However, the glycocalyx of the enterocytes also provides many potential binding sites for microorganisms. Caco-2, HT-29, and HT-29 MTX (mucus producing) tissue culture cells have been used as models for enterocytes. The Caco-2 cell line was originally isolated from human colon adenocarcinoma.[5] Although the cell line originates from the colon, the differentiated cells more closely resemble small intestinal enterocytes than colonocytes. The differentiation and polarization of Caco-2 cells is spontaneous.[6] When the cells in culture reach confluence they cease proliferation and start to differentiate. During the differentiation process the cells become columnar in shape, polarize to have an apical and a basolateral side, and develop microvilli on the apical side. The differentiated cells express apical brush border hydrolases, e.g., disaccharidases and aminopeptidases, normally present on the small intestinal brush border microvilli.[6,7] The apical hydrolases can be used as markers for Caco-2 differentiation, e.g., activity of disaccharidases can be tested by the method of Dahlqvist.[8] The other option is to use indirect immunofluorescence in which antibodies raised against apical hydrolases are utilized to detect the enzymes.[9]

Although intestinal mucus and tissue culture cells provide convenient models for different parts of the mucosa, whole intestinal tissue probably provides the most realistic substratum for the study of adhesion to the intestinal mucosa. Tissue pieces can also be obtained from different parts of the intestine, which allows for

[3] A. A. Weiss, M. W. Babyatsky, S. Ogata, A. Chen, and S. H. Itzkowitz, *J. Histochem. Cytochem.* **44,** 1161 (1996).

[4] P. A. Sanford, "Digestive System Physiology." Edward Arnold, London, 1992.

[5] J. Fogh, W. C. Wright, and J. D. Loveless, *J. Natl. Cancer Inst.* **58,** 209 (1977).

[6] M. Pinto, S. Robine-Leon, M.-D. Appay, M. Kedinger, N. Triadou, E. Dussaulx, B. Lacroix, P. Simon-Assmann, K. Haffen, J. Fogh, and A. Zweibaum, *Biol. Cell* **47,** 323 (1983).

[7] H.-P. Hauri, E. E. Sterchi, D Bienz, J. A. M. Fransen, and A. Marxer, *J. Cell Biol.* **101,** 838 (1985).

[8] A. Dahlqvist, *Anal. Biochem.* **22,** 99 (1968).

[9] G. Chauvière, M.-H. Coconnier, S. Kernéis, J. Fourniat, and A. L. Servin, *J. Gen. Microbiol.* **138,** 1689 (1992).

studying possible site specific adhesion. In the case of tissue from human origin, it will usually be diseased tissue, though some healthy tissue is often resected together with this. Thus, the effect of different diseases on adhesion can be studied as well. Moreover, the effect of the normal microbiota present on the intestinal mucosa can be taken into account as well. A potential disadvantage with the use of tissue pieces, especially from human origin, may be the restricted availability and the requirement of immediate processing when available.

This chapter will describe methods for studying the adhesion of microorganisms to intestinal mucus, intestinal tissue culture cells, and intestinal mucosal tissue pieces. Methods will also be discussed on how to determine the affinity of microorganisms for substrata and the maximum number of binding sites available on a substratum.

Preparation of the Substrata

Preparation of Mucus from Tissue

Mucus can be prepared from different parts of the intestine by a method described earlier in this series,[10,11] but will be repeated briefly for completeness. The same method can also be used to collect mucus from other mucosal surfaces, e.g., fish skin mucus.[12] In the case of experimental animals, the intestines are removed immediately after the animal is sacrificed. When human tissue is concerned, the tissue should be processed as soon as possible after resection and transported on ice. The parts of interest are cut into 4–5 cm long sections. The sections are cut open and gently washed in ice-cold HEPES (N-2-hydroxyethylpiperazine-N'-2-ethanesulfonic acid) buffered Hank's balanced salt solution (HH) to remove intestinal contents. The mucus is then gently scraped with a rubber policeman into a small amount of HH-buffer and centrifuged once at 12,000 g (10 min) and once at 27,000 g (15 min) to remove cell debris and bacteria. The mucus can be stored at $-70°$ until use.

Preparation of Mucus from Feces and Ileostomy Effluent

Isolating mucus from feces has the advantage that the effects of different host factors—age, health, etc.[13]—on the adhesion can be studied. Also, differences between individuals can be studied this way. Fecal mucus samples or ileostomy

[10] P. S. Cohen and D. C. Laux, *Methods Enzymol.* **253**, 309 (1995).

[11] D. C. Laux, E. F. McSweegan, and P. S. Cohen, *J. Microbiol. Meth.* **2**, 27 (1984).

[12] J. C. Olsson, A. Westerdahl, P. L. Conway, and S. Kjelleberg, *Appl. Environ. Microbiol.* **58**, 551 (1992).

[13] A. C. Ouwehand, P. Niemi, and S. J. Salminen, *FEMS Microbiol. Lett.* **177**, 35 (1999).

effluent are suspended in ice-cold phosphate buffered saline (PBS; 10 mM phosphate, pH 7.2) containing 0.5 g/liter NaN$_3$ to prevent bacterial growth, 1 mM phenylmethylsulfonyl fluoride to inhibit serine proteases, 2 mM iodoacetamide to inhibit cysteine containing enzymes, and 10 mM EDTA to inhibit metalloproteases. The suspension has to be thoroughly mixed and subsequently centrifuged 30 min at 15,000 g. From the clear supernatant, the mucus is precipitated twice with ice-cold ($-18°$) ethanol, final concentration 60% (v/v), and dissolved in PBS.[14] The crude mucus can be further purified by applying it to a Sepharose CL-4B (Pharmacia, Uppsala, Sweden) or equivalent column. The void volume ($M_1 > 20 \times 10^6$ Da) is collected and used in the adhesion assay as mentioned below.

Characterization of Mucus

The protein content of the mucus can be determined by the method of Lowry and co-workers[15] after a modification of Miller and Hoskins[16] using bovine serum albumin as standard. This method will hydrolyze the carbohydrate moieties that otherwise might obstruct the protein core of the mucin and give an underestimation of the protein content of the sample.

The carbohydrate content can be determined using the phenol–sulfuric acid method with glucose as standard.[17]

The sulfate content can be determined using a turbidimetric method with BaCl$_2$. In short, mucus suspensions are boiled 30 min in the presence of 0.1 M HCl. Any precipitate has to be removed by centrifugation (5 min, 10,000 g). An equal volume of 10% (w/v) BaCl$_2$ is added to the supernatant and mixed vigorously. The absorbance is read at 550 nm; Na$_2$SO$_4$ can be used as standard.[14] The mucin content can be determined by the alcian blue method using bovine submaxillary mucin (Sigma, St. Louis, MO) as a standard.[18,19]

In order to determine how intact the isolated mucin is, the blood-group antigenicity can be determined. Blood-group determinants are present on the nonreducing termini of the carbohydrate chains that line the mucin protein core. Because blood-group degrading enzymes are mainly exoglycosidases, the blood-group determinants are likely to be the first to be removed upon bacterial degradation of the mucin.[20] Their presence or absence therefore indicates the level of

[14] A. C. Ouwehand, P. V. Kirjavainen, M.-M. Grönlund, E. Isolauri, and S. J. Salminen, *Int. Dairy J.* **9**, 623 (1999).

[15] O. H. Lowry, N. J. Rosebrough, A. L. Farr, and R. J. Randall, *J. Biol. Chem.* **193**, 265 (1951).

[16] M. R. Miller and L. C. Hoskins, *Gastroenterology* **81**, 759 (1981).

[17] M. Dubois, K. A. Gilles, J. K. Hamilton, P. A. Rebers, and F. Smith, *Anal. Chem.* **28**, 350 (1956).

[18] R. L. Hall, R. J. Miller, A. C. Peatfield, P. S. Richardson, I. Williams, and I. Lampert, *Biochem. Soc. Trans.* **8**, 72 (1980).

[19] N. Fontaine and J. C. Meslin, *Reprod. Nutr. Dev.* **34**, 237 (1994).

[20] L. C. Hoskins, *in* "Attachment of Organisms to the Gut Mucosa" (E. C. Boedeker, ed.), p. 51. CRC Press, Boca Raton, FL, 1984.

degradation of the mucin. The blood-group antigenicity can be determined with A and B antisera, A, B, and O erythrocytes, which usually can be obtained from a local blood transfusion service, and anti-H lectin, *Tetragonolobus purpureas* (Sigma).[14,16]

Tissue Samples

Tissue obtained from experimental animals or from operations on humans should be placed on ice and processed immediately. The tissue is gently washed with PBS containing 0.01% gelatin until all contents are removed. The attached biofilm of the normal microbiota should, however, not be disturbed. Circular samples (up to 9 mm diameter) can be stamped out, e.g., with a cork drill.[21] The samples are preferably used immediately after resection. However, if necessary they can be frozen at $-70°$ in PBS with 40% glycerol and washed gently in HH prior to use. This procedure may affect the adhesion to the sample.

Culturing of Caco-2 Cells

For microbial interaction studies Caco-2 cells can be cultured in standard tissue culture plates (24-well or 6-well plates). The standard tissue culture plates are suitable for adhesion and invasion studies in which epithelial cells are lysed to enumerate the attached bacteria. If microscopic enumerations are to be made, Caco-2 monolayers are prepared on glass coverslips placed in tissue culture plates.

Additionally, Caco-2 cells can be cultured on permeable filters to obtain fully polarized monolayers. The filter units are available in different area and pore sizes (Transwell; Costar, Cambridge, MA). The advantage of these filters is that the differentiation process can be followed by measuring the transepithelial electrical resistance using a Millicell-ERS resistance meter (Millipore, Bedford, MA). In addition to adhesion and invasion studies, the Transwell units offer the possibility to study bacterial penetration through the epithelial cell layer. Detailed instructions for use of the Transwell units for microbial interaction studies have been given previously in this series.[22]

The Caco-2 cell line (ATCC HTB 37) can be purchased from the American Type Culture Collection (ATCC, Rockville, USA). The cells are maintained in Dulbecco's modified Eagle's minimal essential medium (DMEM) supplemented with 10% (v/v) heat-inactivated (30 min, 56°) fetal calf serum, 2 mM L-glutamine, 100 U/ml penicillin, and 100 μg/ml streptomycin at 37° in an atmosphere of 10% CO_2/90% air. Other possible supplements include nonessential amino acids and human transferrin. The use of antibiotics is optional. The cells are cultured in tissue

---

[21] A. Henriksson, R. Szewzyk, and P. L. Conway, *Appl. Environ. Microbiol.* **57**, 499 (1991).
[22] M. G. Puccianelli and B. B. Finlay, *Methods Enzymol.* **236**, 438 (1994).

culture flasks (T-25 cm^2, T-175 cm^2, or T-175 cm^2) from which they are removed with trypsin–EDTA treatment once a week. The harvested cells are then seeded as a 1 : 10 dilution in a new flask.

To culture Caco-2 cell monolayers for microbial interaction assays trypsinized cells are counted in a hemocytometer and seeded at densities of 2.5–3 × 10^5 per cm^2 to obtain confluence. The cell culture medium is replaced every 2–3 days and the cultures are maintained for 2 weeks prior to use in microbial interaction assays, because in most studies the cultures are considered fully differentiated about 15 days after confluence.

Adhesion Assays

Bacterial Growth Conditions

In order to be able to quantify the adherent bacteria, the cells are metabolically labeled with methyl-1,2[^{3}H]thymidine (4.40 TBq/mmol, Amersham, UK) using 10 μl/ml of appropriate medium. Thymidine has the advantage that it is almost exclusively used for DNA synthesis and is not used for biosynthesis of other macromolecules without first being degraded. The radioactivity that will subsequently be measured therefore represents bacterial cells and not excreted products.[23] The adhesive ability of the bacteria can be affected by the growth phase and the growth medium and should therefore be carefully selected and standardized.[24] After growth, the bacteria are washed twice with and resuspended in PBS to remove residual radiolable. It is important to determine the optimal concentration of the bacterial suspension. When the numbers of bacteria are too high, the potential binding sites will be saturated and a low percentage of bacteria will be observed to bind.[25] The optimal density of the bacterial suspension should therefore be determined by testing different amounts of each strain. It is important to use the same density of a strain in a series of experiments. Therefore, the optical density of the bacterial suspension is adjusted to obtain a known number of bacteria (determined by plating serial dilutions of the suspension on appropriate solid media or by flow cytometry[26]).

It should be noted that alternative methods of quantification are possible: e.g., staining with crystal violet[27,28] followed by determination of the absorbance,

[23] D. W. J. Moriarty, *Meth. Microbiol.* **22**, 211 (1990).
[24] E. M. Tuomola, University of Turku, Turku (1999).
[25] E. M. Tuomola and S. J. Salminen, *Int. J. Food Microbiol.* **41**, 45 (1998).
[26] M. Virta, S. Lineri, P. Kankaanpää, M. Karp, K. Peltonen, J. Nuutila, and E.-M. Lilius, *Appl. Environ. Microbiol.* **64**, 515 (1998).
[27] D. L. Brasaemle and A. D. Attie, *Biotechniques* **6**, 418 (1988).
[28] D. R. Mack and P. L. Blain-Nelson, *Pediatr. Res.* **37**, 75 (1995).

releasing the bound bacteria and microscopic counting,[29] or detection with antibodies as with an ELISA.[30]

Adhesion to Mucus

Mucus is passively immobilized to polystyrene Maxisorp microtiter plate wells (Nunc, Roskilde, Denmark) by overnight incubation at 4°, 100 μl per well. The optimal mucus concentration for the immobilization should be determined; in general 500 μg/ml has proved sufficient. However, this can be determined by immobilizing a mucus dilution series and determining the adhesion of a strain that exhibits low adhesion to mucus and has high affinity for polystyrene, e.g., *Lactobacillus plantarum* ATCC 8014.[31]

After the mucus has been immobilized, it is washed twice with HH buffer and 100 μl of the radioactively labeled bacteria is added to the wells. Also, 100 μl of the bacterial suspension is added to scintillation vials to be used as a measure of the bacteria added. After 1 hr incubation at 37°, the wells are carefully washed twice to remove nonbound bacteria. Subsequently 250 μl of 1% SDS in 0.1 *M* NaOH is added and incubated 1 hr at 60° to release and lyse the bacteria. The radioactivity of the suspension can then be measured and compared to the radioactivity added to the wells.[10] Each experiment should be performed with three or four parallels in order to correct for intra-assay variation. The level of adhesion can vary significantly between different strains, from 2–3% of the applied bacteria to around 40% (Fig. 1).

Adhesion to Caco-2 Cell Cultures

The following assay protocol is developed to measure adhesion of lactic acid bacteria to differentiated Caco-2 cell cultures, but the method is readily adaptable to other noninvasive bacteria. The volumes given are for Caco-2 cell cultures in 24-well plates; for 6-well plates larger volumes of medium and bacteria have to be used.

The differentiated Caco-2 monolayers are washed with DMEM without any supplements or with PBS (pH 7.3). After washing, 1 ml of the nonsupplemented DMEM is added to the wells containing the Caco-2 cell monolayer followed by the bacterial suspension in PBS.

The attachment of radiolabeled bacteria to Caco-2 cell cultures is examined by adding 50 μl of a radiolabeled bacterial suspension to three wells containing the

[29] M. A. Bordas, M. C. Balebona, I. Zorrilla, J. J. Borrego, and M. A. Morinigo, *Appl. Environ. Microbiol.* **62**, 3650 (1996).

[30] P. Doig and T. J. Trust, *J. Microbiol. Meth.* **18**, 167 (1993).

[31] E. M. Tuomola, A. C. Ouwehand, and S. J. Salminen, *Lett. Appl. Microbiol.* **28**, 159 (1999).

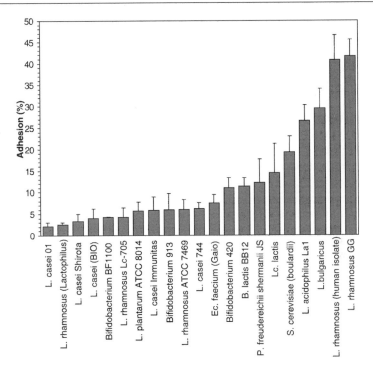

FIG. 1. Adhesion of some lactic acid bacteria to mucus isolated from adult feces. Adhesion is expressed as the percentage of bacteria binding relative to the amount of bacteria added to the immobilized mucus (after Ouwehand et al.[13]).

Caco-2 monolayer (1 ml of nonsupplemented DMEM in the well) or the bacteria are first mixed with 1 ml of medium and added to the wells. After incubation at 37° for 1.5 hr the Caco-2 cell cultures are washed three times with 1 ml of PBS and treated with 100 µl of 0.9 M NaOH at 37° overnight to lyse both the Caco-2 cells and bacteria. The lysed cells are then mixed with scintillation liquid and the radioactivity is measured by liquid scintillation. The radioactivity measured is then compared to the radioactivity added to the wells. This assay has been used to compare the adhesive properties of lactic acid bacteria.[25] With the most adherent strain approximately 15% of added radioactivity was bound to the monolayer. It is important to test a bacterial strain known to adhere poorly and a strain known to adhere effectively to Caco-2 cell cultures in order to control the effectiveness of the washing steps and the whole assay protocol. In our study[25] bovine enterotoxigenic *Escherichia coli* B44 and human enterotoxigenic *E. coli* H10407 were used as nonadhesive and adhesive control strains, respectively. The corresponding values for binding were 4% and 14%.

For light microscopy the nonradiolabeled bacterial suspensions are prepared and used in adhesion assays exactly like the suspensions containing radiolabeled bacteria. After adhesion the Caco-2 monolayers (on glass coverslips) are washed three times with PBS and fixed for 30 min at room temperature with 3% (w/v) paraformaldehyde in PBS. After fixing, the Caco-2 monolayers are washed three times with PBS, removed from the wells, dried in air, and gram stained. Gram staining of the coverslip facilitates distinguishing the bacteria from the background. Adherent bacteria are detected microscopically by counting 15–20 randomized fields per coverslip.

In order to enumerate the attached bacteria by plate counting, the epithelial cells are lysed with a mild detergent (e.g., 1% Triton-X) for 5–10 min to liberate the attached bacteria. The suspension is then diluted with PBS to prevent the effect of the detergent on the bacteria. Serial dilutions of the lysed suspensions are plated on appropriate media. The advantage of plate counting over the microscopic enumeration is that all the attached bacteria can be enumerated instead of counting microscopic fields. A potential disadvantage is that only viable organisms are enumerated.

Invasion to Caco-2 Cell Cultures

None of the above-mentioned enumeration methods distinguish between the adherent and invaded bacteria. Therefore, if the bacteria studied are of an invasive nature, the plate counting and the use of radiolabeled bacteria described above measure both the adherent and invaded bacteria. To enumerate the invaded bacteria, the adherent bacteria can be killed with gentamicin (100 μg/ml; 1 hr treatment), which does not affect the internalized bacteria. After gentamicin treatment the internalized bacteria are enumerated as described above.

Bacterial invasion is an event requiring metabolically active bacteria, and therefore the bacteria can be treated with sodium azide to prevent them from invading the cell without affecting their ability to bind to the cell surface. Another method to prevent the invasion is to fix the Caco-2 cell monolayer with glutaraldehyde before performing the assay.[32]

Adhesion to Tissue Samples

Tissue pieces are placed in the wells of tissue culture plates with the epithelium facing up and 0.5 ml radiolabeled bacterial suspension, of appropriate dilution (see above), is added. The pieces of tissue with the bacterial suspension are incubated 60 min in a 37° water bath. Each well is subsequently washed twice with 1 ml

[32] B. B. Finlay, M. N. Starnbach, C. L. Francis, B. A. D. Stocker, S. Chatfield, G. Dougan, and S. Falkow, *Mol. Microbiol.* **2**, 757 (1988).

PBS and the tissue pieces are fixed with 3% glutaraldehyde in PBS for 45 min at room temperature. The mucosa is then separated from the muscle tissue and digested overnight with 300 μl 30% H_2O_2 and 150 μl 70% perchloric acid in glass scintillation vials at 60–70°. After digestion, 4 ml scintillant is added and the radioactivity determined by liquid scintillation.[21] The percentage adhesion can be determined from the radioactivity in 0.5 ml of the original bacterial suspension as mentioned above for intestinal mucus. Alternatively, the adhesion can be expressed as CFU/surface area tissue.

Binding Kinetics

In the preceding sections how to determine the relative fraction of an applied bacterial suspension that binds to different mucosal models and tissue pieces has been described. It is, however, also possible to determine the affinity of the bacteria for the substrata tested. High affinity for a substratum does not automatically imply a high percentage of adhesion, or vice versa. In addition, the maximum number of binding sites available on a certain substratum can be estimated.

Theory

The relationship between the adhesion of bacteria on the intestinal epithelial layer (density of bound cells) and the numbers of bacteria in suspension (density of added cells) is often found to be a section of a rectangular hyperbola. Such a relationship implies a process of simple dissociation. That is,

$$\text{Bacteria} + \text{enterocyte} \xrightleftharpoons[k-1]{k+1} \text{Bacteria–enterocyte complex} \qquad (1)$$

where $k + 1$ and $k - 1$ represent the dissociation constants for the reaction. The process is similar to the reaction between a substrate and an enzyme to form a substrate–enzyme complex, in this case without the formation of a product. The relationship is based on the assumption that the interaction between the bacterial cells and the epithelial layer remains in equilibrium. This condition should be achieved if the bacterial cells do not embed in the mucus layer or penetrate the intestinal cells. It is also assumed that the density of the bacterial suspension remains essentially unchanged throughout the study, so that the density of the bacterial suspension can be taken as being equal to the initial bacterial numbers. This condition is usually achieved when the total number of bacterial cells is much greater than the number of cells adherent on the epithelial layer.

In Eq. (1) described above, if x is taken for the density of the bacteria added, e for the enterocyte or intestinal mucus concentration, and e_x for the concentration of the bacteria–enterocyte/mucus complex, then the numbers of free bacterial cells will be $(x - e_x)$.

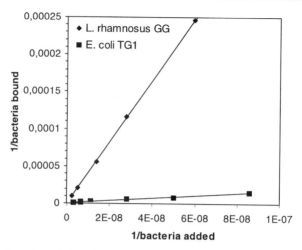

FIG. 2. Double reciprocal plot of the adhesion of *Lactobacillus rhamnosus* GG and *Escherichia coli* TG1 to immobilized human intestinal mucus. The lines indicate the linear fit according to the least squares method (after Lee *et al.*[33]). Reproduced with permission from ASM Press.

Since the process is in equilibrium, the dissociation constant for the process (k_x) can be defined as shown in Eq. (2):

$$k_x = (k-1)/(k+1) = (x - e_x)x/e_x \tag{2}$$

Equation (2) can be rearranged to give an expression for the concentration of the bacteria–enterocyte/mucus complex:

$$e_x = ex/(k_x + x) \tag{3}$$

When x is very much larger than k_x, e_x is approaching e. This maximum value of e_x obtained when the enterocytes or mucus is saturated with bacteria as e_m may be written so that

$$e_x = e_m x/(k_x + x) \tag{4}$$

It may be noted that when x is equal to k_x, e_x is equal to $e_m/2$; thus, the value of k_x could be experimentally obtained from the value of x, which gives half the maximum e_x (i.e., $e_m/2$).

Equation (4) can be rearranged to give a linear relation,

$$1/e_x = 1/e_m + k_x/e_m \cdot x \tag{5}$$

Hence, a plot of $1/e_x$ against $1/x$ will give a straight line, where the intercept on the ordinate gives the value of $1/e_m$ (i.e., 1/maximum number of binding sites), and that on the abscissa gives the value of $-1/k_x$ (i.e., -1/dissociation constant).

[33] Y. K. Lee, C. Y. Lim, W. L. Teng, A. C. Ouwehand, E. M. Tuomola, and S. Salminen, *Appl. Environ. Microbiol.* **66**, 3692 (2000).

Practical

Dilution series of bacterial suspensions are prepared. The range for the series has to be determined empirically for each strain tested, but 10^7–10^8 CFU/ml can serve as an indication. Radioactively labeled bacteria can be used; however, the relation between the radioactivity measured and the number of bacteria needs to be determined by, e.g., plate counting or flow cytometry.[26] Alternatively, especially when using Caco-2 tissue culture cells, microscopic counts can be performed. The adhesion assays are performed as described above for mucus or tissue culture cells.

From the results, double reciprocal plots are prepared, with 1/bacteria added (expressed as CFU/ml) on the x-axis and 1/bacteria bound (expressed as bacteria per Caco-2 cell or per microtiterplate well) on the y-axis. After curve fitting with the least squares method, the intercepts with the ordinate and abscissa can be calculated, which give the value of 1/(maximum number of binding sites) and −1/(dissociation constant).

We have confirmed the relationship between $1/e_m$ and $-1/k_x$ in a study involving *Lactobacillus rhamnosus* GG and *Escherichia coli* TG1 (Fig. 2). The values of e_x and k_x for *L. rhamnosus* GG and *E. coli* are 2×10^7 CFU/well, 8.28×10^{10} CFU/ml, and 1.4×10^6 CFU/well, 2.26×10^8 CFU/ml, respectively. The values suggest that about 10 times more of *L. rhamnosus* GG could adhere onto an unit surface area of epithelium than that of *E. coli* TG1 cells, but the *E. coli* TG1 has higher affinity (dissociates less easily) than the former. The knowledge of the maximum number of cell adhesion and dissociation constants allows better understanding of the mechanism and kinetics of the adhesion process of bacteria on the intestinal mucus layer or to enterocytes, and the competition between different bacteria for adhesion.

In the situation where two types of bacteria are competing for the same receptor, the competition for adhesion is determined by the affinity of the bacteria to the intestinal mucus (k_x) and the density of the bacterial suspension (x). Thus, the ratio

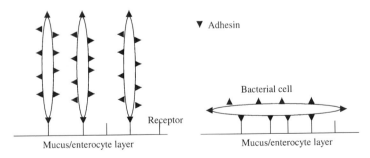

FIG. 3. Schematic representation of adhesion of bacterial cells at high and low densities on a mucus layer.

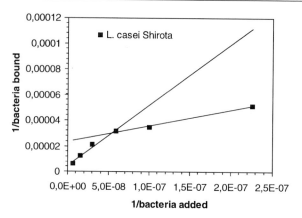

FIG. 4. Double reciprocal plot of the adhesion of *Lactobacillus casei* Shirota to immobilized human intestinal mucus. The lines indicate the linear fit according to the least squares method (after Lee *et al.*[33]). Reproduced with permission from ASM Press.

of e_x for bacterium 1 and bacterium 2 can be described as

$$e_{x1}/e_{x2} = e_{m1}/e_{m2} \cdot x_1/x_2 \cdot (k_{x2} + x_2)/(k_{x1} + x_1) \tag{6}$$

Thus, if *L. rhamnosus* GG at a low density is competing with *E. coli* TG1 for the same adhesion receptor on an intestinal mucus surface, *L. rhamnosus* GG would not be able to prevent the adhesion of *E. coli*, as the former has a higher dissociation constant than the latter. In order to outcompete the *E. coli*, the density of *L. rhamnosus* should approach saturation, i.e., $> 100\,k_x$ or 2.8×10^{13} cells/ml.

In the case where the adhesion of a bacterial cell on the intestinal mucus layer involves more than one adhesion receptor, it could be envisaged that the kinetic constants determined at high and low cell densities would be different. At low cell numbers, the adhesion of lactobacilli on the mucus layer involves the maximum number of adhesion receptors, as illustrated in Fig. 3. At high cell numbers, the minimum number of receptors is involved. Thus, in the plot of $1/e_x$ vs $1/x$, two linear plots are expected at high and low lactobacillus densities, with a curve in between, representing the involvement of various receptors. In a study involving the adhesion of *L. casei* strain Shirota on a mucus layer, two straight lines were observed in the plot of $1/e_x$ vs $1/x$ (Fig. 4). The higher dissociation constant ($k_x = 9.47 \times 10^7$ cells/ml) measured at high *Lactobacillus* numbers compared to that observed at low cell densities ($k_x = 6.10 \times 10^6$ cells/ml) is likely to be due to a 4 times lower number of receptors ($5 \times 10^4 / 20 \times 10^4$) being involved.

Acknowledgments

Financial support from the Academy of Finland is greatfully acknowledged.

[14] Analysis of Microbial Structure and Function of Nitrifying Biofilms

By Satoshi Okabe, Hisashi Satoh, and Yoshimasa Watanabe

Introduction

Wastewater biofilms are very complex multispecies biofilms, displaying considerable heterogeneity with respect to both the microorganisms present and their physicochemical microenvironments. High metabolic rates of the dense microbial population and molecular diffusion create steep microgradients of physicochemical parameters and a successive vertical profile of more energetically favorable electron acceptors in biofilms. Therefore, vertical zonations of predominant respiratory processes occurring simultaneously in close proximity have been found in wastewater biofilms with a typical thickness of only a few millimeters.[1-4] Changes in local microenvironments affect *in situ* microbial activity and consequently arrange microbial populations or vice versa. Such microbial communities have their own particular activities that cannot be achieved by individual microorganisms.

We cannot rely on conventional microbial techniques such as isolation and cultivation to characterize microorganisms and their activities, because many microorganisms cannot be cultivated and microorganisms in pure culture do not necessarily reflect their activities in natural environments. Especially, growth conditions in biofilms are different from those in the bulk medium and thus are very difficult to imitate in artificial cultures. Without knowing the microenvironments in the biofilm, it is almost impossible to understand the relationship between the microbial community structure and its function. Therefore, techniques and tools with a high spatial and temporal resolution are required for direct detection of the spatial distributions of microbial species and their activities in minimally disturbed natural habitats (i.e., biofilms).

At present, microelectrode measurements are the most reliable way of measuring the microenvironments and microbial activities in biofilms with minimal disturbance and have been successfully used for nitrogen cycles[1,3,5,6] and sulfur

[1] D. deBeer, A. Schramm, C. M. Santegoeds, and M. Kuhl, *Appl. Environ. Microbiol.* **63**, 973 (1997).

[2] S. Okabe, T. Itoh, H. Satoh, and Y. Watanabe, *Appl. Environ. Microbiol.* **65**, 5107 (1999).

[3] S. Okabe, H. Satoh, and Y. Watanabe, *Appl. Environ. Microbiol.* **65**, 3182 (1999).

[4] N. B. Ramsing, M. Kuhl, and B. B. Jorgensen, *Appl. Environ. Microbiol.* **59**, 3840 (1993).

[5] A. Schramm, L. H. Larsen, N. P. Revsbech, N. B. Ramsing, R. Amann, and K.-H. Schleifer, *Appl. Environ. Microbiol.* **62**, 4641 (1996).

[6] A. Schramm, D. deBeer, M. Wagner, and R. Amann, *Appl. Environ. Microbiol.* **64**, 3480 (1998).

cycles[2,4,7] in various environmental samples. From the measured concentration profiles, the vertical zonation of respiratory processes and even their rates can be estimated.[2-4]

Specific analyses for microbial species distributions (or for *in situ* identification of microorganisms) in biofilms can only be obtained by techniques such as immunofluorescence[8] or *in situ* whole-cell hybridization[9] without the biases of cultivation. The immunofluorescence approach, however, requires the prior isolation of pure cultures for synthesis of specific antibodies and is therefore not really cultivation independent. Therefore, we will focus here on the use of *in situ* whole-cell hybridization with fluorescently labeled 16S rRNA-targeted oligonucleotide probes.[9] Obviously, analysis at single-cell level by fluorescent *in situ* hybridization (FISH) can provide a more detailed picture of spatial distributions and abundance of so far uncultured microorganisms.[10] FISH has been combined successfully with microelectrode measurements to study nitrification and sulfate reduction in various biofilms.[2-5]

In this chapter, we will address the great potential of the combined use of the current FISH technique and microelectrodes to study the microbial ecology of complex microbial communities such as biofilms. The combination of these two techniques will provide reliable and direct information about relationships between *in situ* microbial activity and the occurrence of specific microorganisms in biofilms. As an example of the combined study, we will illustrate the *in situ* spatial organization of ammonia-oxidizing and nitrite-oxidizing bacteria on fine scale in autotrophic nitrifying biofilms by applying FISH and will correlate to their activity distributions at a similar resolution determined by use of microelectrodes.

Methods

Microelectrodes

There are principally three types of electrochemical microsensors that are most often used in environmental applications: potentiometric microsensors, amperometric microsensors, and optical microsensors. It is not possible to outline all the details for the principle and the construction of each microelectrode here. Since a good introduction to the construction of microelectrodes is given by Revsbech and Jorgensen,[11] we will focus on applications of microelectrodes in biofilm studies.

[7] M. Kuhl and B. B. Jorgensen, *Appl. Environ. Microbiol.* **58**, 1164 (1992).
[8] B. B. Bohlool and E. L. Schmidt, *Adv. Microb. Ecol.* **4**, 203 (1980).
[9] R. I. Amann, W. Ludwig, and K.-H. Schleifer, *Microbiol. Rev.* **59**, 143 (1995).
[10] R. I. Amann, N. Springer, W. Ludwig, H.-D. Gortz, and K.-H. Schleifer, *Nature (London)* **351**, 161 (1991).
[11] N. P. Revsbech and B. B. Jorgensen, *Adv. Microb. Ecol.* **9**, 293 (1986).

O_2 Microelectrodes

Although there are two types of O_2 microelectrodes, a cathode-type and a Clark-type, we will describe only a simpler cathode-type O_2 microelectrode in this chapter. The construction of the cathode-type O_2 microelectrode is easier than that of the Clark-type. However, the cathode-type O_2 microelectrode, in general, has a less stable signal, is relatively sensitive for Ca^{2+} and Mg^{2+}, and is pH dependent (pH should be above 6), which limits the use of this microelectrode in sea water and acidic environments.[12,13] This microelectrode consists of an etched platinum wire covered with a thin glass wall, and the tip is coated with gold (namely, "a tiny gold cathode"). This simple cathode should be coated with DePeX, an oxygen permeable resin. The microelectrode is charged at about -0.75 V vs a standard calomel reference electrode, and the current originating from the oxygen reduction at the surface of the gold cathode is proportional to the oxygen partial pressure in the surrounding medium, which is measured by a sensitive picoampere meter (DC microvolt AM meter PM-18U; TOA). A two-point calibration is usually made by measuring the current when the microelectrode is placed in water saturated with air (21% O_2) and in water sparged with 100% N_2 gas (0% O_2).

Liquid Ion-Exchange Membrane Microelectrodes

The principle of the liquid ion-exchange membrane (LIX) microelectrodes is the same as that of the potentiometric microelectrodes. The electrical potential difference across the liquid ion-exchange membrane becomes proportional to the logarithm of the ion activity in the solution when the ion activity in the electrolyte solution is constant.[14] The generated electrical potential can be measured with a high-impedance voltmeter (e.g., pH/ion Meter F-23; Horiba or Keithley 6512). The detailed principle, design, and application of ion-selective microelectrodes can be found elsewhere.[14] The preparation of LIX microelectrodes for NH_4^+, NO_2^-, and NO_3^- has been described in detail by deBeer et al.[1]; thus the fabrication steps will be discussed briefly. The LIX membrane can be gelled with high-molecular-weight polyvinyl chloride (PVC), and the tip should be coated with 10% (wt/vol) bovine serum albumin (BSA) to improve the stability and performance of the LIX microelectrodes. Finally, the microelectrode is shielded with an outer casing containing 1 M KCL to reduce electrical noise.[1] All liquid ion exchangers and other components are commercially available (e.g., from Fluka). A calibration of LIX microelectrodes can usually be performed in dilution series (10^{-3}–10^{-6} M) of each individual ion in the same medium used for the microelectrode measurements.

[12] H. Baumgartl and D. W. Lubbers, in "Polarographic Oxygen Sensors: Aquatic and Physiological Applications" (E. Gnaiger and H. Forstner, eds.), p. 37. Springer-Verlag, Heidelberg, 1983.

[13] N. P. Revsbech, in "Polarographic Oxygen Sensors: Aquatic and Physiological Applications" (E. Gnaiger and H. Forstner, eds.), p. 265. Springer-Verlag, Heidelberg, 1983.

[14] D. Ammann, "Ion-Selective Microelectrodes." Springer-Verlag, Berlin, 1986.

Microelectrode Measurements

The biofilm samples taken from the reactor should be acclimated in a medium a few hours before the microelectrode measurement to ensure that steady state profiles are obtained. Concentration profiles in the biofilm are measured using a motor-driven micromanipulator (e.g., ACV-104-HP; Chuo Precision Industrial Co. Ltd., Japan) in increments of 25–100 μm from the bulk liquid into the biofilm. The biofilm liquid interface is determined by viewing through a dissection microscope (e.g., Stemi 2000, Carl Zeiss). Microprofiles should be measured several times at different positions in the biofilms for each species and condition because of biofilm heterogeneity. A more detailed discussion of microelectrodes and their use in microbial ecology can be found in the recent literature.[11,15,16]

Fluorescent in Situ Hybridization

Over the past years numerous applications of fluorescent *in situ* hybridization (FISH) have been reported for *in situ* identification and localization of certain bacterial cells in various samples (for a review see Ref. 9). Several studies have achieved *in situ* identification of so far uncultured bacteria based on 16S rRNA sequences directly retrieved from samples.[10] Accordingly, a number of 16S rRNA-targeted oligonucleotide probes used successfully for whole-cell hybridization have been developed and become available. If one uses such probes, fine adjustment and optimization of the hybridization conditions are recommended. If one wants to design new probes and apply them to biofilm samples, many steps from the initial extraction of the nucleic acids to the final optimization of the probe specificity and assay sensitivity must be performed. The design of new rRNA-targeted, group- or species-specific probes by comparative sequence analysis is beyond the scope of this chapter. The principal steps for designing new probes have been reviewed in detail.[17] In the following section, we focus on general procedures of biofilm sample fixation, immobilization on glass slides, and *in situ* whole cell hybridization with probes that have been successfully used previously. One should be aware that different specimens may require adjustment both in sample preparation and hybridization.

Fixation and Cryosectioning of Biofilm Samples for FISH

Immediately after the microelectrode measurements, the biofilms attached on the substratum must be fixed with freshly prepared paraformaldehyde solution

---

[15] M. Kuhl and N. P. Revsbech, *in* "Benthic Boundary Layer: Transport Processes and Biogeochemistry" (B. P. Boudreau and B. B. Jorgensen, eds.), Oxford University Press, Oxford, 2001.

[16] R. Amann and M. Kuhl, *Curr. Opinion Microbiol.* **1,** 352 (1998).

[17] D. A. Stahl and R. Amann, *in* "Nucleic Acid Techniques in Bacterial Systematics" (E. Stackebrandt and M. Goodfellow, eds.), p. 205. John Wiley & Sons Ltd., Chichester, 1991.

[4% in phosphate buffered saline (PBS); 130 mM sodium chloride, 10 mM sodium phosphate buffer, pH 7.2] for 4 to 8 hr at 4° to maintain the morphological integrity of the cells. Paraformaldehyde fixation also minimizes autofluorescence. Thereafter, the biofilms are rinsed twice with PBS solution. The fixed biofilm is then placed in a small aluminum container with the biofilm side up, embedded in Tissue-Tek OCT compound (Miles, Elkhart, IN) overnight to infiltrate the OCT compound into the biofilm, and subsequently frozen at −20°. The embedded biofilm is then separated from the substratum and turned over. The second embedding is performed on the other side, which is previously attached to the substratum. The frozen biofilms are completely surrounded by OCT compound and are finally cut into 10- to 20-μm-thick vertical slices with a cryostat (Reichert-Jung Cryocut 1800, Leica) at −20°. Each sectioned specimen is immobilized on a gelatin-coated (0.1% gelatin and 0.01% chromium potassium sulfate) slide on which a hydrophobic coating separates six glass surface windows (e.g., Cel-Line Associates, Inc., Newfield, NJ) and air dried overnight. The specimen is finally dehydrated by successive passages through 50, 80, and 98% ethanol washes (for 3 min each), air dried, and stored at room temperature. The ethanol dehydration substantially reduces the inherent fluorescence, removes the OCT compound, and also increases probe penetration through cell walls. More detailed procedures can be found elsewhere.[18]

In Situ Hybridization

Hybridization buffer consists of 0.9 M NaCl, 20 mM Tris hydrochloride (pH 7.2), 0.01% sodium dodecyl sulfate (SDS), and a varied formamide concentration. The formamide concentration should be determined experimentally to obtain the appropriate specificity for each probe.[17] Hybridization buffer (8 μl) is spotted on each fixed specimen in a glass window, and then 1 μl of each fluorescent probe is mixed. The final probe concentration is approximately 5 ng μl^{-1}. The glass slide should be transferred to a prewarmed airtight box containing a paper tissue soaked in the hybridization buffer to prevent evaporative concentration of hybridization buffer, which might result in nonspecific hybridization. We usually use 50 ml polypropylene centrifuge tubes with double seal caps (e.g., No. 2341-050, IWAKI). Then hybridization is performed at 46° for 2–3 hr. Subsequently, a stringent wash step is performed at 48° for 20 min in 50 ml of prewarmed washing solution containing 20 mM Tris hydrochloride (pH 7.2), 0.01% SDS, and a varied concentration of NaCl. The NaCl concentration in the washing solution is dependent on the stringency required to achieve the appropriate specificity for each probe.[17] The slides are then rinsed briefly with ddH$_2$O and allowed to air

[18] R. Amann, *in* "Molecular Microbial Ecology Manual" (A. D. L. Akkerman, J. D. van Elsas, and F. J. de Bruijn, eds.), 336, p. 1. Kluwer Academic Publishers, Dordrecht, The Netherlands, 1995.

dry. One should be very careful not to wash the biofilm samples away during the washing and rinsing steps. Simultaneous hybridization with probes requiring different stringency can be performed by a successive hybridization procedure: hybridization with the probe requiring higher stringency is performed first and followed by hybridization with the probe requiring lower stringency. Slides are finally mounted in antifading agents. Several antifading agents are commercially available. We usually use a SlowFade-light antifade kit (Molecular Probes, Eugene, OR) or Citifluor (Citifluor Ltd., Canterbury, UK).

Microscopic Observation

Obviously, successful *in situ* hybridization will also depend on the instrumentation used for final detection. The detectability of probe-conferred fluorescence signals can be significantly improved by reducing background noise caused by broad inherent autofluorescence and out-of-focus fluorescence. Especially, wastewater biofilm samples generally contain detrital matter and mineral grains, which usually exhibit a wide range of emission spectra and thus generate problems with autofluorescence. For this reason, the scanning confocal laser microscope (SCLM) is an ideal tool for analysis of microscale spatial organizations of microbial communities in biofilms. With this SCLM, nondestructive optical sections of a sample can be obtained, which removes out-of-focus fluorescence and facilitates sharper and clearer imaging. In addition, three-dimensional images can be reconstructed from the data obtained by serial optical sectioning.

Representative images shown in this chapter were taken by a Zeiss LSM 510 scanning confocal laser microscope (Zeiss) supplying excitation wavelengths at 488 (argon laser) and 543 nm (HeNe laser). Zeiss filter sets of 09, 10, and 15 and object lenses of 20×, 40×, and 63× oil immersion were used. All image combining and processing were performed with the standard software package provided by Zeiss.

Applications

Microbial Nitrification in Biofilms

The microbial ecology of nitrifying bacteria in an autotrophic nitrifying biofilm will be illustrated below as an example of the combined use of the above-mentioned microelectrode and *in situ* hybridization techniques. Although microbial nitrification processes for nitrogen removal are becoming more important because of strict regulations on nitrogen discharge, nitrification is recognized as being difficult to maintain in wastewater treatment systems. The process of nitrification is carried out by two phylogenetically unrelated groups of lithoautotrophic bacteria, the ammonia-oxidizing bacteria and nitrite-oxidizing bacteria. A better understanding of microbial ecology (i.e., diversity, abundance, localization, and activity) of nitrifying bacteria in wastewater biofilms is essential for improving process performance and control. However, investigation of the microbial ecology of nitrifying bacteria by conventional cultivation techniques has been hampered by their slow growth rates and by the biases inherent in all culture-based techniques. Therefore, the *in situ* detection of nitrifying bacteria and their activity in biofilms is of great practical and scientific relevance.

Microbial Structure and Function of Autotrophic Nitrifying Biofilms

An autotrophic nitrifying biofilm was cultured in partially submerged rotating disk reactors (RDR) consisting of five disks with removable slides (1 × 6 cm) for sampling biofilms.[19] The autotrophic nitrifying biofilms were first cultured with the primary settling tank effluent from the Shoseigawa municipal wastewater treatment plant (Sapporo, Japan) for a few days and then were cultured with synthetic nutrient.[3]

Microelectrode Measurements

Steady-state concentration profiles of O_2, NH_4^+, NO_2^-, and NO_3^- within the nitrifying biofilm were measured by microelectrodes (Fig. 1A, see color insert). All measurements were performed in a flow cell reactor at $20°$, with an average liquid velocity of 2–3 cm sec^{-1} by blowing air on the water surface. The medium used for microprofile measurements contained the following ingredients: 200 μM NH_4Cl, 50 μM $NaNO_2$, 300 μM $NaNO_3$, 570 μM Na_2HPO_4, 84 μM $MgCl_2 \cdot 6H_2O$, 200 μM $CaCl_2$, and 270 μM EDTA (pH 7.0). Oxygen penetrated approximately 200 μm into the biofilm (thickness ca. 250 μm). The NH_4^+, NO_2^-, and NO_3^- profiles showed that the consumed NH_4^+ was primarily converted to NO_2^- in the upper 75 μm with a NO_2^- peak of 64–73 μM at 50–75 μm and no significant NO_3^- production in this zone. The produced NO_2^- was eventually converted to NO_3^- in the deeper oxic layer (75–150 μm). This result demonstrates the sequential oxidation of NH_4^+ and NO_2^- in the oxic biofilm strata, that is, the active NH_4^+-oxidizing zone is located in the outer part of the oxic biofilm, whereas the active NO_2^--oxidizing zone is located just below the NH_4^+-oxidizing zone. Obviously, this sequential oxidation of NH_4^+ and NO_2^- on such a micro scale can only be observed with microelectrodes. This characteristic activity distribution will be correlated to the vertical distribution of NH_4^+- and NO_2^--oxidizing bacteria in the corresponding biofilm determined with the use of the FISH technique in the following section.

[19] S. Okabe, K. Hirata, Y. Ozawa, and Y. Watanabe, *Biotechnol. Bioeng.* **50**, 24 (1996).

TABLE I
LIST OF 16S rRNA TARGETED-OLIGONUCLEOTIDE PROBES USED IN THIS STUDY

Probe	Specificity	Sequence of probe (5′–3′)	Target site[a]	FA[b] (%)	NaCl[c] (mM)	Ref.
EUB338	Domain *Bacteria*	GCTGCCTCCCGTAGGAGT	338–355	20	0.225	20
Nso190	Ammonia-oxidizing *beta*-proteobacteria	CGATCCCCTGCTTTTCTCC	190–208	20	0.020	21
NIT2	*Nitrobacter* spp.	CGGGTTAGCGCACCGCCT	1433–1450	40	0.056	22
NIT3	*Nitrobacter* spp.	CCTGTGCTCCATGCTCCG	1035–1048	40	0.056	22
CNIT3[d]	Competitor for NIT3	CCTGTGCTCCAGGCTCCG	1035–1048	—		22
Ntspa454	*Nitrospira moscoviensis* aquarium clone 710-9	TCCATCTTCCCTCCCGAAAA	435–454	20	0.225	23
Ntspa1026	*Nitrospira moscoviensis* activated sludge clones A-4, A-11	AGCACGCTGGTATTGCTA	1026–1043	20	0.225	24

[a] 16S rRNA position according to *Escherichia coli* numbering.
[b] Formamide concentration in the hybridization buffer.
[c] Sodium chloride concentration in the washing buffer.
[d] Used as unlabeled competitor probe together with probe NIT3.

In Situ Hybridization

To investigate the spatial distributions of NH_4^+- and NO_2^--oxidizing bacteria within the biofilm, simultaneous *in situ* hybridization with specific probes was performed for the entire vertical biofilm sections (Fig. 1B). All oligonucleotide probes used in this study, their sequences, specificity, formamide concentrations in hybridization buffer, NaCl concentrations in washing buffer, and references are given in Table I. Probes were labeled with fluorescein isothiocyanate (FITC) or tetramethylrhodamine 5-isothiocyanate (TRITC).

Spatial Distributions of Nitrifying Bacteria

Spherical clusters of densely packed probe Nso190-stained NH_4^+-oxidizing bacterial cells were detected throughout the oxic biofilm strata, indicating more or less a homogeneous spatial distribution of NH_4^+-oxidizing bacteria (Fig. 1B). In

[20] R. I. Amann, L. Krumholz, and D. A. Stahl, *J. Bacteriol.* **172,** 762 (1990).
[21] B. K. Mobarry, M. Wagner, V. Urbain, B. E. Rittmann, and D. A. Stahl, *Appl. Environ. Microbiol.* **62,** 2156 (1996).
[22] M. Wagner, G. Rath, H.-P. Koops, J. Flood, and R. I. Amann, *Water Sci. Technol.* **34,** 237 (1996).
[23] T. A. Hovanec, L. T. Taylar, A. Blakis, and E. F. Delong, *Appl. Environ. Microbiol.* **64,** 258 (1998).
[24] S. Juretschko, G. Timmermann, M. Schmid, K.-H. Schleifer, A. Pommerening-Roser, H.-P. Koops, and M. Wagner, *Appl. Environ. Microbiol.* **64,** 3042 (1998).

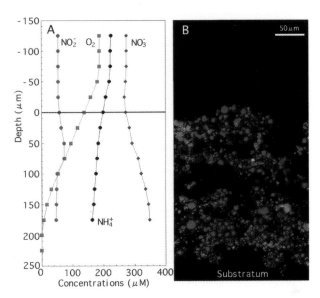

FIG. 1. An example of the combined use of microelectrode measurement and fluorescent *in situ* hybridization (FISH). (**A**) Steady-state microprofiles of O_2, NH_4^+, NO_2^-, and NO_3^- in an autotrophic nitrifying biofilm measured by microelectrodes. The points are means of three to five measurements. The biofilm surface was at a depth of zero. (**B**) A composite CSLM projection image of the entire vertical section of the same biofilm after simultaneous *in situ* hybridization with TRITC-labeled Nso190 probe for NH_4^+-oxidizing bacteria (red) and FITC-labeled Ntsp454 probe for *Nitrospira*-like NO_2^--oxidizing bacteria (green). The biofilm surface was at the top. The biofilm thickness is about 250 μm. Panel B is reproduced from Okabe *et al., Appl. Environ. Microbiol.* **65**, 3182 (1999).[3]

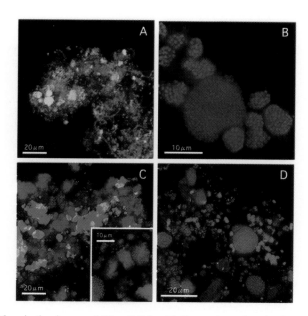

Fig. 2. CLSM projection images of thin nitrifying biofilm sections after simultaneous *in situ* hybridization with TRITC-labeled Nso190 probe for NH_4^+-oxidizing bacteria (yellow) and FITC-labeled Eub338 probe for domain bacteria (green) (A). Cells of NH_4^+-oxidizing bacteria appear to be yellow because of binding of both probes. A magnification of Nso190 probe-stained NH_4^+-oxidizing bacterial clusters (B). Cells were densely packed and formed spherical clusters. Close association of TRITC-labeled Nso190 probe-stained NH_4^+-oxidizing bacteria (red) with FITC-labeled Ntsp454 probe-stained *Nitrospira*-like NO_2^--oxidizing bacteria (green) (C). The inset in panel C is a magnification of both microbial clusters. Similarly, close association of NH_4^+-oxidizing bacteria (red) with *Nitrospira*-like NO_2^--oxidizing bacteria (green) (D). Ntsp1026 probe was used instead of Ntsp454 probe. Panel D is reproduced from Okabe *et al.*, *Appl. Environ. Microbiol.* **65**, 3182 (1999).[3]

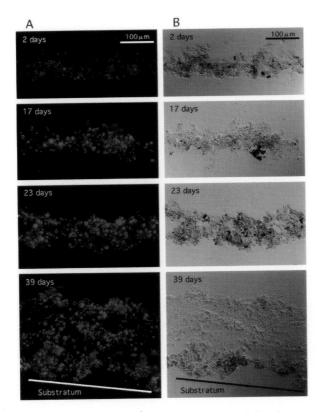

Fig. 3. Time-dependent development of NH_4^+-oxidizing bacteria populations in an autotrophic nitrifying biofilm cultured with a synthetic medium (the substrate C/N ratio is zero). CSLM images after *in situ* hybridization of the vertical biofilm sections with TRITC-labeled Nso190 probe (red). Biofilm sections were taken at 2-, 17-, 23-, and 39-day cultivations, respectively (A). Differential interference contrast (DIC) images of the same microscopic fields (B), showing the development of the entire biofilm structure. The biofilm surface was at the top in all panels.

contrast, clusters of the Ntspa 454 probe-stained *Nitrospira*-like NO_2^--oxidizing bacteria were primarily detected in the deeper part of the oxic region. No hybridization signal was observed when *Nitrobacter*-specific probes NIT2 and NIT3 were used with any of the samples.

The sequential oxidation of NH_4^+ and NO_2^- found by microelectrode measurements coincides with locations where higher abundance of NH_4^+- and NO_2^--oxidizing bacteria were detected. The lower NO_2^- oxidation activity in the surface zone can be explained by the absence (or lower abundance) of NO_2^--oxidizing bacteria. This is a good example of a correlation between the distribution of microbial species and accompanying expected activity. However, it is not always possible to find such a correlation, because the spatial distributions of microbial species can be a result of previous stages of the biofilm development, rather than an optimal adaptation to the actual microenvironments.

Micro-Scale Spatial Organization of Nitrifying Bacteria

Although the biofilm was cultured in the synthetic medium containing no organic carbon, the NH_4^+-oxidizing bacterial clusters were surrounded by a number of heterotrophs, including filamentous bacteria (Fig. 2A, see color insert), which may suggest these heterotrophs could utilize soluble organic compounds secreted from NH_4^+-oxidizing bacteria. NH_4^+-oxidizing bacteria formed densely packed spherical clusters (Fig. 2B) and closely associated with NO_2^--oxidizing bacteria (Figs. 2C and 2D), demonstrating the sequential metabolism of NH_4^+ via NO_2^- to NO_3^-. By such close association, the diffusion path from NH_4^+-oxidizing bacterial clusters to the surrounding NO_2^--oxidizing bacteria is short and facilitates an efficient transfer of the intermediate NO_2^-. It should be noted that these micro-scale spatial organizations of two phylogenetically unrelated species can only be obtained by use of the FISH technique.

Development of NH_4^+-Oxidizing Bacterial Populations in Biofilms

To visualize time-dependent development of NH_4^+-oxidizing bacteria communities in an autotrophic nitrifying biofilm, vertical sections of the biofilms taken at different developmental stages were hybridized with TRITC-labeled Nso190 probe (Fig. 3A, see color insert). Differential interference contrast (DIC) images of the same microscopic fields were presented to illustrate development of the entire biofilm structure (Fig. 3B). The Nso190 probe-stained NH_4^+-oxidizing bacteria were barely found and were initially present in the forms of single scattered cells or small clusters just 2 days after switching to the synthetic medium from the primary settling tank effluent. However, these small microbial clusters gradually became bigger spherical clusters, and the number of clusters increased with time.

After 39-day cultivation, the biofilm thickness reached approximately 250 μm, and a relatively homogeneous spatial distribution of NH_4^+-oxidizing bacteria was developed throughout the biofilm. This observation clearly reveals a characteristic growth pattern of NH_4^+-oxidizing bacteria within the biofilm, which had not been demonstrated by any culture-based techniques so far.

Discussion

Selectivity and Detection Limit of LIX Microelectrodes

For LIX microelectrodes, detection limit and ion selectivity of microelectrodes are crucial points and thus will be discussed below. Figure 4 shows typical calibration curves of nitrite and nitrate microelectrodes in various solutions. The responses of the nitrite microelectrodes were log-linear down to a nitrite concentration of 10^{-4} M in all solutions including the medium used for the measurement, and then went off slightly. The selectivity constants for Cl^-, NO_3^-, and HCO_3^- were 0.02, 0.02, and 0.08, respectively. The selectivity of this microelectrode could be improved by adjusting the tip diameter to approximately 10 μm.[1] Similar results were obtained for nitrate microelectrodes. The nitrate microelectrodes showed a log-linear response down to a nitrate concentration of 10^{-5} M in pure water. However, the response slightly deteriorated below a nitrate concentration of 10^{-4} M in the medium and in the presence of 10 mM of NO_2^- and Cl^-. The selectivity

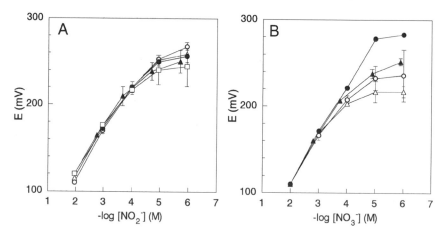

FIG. 4. (A) Calibration curves of nitrite microelectrodes in pure water (●), in the medium used for measurements (▲), in 10 mM sodium chloride solution (○), in 10 mM nitrate solution (△), and in 10 mM sodium bicarbonate solution (□). (B) Calibration curves of nitrate microelectrodes in pure water (●), in the medium used for measurements (▲), in 10 mM sodium chloride solution (○), and in 10 mM nitrite solution (△).

constants of the nitrate microelectrodes for Cl^- and NO_2^- were 0.03 and 0.05, respectively. The selectivity of NH_4^+ microelectrodes for Na^+ and K^+ was relatively low (data not shown).

Limitation of Microelectrode Measurements

Microelectrode measurement is, of course, not perfect, and there are some limitations. First, it should be noted that the microprofiles presented in this chapter are artificial results under the conditions applied during microelectrode measurement and are not the profiles that actually occurred under growth conditions in the biofilm reactor, because the reactor hydrodynamics are different. The lack of stirring causes a thickening of the diffusive boundary layer above the biofilm, which significantly reduces the substrate concentrations at the biofilm surface and consequently affects substrate consumption rates in the biofilm. Second, biofilms, in general, are not continuum flat films and are structurally very heterogeneous as shown in Fig. 1B. Accordingly, the diffusion coefficient may vary with depth. However, the influence of the biofilm heterogeneity on diffusion coefficients is not presently well known. Third, when biofilm thickness is thin, it is difficult to completely measure microprofiles until the curvature ends (e.g., NH_4^+ and NO_3^- profiles in Fig. 1A), which leads to the uncertainties of quantitative activity analyses. Finally, the lifetime of LIX microelectrodes is usually a few days, and thus the fabrication of fresh LIX microelectrodes is laborious.

Obstacles to General Application of FISH to Biofilm Samples

Although fluorescent *in situ* hybridization is a powerful tool for identifying and counting microorganisms in complex microbial communities, application of this technique to natural biofilm systems is sometimes limited for several reasons. First, since the amount of rRNA in cells is directly correlated to the fluorescence conferred by rRNA-targeted probes, quiescent cells with a low rRNA content may result in a weak or even undetectable hybridization signal. It is most likely that many microorganisms present in natural biofilms are in such a quiescent state and thus difficult to detect by FISH. Second, the detection and counting of target cells at low concentrations (approximately 10^3–10^4 cells cm^{-2}) in a high background of nontarget cells becomes very tedious and not reliable. Third, wastewater or natural biofilms generally contain detrital matter and mineral grains, which generate problems with autofluorescence and unspecific staining. Finally, microscopic observation of an exact location of the targeted cells in thick samples remains difficult because of problems with limited depth of focus. The last two problems can be overcome with the use of a scanning confocal laser microscope (SCLM) as mentioned above. Another frequently encountered difficulty is the enumeration of probe-stained nitrifying bacteria cells in biofilm thin sections because they often form densely packed aggregates (see Fig. 2B) in which a reliable cell count is very difficult. If the sample consists of suspended microorganisms, the enumeration of probe-stained cells and quantification of fluorescence can be achieved with a flow cytometer.[25,26] A possible solution would be to measure the surface fraction of total biomass area and probe-stained cell (cluster) area,[3] which should, of course, be regarded as a semiquantitative analysis. In regard to fundamental problems of *in situ* hybridization such as probe specificity, probe accessibility, and fluorescence signal intensity, the reader is referred to a comprehensive review by Amann *et al.*[9]

Conclusions

Obviously, FISH and microelectrode techniques on their own have a high spatiotemporal resolution and great potential and provide reliable and direct information about the occurrence of specific microorganisms and their *in situ* activity in biofilms, respectively. Such information cannot be obtained by conventional culture-dependent techniques. When these two techniques are used together, the abundance, localization, and activity of targeted microorganisms *in situ* can be directly correlated, and thus the most detailed pictures of microbial ecology of complex microbial communities can be drawn on a micro scale. This is extremely important for an understanding of microbial processes occurring in environmental and engineered biofilm systems, because most microbial transformations are catalyzed by consortia and not by single species of microorganisms.

Acknowledgment

We thank Naoko Norimatsu for technical assistance with FISH. This work was partially supported by the CREST (Core Research for Evolutional Science and Technology), Japan Science and Technology Corporation (JST) and Grant-in Aid (No. 09750627) for Developmental Scientific Research from the Ministry of Education, Science and Culture of Japan.

---

[25] G. Wallner, R. Amann, and W. Beisker, *Cytometry* **14**, 136 (1993).
[26] B. Bertin, O. Broux, and M. van Hoegarden, *J. Microbiol. Methods* **12**, 1 (1990).

[15] Computational and Experimental Approaches to Studying Metal Interactions with Microbial Biofilms

By D. SCOTT SMITH and F. GRANT FERRIS

Introduction

In modeling the fate and transport of metals in aquatic environments it is well appreciated that speciation information is critical. In particular, it is critical to determine whether the metal of interest is free in solution or bound to any available solid substrates. Possible natural substrates include mineral and bacterial surfaces, as well as organic matter such as humic and fulvic acid. Often these three components do not occur as discrete phases, but rather as biofilm systems in which they are combined in a complex mixture that coats some underlying surface. The various organic and inorganic components within biofilms are chemically reactive and afford numerous potential metal binding sites. In addition, metabolic processes that occur within biofilms are known to be very complex, and a function of the underlying substratum as well as the overlying bulk water phase. With the advent of confocal laser scanning microscopy (CLSM), attenuated total reflectance Fourier transform infrared spectroscopy (ATR-FTIR), microelectrodes, and other techniques, a range of chemical gradients can be seen.[1-3] Biofilms are very complicated and are spatially as well as chemically heterogenous. Overall, this complex system makes measuring and modeling metal–biofilm interactions very difficult.

Methods to determine metal binding to natural sorbents, such as minerals, organic matter, and bacteria, have been developed,[4-7] and similar methods would apply for the mixed system that makes up a biofilm. In order to simplify the theoretical framework for modeling metal–biofilm interactions it is useful to start with the assumption that the system can be explained with an equilibrium description. In addition, if additivity is assumed, then results for individual systems can be combined to obtain an overall description of biofilm binding.

In this paper we present methods to investigate metal–biofilm interactions. In particular, we discuss experimental approaches to quantify metal partitioning

[1] G. Silyn-Roberts and G. Lewis, *Wat. Sci. Technol.* **36,** 117 (1997).

[2] J. Schmitt and H. C. Flemming, *Int. Biodeterioration Biodegradation* **41,** 1 (1998).

[3] P. L. Bishop and T. Yu, *Wat. Sci. Technol.* **39,** 179 (1999).

[4] J. S. Cox, D. S. Smith, L. A. Warren, and F. G. Ferris, *Env. Sci. Technol.* **33,** 4514 (1999).

[5] E. Jenne, (ed.) "Adsorption of Metals by Geomedia: Variables, Mechanisms and Model Applications." Academic Press, San Diego, 1998.

[6] J. B. Fein, C. J. Daughney, N. Yee, and T. A. Davis, *Geochim. Cosmochim. Acta* **61,** 3319 (1997).

[7] D. A. Dzombak and F. M. M. Morel, "Surface Complexation Modeling—Hydrous Ferric Oxide." John Wiley and Sons, New York, 1990.

between the aqueous and solid phases as well as modeling approaches to interpret the resultant data. Throughout is a discussion of the limits and advantages of the described techniques. Finally, a consideration of the limits of applying the additivity assumption to investigate total binding in biofilms is presented. Properties of metal ions in bulk solution are considered first, because of the similarity of complexation reactions in aqueous solution to complexation reactions in biofilms.

Metal Ions in Aqueous Solution

The chemical behavior of any metal in aqueous solution depends strongly on the nature of the individual solution components (i.e., metal concentration, the presence of complexing agents, pH, Eh, ionic strength, temperature). An implicit corollary to this realization is that, for the factors controlling reactions of metal ions in aqueous systems with biofilms to be understood, their chemical reactivity and affinity for other solution components must be known.

The identity of individual chemical entities in solution, i.e., speciation, is extremely important for understanding the potential of metal ions to sorb to biofilm surfaces. The sorption of cations or anions may be quite different when they occur as complexed, rather than uncomplexed, ions. For example, cationic U(VI) hydroxo complexes interact strongly with surfaces, whereas anionic U(VI) cabonato complexes interact only weakly with the same solids.[8]

Metal ions tend to form coordination complexes as a result of Lewis acid–base reactions in which the metal ion behaves as an acid and the conjugate base is a complexing agent, i.e., ligand. The formation and dissociation of the complex compounds often involve successive equilibria that depend on the number of metal ions (Me^{n+}) and/or ligands (L^{m-}) participating in the complexation reaction:

$$a Me^{n+} + b L^{m-} \rightleftarrows Me_a L_b^{an-bm} \tag{1}$$

$$\beta_{ba} = \frac{\{Me_a L_b^{an-bm}\}}{\{Me^{n+}\}^a \{L^{m+}\}^b} \tag{2}$$

where the brackets denote the activity of the various species and β_{ba} is the formation constant for complexes comprised of a metal ions of charge n^+ and b ligands of charge m^-. Activities of aqueous species are related to their dissolved concentrations by an activity coefficient (γ); e.g., $\{Me^{n+}\} = \gamma_{Me}[Me^{n+}]$.

Because metals ions are always hydrated or coordinated by water molecules in aqueous solutions, the complex formation constant provides a measure of how effectively a ligand can compete with water in the metal ion coordination sphere. Moreover, equilibria involving ionic species in solution are affected to varying degrees by the presence of all ions in solution, which together make up the ionic

[8] R. T. Padalan, D. R. Turner, F. P. Bertetti, and J. D. Prikryl, *in* "Adsorption of Metals by Geomedia: Variables, Mechanisms and Model Applications." (E. Jenne, ed.), p. 100. Academic Press, San Diego, 1998.

strength (I). Ionic strength is defined as half the sum of the concentration (c_i) of all species in solution multiplied by their charge (Z_i) squared. The expression for ionic strength can be written $I = 1/2 \sum(c_i Z_i^2)$. Specifically, ion activities in concentrated solutions ($I > 0.1\ M$) can be markedly different from corresponding dissolved concentrations. In such situations, activity coefficients can be calculated using various approximations (e.g., the Debye–Hückel equation; see Stumm and Morgan[9]) to estimate activities from known concentration values. For dilute freshwater systems, however, γ is often taken to be unity and activities are assumed to be equal to dissolved concentrations.

The acidity of water molecules coordinated to metal ions is generally much larger than that of uncomplexed water and tends to increase with decreasing radius and increasing charge of the metal ion. In aqueous solution, the loss of a proton from a hydrated metal ion to form hydroxo complexes is particularly important. For simple metal ions, the determination of hydrolysis constants is of fundamental importance because the behavior of metal ions in various chemical reactions is a function of pH and is governed by the value of the hydrolysis constant. As the pH of a solution is changed, hydrolysis can proceed systematically in a stepwise manner to form neutral or even anionic species. For example, the hydrolysis of Fe^{3+} evolves from $Fe(OH)^{2+}$ through an entire series of hydroxo complexes: $Fe(OH)_2^+$, $Fe(OH)_3^0$, and even $Fe(OH)_4^-$. An additional complexity when hydrolyzed ions are present in high concentrations is their tendency to form polymeric ions, such as $Fe_2(OH)_2^{4+}$, which can be considered a dimer of $Fe(OH)^{2+}$. A quantitative discussion of iron hydrolysis is presented in Stumm and Morgan.[9]

During successive hydrolysis steps, the overall charge of the metal complex is reduced concomitantly, and the resulting hydroxo complex ions are very different in terms of chemical properties such as solubility and sorption. The distribution of metal ion species under a variety of conditions can, however, be predicted if the hydrolysis constants and solubility product of the insoluble solid phase are known. Various hydrolysis constants compiled from Stumm and Morgan[9] are given in Table I. Hydrolysis constants are useful because they indicate at what pH it is necessary to start considering hydrolysis species in overall speciation calculations.[10] In the absence of other complexing agents the first hydrolysis product will be equal to the free metal in solution when the pH is equal to $-\log \beta^*$ as given in Table I. Thus, it is necessary to consider Fe^{3+} hydrolysis even in acidic solution, pH ~ 2, but Cd^{2+} and Mg^{2+} will not hydrolyze significantly until the pH is >9.

Species predominance diagrams can be calculated for specific solution conditions using a number of available computer programs, such as EQ3/6[11] or

[9] W. Stumm and J. J. Morgan, "Aquatic Chemistry." John Wiley, New York, 1996.

[10] D. S. Smith, N. W. H. Adams, and J. R. Kramer, *Geochim. Cosmochim. Acta* **63**, 3337 (1999).

[11] T. J. Wolery, "EQ3NR, A Computer Program for Geochemical Aqueous Speciation–Solubility Calculations: Theoretical Manual, User's Guide, and Related Documentation. (Version 7.0)." UCRL-MA-110662-PT-111. Lawerence Livermore Natl. Lab., Livermore, CA, 1992.

TABLE I
METAL ION HYDROLYSIS REACTIONS AND CONSTANTS

Reaction	Symbol	Log β^a
$Fe^{3+} + H_2O = FeOH^{2+} + H^+$	β_1	-2.19
$Fe^{3+} + 2H_2O = Fe(OH)_2^+ + 2H^+$	β_2	-5.67
$Fe^{3+} + 3H_2O = Fe(OH)_3^0(aq) + 3H^+$	β_3	<-12
$Fe^{3+} + 4H_2O = Fe(OH)_4^- + 4H^+$	β_4	-21.6
$2Fe^{3+} + 2H_2O = Fe_2(OH)_2^{4+} + 2H^+$	β_{22}	-2.95
$Al^{3+} + H_2O = AlOH^{2+} + H^+$	β_1	-4.97
$Al^{3+} + 2H_2O = Al(OH)_2^+ + 2H^+$	β_2	-9.3
$Al^{3+} + 3H_2O = Al(OH)_3^0(aq) + 3H^+$	β_3	-15.0
$Al^{3+} + 4H_2O = Al(OH)_4^- + 4H^+$	β_4	-23.0
$2Al^{3+} + 2H_2O = Al_2(OH)_2^{4+} + 2H^+$	β_{22}	-7.7
$3Al^{3+} + 4H_2O = Al_3(OH)_4^{5+} + 4H^+$	β_{43}	-13.9
$Zn^{2+} + H_2O = ZnOH^+ + H^+$	β_1	-8.96
$Cd^{2+} + H_2O = CdOH^+ + H^+$	β_1	-10.1
$Hg^{2+} + H_2O = HgOH^+ + H^+$	β_1	-3.4
$Mg^{2+} + H_2O = MgOH^+ + H^+$	β_1	-11.44
$Ca^{2+} + H_2O = CaOH^+ + H^+$	β_1	-12.85

^{a} At 25° and $I = 0$.

MINEQL.[12] Note, however, that care should be taken to ensure that the results of such calculations are consistent with the principles of aqueous chemistry, and that the thermodynamic database used in the computational exercise is self-consistent, up-to-date, and appropriate for the solution conditions (i.e., T and ionic strength). The methods whereby MINEQL calculates chemical equilibrium problems are outlined below and could be reproduced with any sufficiently flexible optimization software, such as MATLAB (The MathWorks, MA) or Scilab (INRIA, France).

The influence of chemical equilibrium on the progress of chemical reactions determines the abundance and distribution of chemical species in solution. On the other hand, an appreciation of reaction kinetics is important for determining whether a reaction is sufficiently fast and reversible that it can be considered as chemical equilibrium controlled. Once the equilibrium condition is established, measurement of corresponding equilibrium constants unlocks a door to a more complete understanding of the behavior of metal ions in aqueous solution.

[12] J. C. Westall, J. L. Zachary, and F. M. M. Morel, "MINEQL: A Computer Program for the Calculation of Chemical Equilibrium Composition of Aqueous Systems." Ralph M. Parsons Laboratory, Department of Civil Engineering, Massachusetts Institute of Technology, Cambridge, MA, 1976.

Surface Complexation Theory

When a metal ion species partitions from aqueous solution on to a reactive solid surface, this process is termed sorption, and the surface-associated ion is known as a sorbate. The solid phase involved in sorption reactions is referred to as the sorbent. A generalized reaction for the sorption of a metal ion to a reactive site ($\equiv SOH^0$) on a sorbent can be written in an analogous manner to a complexation reaction involving a protonated ligand in solution:

$$Me^{n+} + \equiv SOH^0 \rightleftarrows \equiv SOMe^{n-1} + H^+ \tag{3}$$

The corresponding mass action expression is

$$K_s = \frac{\{\equiv SOMe^{n-1}\}\{H^+\}}{\{\equiv SOH^0\}\{Me^{n+}\}} \tag{4}$$

These relationships emphasize that metal ion sorption reactions are intrinsically sensitive to pH and are enhanced by a low equilibrium proton condition (i.e., elevated pH). At low ionic strength, proton and metal ion activities can be taken to be equal to dissolved concentrations; otherwise they may be calculated by determining appropriate activity coefficients. The activities of surface species (i.e., $\{\equiv SOH^0\}$ and $\{\equiv SOMe^{n-1}\}$) are, by convention, assumed to be proportional to the electrostatic potential that arises in response to charge development caused by the sorbed metal ion on the solid surface. Thus, the mass action relationship can be rewritten in terms of an apparent concentration sorption constant K_s^* that is a function of the intrinsic sorption constant K_s, the surface potential (Ψ) of the sorbent, and the change in surface charge (Δz), as well as the gas constant (R) and the absolute temperature (T):

$$K_s^* = \left(\frac{\gamma_{Me}}{\gamma_H}\right) K_s \exp\left(\frac{-\Delta z F \Psi}{RT}\right) = \frac{[\equiv SOMe^{n-1}][H^+]}{[\equiv SOH^0][Me^{n+}]} \tag{5}$$

where the term $\exp(-\Delta z F \Psi / RT)$ is the Boltzmann correction factor resulting from the electrostatic contribution to the total energy of reaction. The intrinsic energy component corresponds to the energy released on binding the metal to the surface and the electrostatic component is the energy necessary to bring the metal ion from an infinite distance to the surface. The total energy is the sum of these two contributions. The assumptions involved in formulating the Boltzmann correction factor are detailed elsewhere.[7,9]

The surface potential, Ψ, is not a measurable quantity, but must be approximated in some consistent way for the system. The most commonly used models are the constant capacitance,[13] the diffuse double layer model of Dzombak and Morel,[7] and the triple layer models.[14] The diffuse double layer model has the advantage of

[13] P. W. Schindler and H. Gamsjager, *Kolloid Z. Z. Pollymere* **250,** 759 (1972).
[14] J. A. Davis, R. O. James, and J. O. Leckie, *J. Colloid Interface Sci.* **63,** 480 (1978).

not requiring any additional parameters other than those specified in Eq. (5) and is probably the most widely used model. Exactly how the diffuse double layer model is incorporated into speciation determinations is discussed further below.

In addition, metal binding to solid surfaces at high surface coverage can lead to heterogenous mineral precipitation. This will not be discussed further here, but is explained for hydrous ferric oxide by Dzombak and Morel[7] and bacterial surfaces by Warren and Ferris.[15]

Sorption Sites in Biofilms

Biofilms have strong reactive chemical properties, even in the absence of any metabolic activity, that sustain the uptake and immobilization of cations, anions, and organic molecules, depending on the nature of the biofilm produced (e.g., the different species in the adherent microbial consortium, extracellular polymers, mineral precipitates, etc.). Although biofilms themselves are not sorbents *per se,* they are composite amalgams of living and nonliving sorbent materials. The ability to quantify and predict the degree to which biofilms sorb metals depends on the extent to which the fundamental aspects of sorption are understood, and on the accuracy with which the phenomenon can be characterized and modeled in aqueous environments.

The extracellular polymeric substances (EPS) produced by bacteria in biofilms comprise the most immediate interfacial boundary between the bulk aqueous phase and the bacterial cells. As a largely undifferentiated structural matrix, EPS consist generally of a wide variety of macromolecular compounds including acidic polysaccharides and proteins, as well as lipids. Because these various polymers often contain reactive amphoteric groups such as carboxylate, phosphoryl, and amino groups, EPS behave as potent solid phase sorbents of metal ions.[16]

Bacterial cell wall polymers also contain a variety of amphoteric functional groups that interact with and sorb metal ions (Table II). For example, in gram-negative cells, lipopolysaccharides in the outer membrane may contain carboxylate and phosphoryl as well as amino groups. In gram-positive cells, teichoic acids contain phosphates in both terminal phosphoryl groups and phosphodiester bonds, whereas carboxylate groups are found in teichuronic acids. The peptidoglycan component of gram-positive, and gram-negative, bacteria also contains carboxylate and amino groups.

Mineral grains that accumulate in biofilms owing to *in situ* precipitation or physical entrapment comprise yet another class of reactive solid phase sorbents. The metal retention properties of clays, for example, are well known; however, biofilms in aqueous environments often contain hydrous Fe(III) and Mn(III/VI)

[15] L. A. Warren and F. G. Ferris, *Env. Sci. Technol.* **32,** 2331 (1998).
[16] H. C. Flemming, *Wat. Sci. Technol.* **32,** 27 (1995).

TABLE II
TYPES OF FUNCTIONAL GROUPS AVAILABLE IN BIOFILMS

Site	Portion of the bacteria cell walls
Carboxylate	Peptidoglycan (peptide and muramic acid residues), teichuronic acids
Phosphodiesters	Teichoic and linkage of teichuronic acid to peptidoglycan
Phosphoric	Teichoic acid
Amines	Peptidoglycan (peptide part)
Hydroxyl	Peptidoglycan (muramic acid residue and possibly on peptide part) as well as teichoic and teichuronic acids

Site	Mineral components
Amphoteric	Mostly surface sites on hydrous Mn and Fe oxides and clay minerals
Hydroxyl	

Site	Humic and fulvic acid components
Carboxylic	Salicylic acid–like binding sites
Phenolic	Hydroxyl groups attached to an aromatic ring

oxides that are hailed as dominant inorganic sorbents of dissolved metals in both pristine and contaminated systems. The reactivity of these oxides also stems directly from the presence of amphoteric hydroxyl groups at the mineral surface.

Most studies on the immobilization of metals by biofilms do not distinguish between the myriad different sorption sites that are likely to be available. Usually sorption is assumed to occur predominantly in association with EPS and the cell walls of bacteria; however, a number of recent studies have established that inorganic mineral sorbents accumulating in biofilms have a profound influence on the solid phase partitioning of metal ions.[17,18] The situation is complicated further by the existence of multiple sorption sites on each of the individual sorbent solids that may occur with a biofilm.

Chemical Equilibrium Modeling of Metal–Biofilm Interactions

In order to determine metal speciation in fate/transport modeling it is necessary to experimentally determine predictive parameters. In this discussion we will assume equilibrium although in a more complete description kinetic considerations must be included. Possible equilibrium parameters include the well known distribution coefficient, K_D, as the simplest description of metal–biofilm interactions,

[17] T. D. Small, L. A. Warren, E. E. Roden, and F. G. Ferris, *Env. Sci. Technol.* **33**, 4465 (1999).
[18] Y. M. Nelson, L. W. Lion, M. L. Shuler, and W. C. Ghiorse, *Env. Sci. Technol.* **30**, 2027 (1996).

where K_D is the ratio of the concentrations of bound and free metal. The free metal is defined as all the metal that is not bound to the solid. The magnitude of the distribution coefficient is dependent on the conditions, especially pH and ionic strength, at which it was measured. Individual site interactions, however, may include any proposed reaction stoichiometries and mixed-metal–mixed-ligand complexes. The complicated nature of the potential metal–biofilm intereactions highlights the importance of system definition in studying natural surfaces, which is discussed by Smith et al.[10] Additional modeling approaches that consider the sorbent as a heterogenous gel phase using Donnan potentials[19] or as a continuous distribution function[20] have been advanced, but are not discussed here. Instead, we focus on modeling the biofilm as a mixture of simultaneous equilibrium reactions. The advantage of modeling biofilm surfaces as a set of simultaneous equilibria is that available geochemical code, such as MINEQL, can be used to apply speciation parameters in modeling natural systems.

In terms of relating biofilm structure and composition to surface complexation theory, it is notable that the diffuse double layer of interest in modeling solid water interfaces is estimated to be of the order of 0.4 nm, whereas the solvent layer of interest may extend several nanometers from the surface. In contrast, an average microbial cell is approximately 500 nm in diameter. This, coupled with the fact that individual cells in biofilms are often embedded in a dense matrix of EPS and mineral precipitates, creates a particularly demanding theoretical and experimental problem.

Experimental Approach

Speciation determinations have been dominated by the importance of determination of the concentration of the free aquo ions of the metal of interest because of the importance of the free ion model in fate transport and toxicity models.[21] Most experiments measure the partitioning of the metal between the free and bound forms, using standard analytical techniques, such as flame atomic adsorption or inductively coupled plasma atomic emission spectroscopy. The simplest experiments involve measuring the free metal by using filtration to remove all forms of bound metal. Alternatively, the metal can be measured in the bound phase after filtration. Also, ion selective electrodes (ISE) can be used to determine the concentration of the free metal, as Cabaniss and Shuman[22] have done for Cu. It is also possible to use spectroscopic methods to look at the metals, as for U[23] or to

[19] M. F. Benedetti, W. H. van Riemsdijk, and L. K. Koopal, Env. Sci. Technol. 30, 1805 (1996).
[20] J. Buffle, R. S. Altmann, M. Filella, and A. Tessier, Geochim. Cosmochim. Acta 54, 1535 (1990).
[21] L. Parent, M. R. Twiss, and P. G. Campbell, Env. Sci. Technol. 30, 1713 (1996).
[22] S. E. Cabaniss and M. S. Shuman, Geochim. Cosmochim. Acta 52, 185 (1988).
[23] J. R. Bargar, R. Reitmeyer, and J. A. Davis, Env. Sci. Technol. 33, 2481 (1999).

measure ligands, as Smith and Kramer [24] have done using fluorescence to measure specific sites in natural organic matter.

In order to quantitatively understand the binding of metals to biofilms it is necessary to first characterize the bulk parameters of the biofilm system. This is in fact a very difficult problem and becomes an operationally defined step of the procedure. The types of information required would be the proportion of the system that is organic versus inorganic, which mineral phases are present, moisture content, etc. Another important parameter that is of crucial importance is dry mass concentration of biofilms in contact with solution. Using the dry mass concentration, all the measured sorption isotherms can be normalized per mass of solid. This is vital in order to be able to compare results with the literature.

To characterize the binding of metals in aqueous biofilm systems it is necessary to start from some known reference state. The most obvious and easiest starting point is a total concentration of zero for the metal of interest. Then the system can be perturbed with additional metal and measurements made on the reequilibrated system. In practice, this is a metal titration. For each addition of total metal it is possible to measure the pH and free metal (by filtration) and check the total metal (e.g., by strong acid digestion). Generally the pH is held constant in a metal titration by small additions of strong acid or base. Titrations can be performed in batch mode where each titration point corresponds to an individual reaction vessel, or continuous mode where subsamples are taken from a single reaction vessel after each addition of titrant.

To investigate the effect of pH on metal binding it is possible to hold the total metal concentration constant and to vary the pH. It is important in this type of study to keep the total metal constant in excess of the total site density, or estimates of binding capacity will be in error. For metal titrations at fixed pH, the titration should proceed until there is no further sorption. Also, in natural systems, there are "analytical window" considerations as well. The concept of an analytical window is discussed best by Buffle et al. [20] and emphasizes the fact that if the concentrations of sorption sites are far below the metal concentration range used in the study, then they will not be quantifiable. The same holds true for sites above the titration range.

Fitting Approaches

After obtaining experimental data it is necessary to fit it to some mathematical model of the system. This fitting can be motivated by a need for predictive parameters, such as for use in fate and transport model code calculations, [25] or a need for fundamental understanding of the nature of metal binding sites in biofilms

[24] D. S. Smith and J. R. Kramer, *Anal. Chim. Acta,* in press (2000).
[25] C. M. Bethke, "Geochemical Reaction Modeling Concepts and Applications." Oxford University Press, New York and Oxford, 1996.

(e.g., sorption stoichiometry). Structural information about biofilm binding sites is indirectly contained in the determined stability constants, as it is possible to constrain the possible functional identity of sorption sites by comparison to known molecular analogs. A note of caution: while there is a commonly held view that a good fit to the data implies that a model is true, it should be emphasized this is not always the case. A good fit does not confirm that the model is true, but only that it is possible.

There are many modeling approaches used in the literature. If the system is assumed to be represented sufficiently well by a one-site model, then isotherm linearization approaches, such as that of Langmuir and Freundlich, could be used.[26] In general, a single site is not sufficiently complex to represent natural systems over a range of conditions, and so we will focus on two widely used multisite approaches here: the surface complexation modeling approach typified by the program FITEQL[27] and the pK spectrum approach.[28] In addition, there are other modeling approaches such as the site occupation distribution function,[20] and NICA–Donnan[19] that will not be discussed further here. An approach that has received significant interest is the CD-MUSIC model developed for multisite sorption on mineral surfaces.[29] This approach is of great theoretical interest, but is not presently useful for biofilms because the modeling method requires information about the surface structure (e.g., types and nature of surface functional groups) that is not precisely known for natural biofilms.

Surface Complexation Modeling

Surface complexation modeling (SCM) is often performed using FITEQL as developed by Westall,[27] but the theory could be applied and the data fit using any available, sufficiently flexible optimization software, such as MATLAB, SIMU-SOLV (Dow Chemical Company, Midland, MI), or Scilab. FITEQL is a general purpose fitting program that can be used to determine stability constants and site densities from experimental observations. Typical experimental observations would be free metal versus pH for a fixed total metal concentration, or free metal versus total metal for fixed pH. The site densities and stability constants are determined by minimizing the error between the predicted and measured free metal. What is necessary is first to assume the relevant reactions and then to minimize a weighted sum of squares function to determine the speciation parameters. In

[26] I. Ružić, *Marine Chem.* **53,** 1 (1996).
[27] J. C. Westall, "FITEQL, a Computer Program for Determination of Chemical Equilibrium Constants from Experimental Data. Version 2.0." Report 82-02, Dept. Chem., Oregon St. Univ., Corvallis, OR, 1982.
[28] P. Brassard, J. R. Kramer, and P. V. Collins, *Env. Sci. Technol.* **24,** 195 (1990).
[29] P. Venema, T. Hiemstra, P. G. Weidler, and W. H. van Riemsdijk, *J. Colloid Interface Sci.* **198,** 282 (1999).

practice, the number of sites is increased until a sufficiently good description of the data is obtained.

Models are developed conveniently using the tableau notation to compile chemical equilibrium problems.[12,30] This method is a quick and efficient method of writing chemical equilibrium problems and for solving them based on mass conservation and mass action expressions. Implicit in the approach is charge and proton balance conditions. Computational details for calculating speciation given a fixed tableau are given below. In fitting data to a model (tableau) for the system typically some or all of the stability constants and/or the total site densities (S_T) are taken as unknowns for a prescribed total number of sites. The values of these parameters are varied in computational runs to minimize a weighted sum of squares function. Usually the sum of squares is determined from the difference between calculated and observed pH, or calculated and observed free metal, or both.

Using surface complexation modeling many different tableaus (i.e., reaction stoichiometeries) can be considered and many may yield an equally good description of the data. For practical purposes the simplest model that adequately describes the data should be taken as the working model. For better model constraints, it is necessary to have independent information on the structure of the surface complex as can be obtained by EXAFS[31] or other spectroscopic techniques.[24]

Overall the procedure for fitting titration data to speciation parameters is as follows:

1. Define a tableau, as described further below, and decide which entries to make unknowns. Usually one or more stability constants and/or site densites are unknowns. These unknowns are the parameters that must be fit to the data. In certain cases it is possible to fit the acidity constants and site densities if independent acid–base titrations are performed to fit these parameters.

2. Select an electrostatic model to use, e.g., constant capacitance or diffuse double layer. A possible choice is no electrostatic model, but then the resultant parameters must be termed conditional, because they are dependent on the ionic strength of the titration.

3. Make initial guesses for the parameters; this step is critical. For example, in FITEQL the final answer can be very dependent on the initial guess.[32,33] This dependence on the initial guess is a minimization problem. The error surface has numerous local minima, but only the global minimum corresponds to the "best" answer.

4. With the initial guess calculate the speciation using the tableau solution method, as outlined below.

[30] F. Morel and J. Morgan, *Env. Sci. Technol.* **6**, 58 (1972).
[31] G. Sarret, A. Manceau, L. Spadini, J. C. Roux, J. L. Hazemann, Y. Soldo, L. Eybert-Bérard, and J. J. Menthonnex, *Env. Sci. Technol.* **32**, 1648 (1998).
[32] J. Lützenkirchen, *J. Colloid Interface Sci.* **217**, 8 (1999).
[33] J. Lützenkirchen, *J. Colloid Interface Sci.* **210**, 384 (1999).

5. Now that the speciation dependence on the current set of parameters has been determined, it is necessary to compare these results to the experimental measurements. At this stage some definition of fitting error is needed. Generally, this corresponds to a weighted sum of squares function. For an equal weighting of data, the error $= \sum$ (calculated $-$ observed)^2.

6. Based on the results from step 5, the initial guess is changed in a direction closer to the minimum on the error surface. There are many methods for revising the initial guess. There are gradient methods that depend on the derivative of the response surface, such as Newton's method,[25] or nongradient methods such as the simplex method.[34]

7. The revised guess is applied for steps 4 to 6 until the error and/or the solution vectors converge on a constant value within some tolerance. When the answer converges the "best-fit" speciation parameters for the tableau defined in step 1 have been determined.

8. Once the "best-fit" solution has been determined for a fixed tableau, the fit of the data must be evaluated to decide if the model fits the data sufficiently well. There are statistical tests,[35] but the best practical indicator is a plot of the residuals (calculated values $-$ observed values). If the residuals are randomly distributed about zero, then the model makes a good fit to the data. If the model does not accurately describe the data, then the tableau in step 1 must be revised and the entire process repeated.

Other than the decision for the form of the tableau, this entire process can be programmed using computer software. Optimization routines to revise the parameter guesses and determine when convergence has been obtained are readily available. In particular, for an interpretive language such as MATLAB, programming this fitting method is relatively easy. Of course, the software FITEQL is also already available to fit experimental data.

Tableau Notation

If there is one divalent metal (Me^{2+}) sorbing to one reactive surface site in a biofilm ($\equiv SOH^0$), where the site can be associated with organic matter, or on a bacterial or mineral surface, then the sorption reaction can be written:

$$Me^{2+} + \equiv SOH^0 \rightleftarrows \equiv SOMe^+ + H^+ \tag{6}$$

with the formation constant defined as

$$K_s = \frac{[\equiv SOMe^+][H^+]}{[\equiv SOH^0][Me^{2+}]} \exp\left(\frac{F\Psi}{RT}\right) \tag{7}$$

[34] J. A. Nedler and R. Mead, *Comp. J.* **7**, 308 (1965).

[35] D. M. Bates and D. G. Watts, "Nonlinear Regression Analysis and Its Applications." John Wiley & Sons, New York, 1988.

The surface site and the metal are not independent of pH; for the sake of simplicity let us assume that the site is diprotic and amphoteric in the pH range studied, and that the metal has a single hydrolysis constant. The three additional simultaneous equilibria are

$$Me^{2+} + H_2O \rightleftarrows \equiv MeOH^+ + H^+ \tag{8}$$

$$\equiv SOH_2^+ \rightleftarrows \equiv SOH^0 + H^+ \tag{9}$$

$$\equiv SOH^0 \rightleftarrows \equiv SO^- + H^+ \tag{10}$$

with associated mass action expressions of

$$K_H = \frac{[MeOH^+][H^+]}{[Me^{2+}]} \tag{11}$$

$$K_{a1} = \frac{[\equiv SOH^0][H^+]}{[\equiv SOH_2^+]} \exp\left(\frac{-F\Psi}{RT}\right) \tag{12}$$

$$K_{a2} = \frac{[\equiv SO^-][H^+]}{[\equiv SOH^0]} \exp\left(\frac{-F\Psi}{RT}\right) \tag{13}$$

In general, the surface sites are considered to be amphoteric, such as on metal oxides, where the reactive group can acquire a positive or negative charge. In addition to the mass action expressions from the simultaneous equilibria given above, there are also mass balance constraints:

$$Me_T = [Me^{2+}] + [MeOH^+] + [\equiv SOMe^+] \tag{14}$$

$$S_T = [\equiv SO^-] + [\equiv SOMe^+] + [\equiv SOH_2^+] + [\equiv SOH^0] \tag{15}$$

$$TOTH = [H^+] + 2[\equiv SOH_2^+] + [\equiv SOH^+] - [OH^-] - [MeOH^+] \tag{16}$$

where TOTH is the mole balance condition for protons, and it corresponds to species containing H minus those containing OH.[30] The proton condition is analogous to the electroneutrality condition, and it is only necessary to have one of them to completely describe the system. In an acid–base titration TOTH is equivalent to the added concentration of acid minus the added concentration of base.

The information contained in the simultaneous set of Eqs. (7) and (11)–(16) is sufficient to define the speciation for the system given fixed values for the formation constants, site density (S_T), total metal concentration (Me_T), and TOTH. Determining the solution to this problem is simplified if a set of components is defined such that each species can be represented by a combination of these components. Any set of components can be selected as long as they are independent of each other. In this example, the components are defined to be $[H^+]$, $[Me^{2+}]$, and $[\equiv SOH^0]$. The problem can now be set up in tableau notation where the components are given across the top, the species are listed along the rows, and each entry corresponds to the stoichiometric coefficient for the formation of the species from the

TABLE III
EXAMPLE TABLEAU FOR SURFACE COMPLEXATION OF A METAL AT AMPHOTERIC SITES

	H^+	$\equiv SOH^0$	Me^{2+}	$P = \exp(-F\Psi/RT)$	K
H^+	1				
$\equiv SOH^0$		1			
Me^{2+}			1		
OH^-	−1				K_w
$\equiv SO^-$	−1	1		−1	K_{a2}
$\equiv SOH_2^+$	1	1		1	$1/(K_{a1})$
$\equiv SOMe^+$	−1	1	1	1	K_S
$MeOH^+$	−1		1		K_H
TOTH	S_T	Me_T	σ/F		

components. The tableau corresponding to this example problem is given in Table III. This is the way problems are input into programs such as FITEQL or MINEQL.

So far a discussion of electrostatic modeling has been omitted. Electrostatic corrections are accomplished using the so-called Boltzmann correction factor. As can be seen in Table III, the correction factor ($P = \exp(-F\Psi/RT)$) has been entered as a component of the problem. It is treated just like any of the other components until a solution to the problem is found (the solution method is detailed below). In calculating the value for this component some electrostatic model must be defined for the surface potential Ψ. The total in this column is the total charge of the surface sites in moles per liter. This is represented as σ/F where σ corresponds to the theoretical surface charge (in coulombs) dependent on the surface potential term and F is Faraday's constant (in coulombs/mol). Often, as in Dzombak and Morel,[7] the charge is normalized by area to express charge density (C/m^2). This is a useful normalization to allow comparison to previous work, but requires a measurement of surface area, which is difficult experimentally as discussed by Dzombak and Morel.[7] Alternatively, it is possible to normalize using the dry mass and thus express the charge in coulombs (or μmol) per mg of solid.[4]

Solving for Speciation Given a Fixed Tableau

This solution method is given in the notation presented by Westall[36] and is a very simple way of expressing the problem. In solving an equilibrium problem, the values of the components are varied until the calculated totals agree within a

---

[36] J. C. Westall, *in* "Particulates in Water" (M. C. Kavanaugh and J. O. Leckie, eds.), p. 33. American Chemical Society, Washington, D.C., 1980.

prescribed tolerance with the actual totals. This problem can be set up in matrix notation for n species and m components. We define an $m \times 1$ vector of component concentrations, $\mathbf{X}$; this is what we want to solve for. Once the concentrations of the components have been determined, the rest of the species are easy to calculate. Also, there is an $n \times 1$ vector of species concentrations ($\mathbf{C}$), a $1 \times m$ vector of total concentrations ($\mathbf{T}$), and an ($n \times 1$) vector of formation constants ($\mathbf{K}$). Finally, we define a matrix $\mathbf{A}$, which is $n \times m$ and contains the stoichiometric coefficients for the species. $\mathbf{A}$ is the center part of the tableau given in Table III. Now we can calculate for a given $\mathbf{A}$, $\mathbf{X}$, and $\mathbf{K}$ the concentrations of the species:

$$\log \mathbf{C} = \log \mathbf{K} + \mathbf{AX} \tag{17}$$

With these concentrations we can calculate a total concentration, $\mathbf{T}_{calc} = \mathbf{A}'\mathbf{C}$, where $\mathbf{A}'$ corresponds to the transpose of $\mathbf{A}$. Now the error is defined as error $= \sum(\mathbf{T}_{calc} - \mathbf{T})^2$. The error is minimized until $\mathbf{X}$ and/or the error converges on a constant value within a prescribed tolerance. The minimization can be performed using the Newton–Raphson method,[36] or using commercial optimization software such as MATLAB.

The only modification to this method necessary to include the Boltzmann correction factor is a calculation of the total charge σ/F entered as the last entry to the $\mathbf{T}$ vector. The calculation continues until the total charge determined from summing the concentrations of the charged sites and the theoretical charge as a function of the surface potential are the same within some tolerance. In order to do this a relation between the surface charge and surface potential must be defined. For the diffuse double layer model[7] this relation is:

$$\sigma = 0.1174 I^{0.5} \sinh(Z\Psi x 19.46) \tag{18}$$

where Z is the charge of the symmetric supporting electrolyte. Equation (18) is based on Gouy–Chapman theory and the derivation is shown in Dzombak and Morel.[7] It is possible to fit metal sorption data without invoking any electrostatic modeling if the experiments were performed at fixed ionic strength. The constants so obtained are apparent stability constants and are dependent on ionic strength.

pK Spectrum Approach

In the pK spectrum approach it is not necessary to assume a number of sites as in the tableau method (e.g., FITEQL), but it is necessary to assume that the sites can be represented as a sum of monoprotic and/or monometallic sites. Thus, diprotic sites would be represented as two monoprotic sites. In addition, it is only possible to include proton competition indirectly, but if both acid–base and metal titrations are performed, then complete information about the system has been obtained as long as the equilibrium assumption holds true.

The method applied to acid–base titrations has been described in detail else-where.[4,28,37] Here we discuss metal binding. Charges are omitted from the notation for simplicity. The first step is to assume the form of the reactions that will describe the binding isotherm, where the isotherms are determined as free metal vs bound metal at fixed pH. For each of $i = 1 \ldots n$ sites:

$$Me + S_i = MeS_i \quad \text{with corresponding } K_i = \frac{[MeS_i]}{[Me][S_i]} \tag{19}$$

From conservation of mass considerations, the amount of bound metal at the ith site can be expressed as

$$[MeS_i] = \frac{[MeS_i]S_{iT}}{[MeS_i] + [S_i]} \tag{20}$$

and by isolating $[S_i]$ from the stability constant expression [Eq. (19)] and substi-tuting back into Eq. (20) it is possible to write

$$[MeS_i] = \alpha_{ij}S_{iT} \quad \text{where } \alpha_{ij} = \frac{K_i[Me_j]}{K_i[Me_j] + 1} \tag{21}$$

for j additions of metal at fixed pH. For fixed metal, variable pH titrations the corresponding reaction of interest is

$$Me + HS_i = MeS_i + H \quad \text{with corresponding } K_i = \frac{[MeS_i][H]}{[Me][S_i]} \tag{22}$$

and the resultant alpha terms are

$$[MeS_i] = \alpha_{ij}S_{iT} \quad \text{where } \alpha_{ij} = \frac{K_i[Me_j]}{K_i[Me_j] + [H]} \tag{23}$$

What is desired now is to solve for all of the S_T, K pairs, but the number of pairs (n) must be determined. This can be done, as in FITEQL, by increasing n until a sufficiently good fit to the data is obtained or by using the pK spectrum approach.

In the pK spectrum approach, a grid of pK values is defined, where p$K_i = -\log K_i$. The pK values generally start at the minimum experimental pMe through to the maximum pMe at fixed pK intervals. Now, what remains is to solve for the site densities (S_{iT} values). For n sites (pK values) and m experimental measurements the site densities can be determined using regularized least squares[37] or linear programming[28] to solve the matrix equation $Ax = b$ for x, where x is an $n \times 1$ vector of site densities (S_{iT}), and where A is an $m \times n$ matrix of alpha terms and b is a $m \times 1$ vector containing the measured bound metal concentrations. Note that A here has a different meaning than A defined above in the discussion of solving speciation from a fixed tableau. Regularized least-squares and linear programming

[37] M. Černík, M. Borkovec, and J. C. Westall, *Env. Sci. Technol.* **29**, 413 (1995).

are discussed and compared by Smith *et al.*[10] Linear programming is advantageous in that it automatically minimizes the number of sites and is a robust regression method because it utilizes the absolute of the error rather than the least-squares. For linear programming, the minimization problem is defined as

$$\text{Minimize} \left(\sum_{i=1}^{n} \left| \sum_{j=1}^{m} A_{ij} x_j - b_i \right| \right) \tag{24}$$

which can be written as a linear programming problem.[38] The exact solution method and the details of linear programming are beyond the scope of this paper, but for practical purposes it is sufficient to set up the matrices as defined in Brassard *et al.*,[28] and solve for the linear programming problem using available software with linear programming capabilities, such as MATLAB, Scilab, or many commercially available spreadsheet programs.

Numerically, pK spectrum approaches are better behaved then FITEQL because correlation problems between the site capacity and stability constant parameters in FITEQL are avoided by fixing the log K values and only solving for the site density values. Linear programming has the added advantage of only having one unique answer, whereas traditional surface complexation modeling results can be plagued by local minimum problems.[32,33] There is less flexibility in the model definition, though, and it is impossible to perform electrostatic corrections, at least in currently available pK spectrum approaches. Thus, pK spectrum parameters are conditional parameters and only apply for the ionic strength at which the experiment was performed.

Additivity

As a first approximation, the reactivity of individual solid phase sorbents comprising the biofilm can be considered separately. The aim of this reductionist approach is to be able to quantify and predict the sorptive behavior of each phase that may be present within a biofilm. For example, if speciation parameters are available for mineral phases and bacterial cell surface present in a biofilm, the total binding could be a sum of contributions from each phase weighted by the mass of that phase in the total biofilm. The implicit assumption (which may not necessarily be true) is that, once the amount of metal sorbed by each phase (*i*) is quantified, summation will yield the total metal sorption for all of the sorbents (*N*) within a biofilm:

$$\text{Me}_{T(\text{Biofilm})} = \sum_{i=1}^{N} (\text{Me}_{\text{sorbed}})_i \tag{25}$$

---

[38] H. M. Wagner, *Am. Statist. Assoc. J.* 206 (1959).

The basic premise on which the principle of additivity for the total amount of metal sorbed by different sorbent phases within biofilms is that the sorbents do not interact chemically. Of course, this is highly unlikely and represents a major challenge for future studies on biofilm–metal interactions.

In an application of additivity modeling, Nelson et al.[18] modeled Pb adsorption as a sum of adsorption onto two components. The components were polymer-bound Pb and cell-bound Pb. It was found that the total predicted binding using this two-component model overestimated bound Pb by as much as 30%. This overestimation was attributed to polymer–cell interactions reducing the concentration of available sorption sites. Small et al.[17] have noted similar nonadditive behavior with Sr^{2+} sorption in bacterial–iron oxide composites. In another study Nelson et al.[39] found much better predictions to the total bound lead in a system containing three components, the components being glass, iron deposits, and bacterial cells comprising a biofilm. In this study they found that the total bound Pb was equivalent to the predicted bound Pb from known amounts of cells and iron colloids. The issue of nonadditivity remains to be addressed in a more rigorous fashion.

Summary

The structural and compositional heterogeneity of biofilms poses unique problems in metal fate and transport. A starting point for quantitative understanding of biofilm–metal interactions is surface complexation theory, with roots in chemical equilibria and thermodynamics. This approach permits fitting of experimental data to a variety of mathematical models from which predictive parameters, such as K, may be extracted. Applications of more sophisticated fitting routines such as tableau (as in FITEQL) or spectra pK methods provide a better measure of the heterogeneity. There remain large theoretical and computational challenges, as there is ample evidence to suggest that the principle of additivity is problematic, owing to chemical interactions between individual sorbent phases within biofilms. And finally, the question of how bacterial metabolic activity is likely to influence metal uptake by biofilms adds yet another layer of complexity for future investigations.

Acknowledgment

This work was supported by a Natural Science and Engineering Research Council (NSERC) of Canada research grant to FGF.

[39] Y. M. Nelson, W. Lo, L. W. Lion, M. L. Shuler, and W. C. Ghiorse, *Wat. Sci. Technol.* **29**, 1934 (1995).

[16] Microscopy Methods to Investigate Structure of Potable Water Biofilms

By JAMES T. WALKER, JOANNE VERRAN, ROBERT D. BOYD, and STEVEN PERCIVAL

Introduction

The structure of microbial growth on surfaces is defined by the conditions in which the biofilm developed. In oligotrophic water systems where nutrients are scarce biofilms are likely to be thin and sporadically spread across the surfaces,[1] whereas, in reverse osmosis systems, biofilm can be compacted onto the filter surface creating layers on layers of cells.[2] In patients with permanent urinary catheters, biofilm can often be relatively thick, partly because of the excess nutrients that are available to it and also from the intermittent flow rates.[3] Biofilm structure has been seen from a different viewpoint by a number of researchers and has been thought of as being a confluent layer across a surface[4] and as a heterogenous mosaic[5] that mushrooms from the surface[6] providing streamers that float in the planktonic phase.[7]

Laboratory models to simulate biofilms provide an opportunity to produce biofilms in a reproducible and defined manner.[8] Flow cells present a situation in which the biofilm can be visualized *in situ* allowing the intact undisturbed mass of cells to be assessed for structure, viability, and the effect of flow rates and disinfectants.[9]

In situ visualization using flow cells cannot always be used, and so other laboratory models, perhaps where the biofilm has to be removed to be examined, have to be considered. In other cases one may want to visualize biofilm from its natural habitat. This paper discusses a number of methods used for the visualization of biofilm that has been developed primarily on plumbing tube materials.

[1] J. T. Walker, A. B. Dowsett, P. J. L. Dennis, and C. W. Keevil, *Int. Biodet.* **27**, 121 (1991).

[2] H.-C. Flemming and G. Schaule, *in* "Microbial Deterioration of Materials" (W. Sand, E. Heitz, and H.-C. Flemming, eds.), p. 121. Springer Verlag, Heidelberg, 1996.

[3] D. J. Stickler, J. B. King, C. Winters, and S. L. Morris, *J. Infect.* **27**, 133 (1993).

[4] W. A. Hamilton, *Annu. Rev. Microbiol.* **39**, 195 (1985).

[5] C. W. Keevil and J. T. Walker, *Binary* **4**, 92 (1992).

[6] J. W. Costerton, Z. Lewandowski, D. DeBeer, D. Caldwell, D. Korber, and G. James, *J. Bacteriol.* **176**, 2137 (1994).

[7] P. Stoodley, F. Jorgensen, P. Williams, and H. M. Lappin-Scott, *in* "Biofilms: the Good, the Bad and the Ugly" (J. Wimpenny, P. Gilbert, J. Walker, M. Brading, and R. Bayston, eds.), p. 223. Bioline, Cardiff, UK, 1999.

[8] J. W. Wimpenny, *in* "Dental Plaque Revisited: Oral Biofilms in Health and Disease" (H. Newman and M. Wilson, eds.), p. 89. Bioline, Cardiff, UK, 1999.

[9] R. J. Palmer, Jr., *Methods Enzymol.* **310**, 160 (1999).

Use of Light Microscopy to Study Biofilm

Differential interference contrast (DIC) and fluorescence microscopy examination was carried out using a Nikon Labophot-2 that combined both epifluorescence and DIC. A B-2A filter block was used for the epifluorescence. This has a 510 dichroic mirror, an excitation filter of 450–490, and a barrier filter of 520 nm. An IGS block was used for DIC. Objective lens were noncontact M Plan Apo lenses of 40×, 100×, and 150×. Light source was a 100 W halogen lamp. Neutral density filters were used on occasion to suppress high background fluorescence. A zoom adapter was fitted above the specimen to double the magnification already available with the microscope. Photography was carried out using a Nikon F-801 35 mm camera and a JVC TK-1085E color video camera head with built in compact power source fitted to a trinocular head. The JVC color video camera had approximately 400,000 effective pixels and was a CCD (charge coupled device) with solid-state pickup. High resolution and high sensitivity allowed for a horizontal resolution of 400 TV lines with a low light sensitivity of 7 lux. Using the video camera, visualization of the subject matter was relayed either to a video monitor for time lapse photography or to the multisync monitor (Taxan) (capable of automatically switching between noninterlaced and interlaced modes when the appropriate signal is recognized) for the image analysis program— both of which increased magnification to greater than 1500 times. The Digithurst Microeye TC image analysis transputer card and software were operated on a PC compatible computer. Specimens were stained with the fluorescent dye acridine orange (0.02% w/v).

Biofilm Viewed using Scanning Confocal Laser Microscopy

A Medical Research Council (MRC)-600 (Bio-pad) fitted with an argon laser with maximum emission at 488 nm and excitation at 514 was mounted in the upright position above an optical light microscope (Microphot SA, Nikon Ltd). The images from two different fluorescent markers can be imaged simultaneously using an excitation filter and dichroic (DC) mirror to direct the image to either photomultiplier (PMT) 1 or 2. These filters could be changed without disrupting the optical alignment of the microscope, thus allowing imaging of a single field of view under different spectral conditions. Facilities included a z-axis stepping motor providing precise increments of $1/1000$ revolutions or multiples, essential for accurate optical sectioning using the aplano oil objective lenses. Images were relayed to a 486 processor in an IBM PC/AT compatible desktop computer where they were integrated and digitized with a Kalman true-running-averaging filter. To visualize the surface of the copper tube, fluorescein (0.1%) was used as a negative stain. Bacteria were stained with either FITC or RITC. For fluorescein imaging optical filters with a 488 nm line were used in conjunction with the 510 nm line, whereas for rhodamine imaging the 514 nm line and 580 nm beam splitter were used. As an indicator of pH, 5-(and-6)-carboxy-2',7'-dichlorofluorescein (Molecular Probes, Eugene, OR) was used, as the excitation and emission wavelengths decrease due to acidification and would therefore be an indicator of metabolic activity within the biofilm. In this procedure RITC was used to stain protein on the bacterial cell wall to complement the corresponding image of the pH indicator. The images were not subjected to any form of restoration, e.g., nearest neighbor, delinear, or nonlinear deblurring methods.

Transmission Electron Microscopy

Internal cross-sectional detail of the individual microorganisms and their relationship to each other, including the overall biofilm, can be visualized using TEM. Biolayer filter supports were fixed in glutaraldehyde (2%)/ruthenium red (2%) in cacodylate buffer (pH 7.2) for 4 hr before being washed in cacodylate buffer (0.2 M) at pH 7.2. The samples were then postfixed in osmium tetroxide (0.5%)/ruthenium red (0.5%) (pH 7.2) for 1–2 hr and then immersed in graded alcohols of 35%, 70%, and 95% for 20 min each. Samples were placed in propylene oxide for 20 min and then into a mixture of propylene oxide/spurr resin for 1 hr. Samples were then placed into Spurr's resin again for 1 hr, and finally into fresh resin overnight at 70°. Ultrathin sections were then prepared for TEM (Philips 300) and the samples stained using an LKB ultrastainer.

Scanning Electron Microscopy

Copper tiles were immersed into 30% absolute alcohol immediately after being taken from the chemostat. After 15 and 30 min they were placed into 50% and 70% absolute alcohol, respectively. Following immersion in 90% absolute alcohol (45 min) the tiles were placed in osmium tetroxide (BDH, UK) for 2 hr before being placed into a desiccator. After transportation to the SEM suite the tiles were gold sputtered before being viewed in a Cambridge SEM.

Environmental Scanning Electron Microscopy

The ESEM is a fully integrated general purpose microscope. A four-stage differential pumping system in the electron-optical column allows the entire specimen chamber to be maintained at gas pressures higher than those permitted in conventional SEM. Chamber pressure was kept steady by a fully automated electronic pressure servo system enabling pressures between 1 and 50 torr.

Within the chamber, low-energy secondary electrons from the beam strike the specimen surface and are accelerated toward the detector electrode by a moderate electric field. The secondary electron detector is based on the principle of gas ionization. Successive collisions in the ambient gas molecules liberate more free electrons, resulting in a proportional cascade of current within the gas phase where positive ions serve to effectively neutralize the destructive buildup of excess electron charge at the specimen.

Images were captured via a JVC TK-1085E color video camera head with built-in compact power source fitted to a trinocular head. Using the video camera, visualization of the subject matter was relayed via a 0.4× wide field lens to the Microeye TC image analysis transputer card and software (Digithurst, Royston) operated under the Windows 3 (Microsoft) environment on a PC 386 compatible computer (Elonex, London). The photographed images were relayed to a multisync monitor (Taxan 775; capable of automatically switching between noninterlaced and interlaced modes when the appropriate signal is recognized) for image analysis.

Atomic Force Microscopy

Two stainless steel substrata were used: 316 cold rolled stainless steel (Avesta Stainless Steel, Sheffield) used as received, and 316 stainless steel that had been abraded unidirectionally using emery paper (average grit size 26 mm). All substrata were ultrasonically cleaned in a 1% detergent solution followed by washing and storing in pure methanol. Surfaces were examined using atomic force microscopy (AFM) prior to and after bacterial adhesion. *Pseudomonas aeruginosa* strains were incubated for 24 hr at 30° in a nutrient growth medium. The cells were harvested by centrifugation and suspended in phosphate buffer to achieve 1×10^8 cells/ml, and 500 ml was poured over the stainless steel samples that were covered before incubating at room temperature for 1 hr. The samples were then rinsed and air-dried before imaging. Atomic force microscopy images were obtained in air using a Resolver instrument manufactured by Quesant Instruments (Agoura Hills, CA). Images were taken using the intermittent contact mode, minimizing the high lateral force of contact method[10] to give topographical information. Silicon nitride tips and cantilevers were used with a spring constant of 0.12 Nm^{-1}.

Visualization of Biofilm using Different Techniques and Their Limitations

The Nikon Labophot was designed and used specifically for the microscopy of the biofilm and was adapted to increase the usability of the instrument to cater for cumbersome or unusual shaped samples such as curved pipe sections. Noncontact M Plan Apo lenses were used to enable specimens to be viewed from a distance of up to 1 cm with the 40× lens and 2 mm with the 100× lens.

---

[10] Q. Zhong, D. Inniss, K. Kjoller, and V. B. Elings, *Surf. Sci. Lett.* **290**, 68 (1993).

With such a long working distance, curved specimens did not present a problem for viewing. Often, the attachment of the specimen to a microscope slide presented the greatest problem and could be overcome by using either blue-tack or strips of tape. Using this setup, biofilms on flat uniform surfaces are the most easily viewed. For example, biofilm generated on glass, a flat and smooth surface (Fig. 1a), allows the DIC to differentiate the bacteria most easily from the background surface. With materials such as stainless steel much of the biofilm and the surface topography is out of focus (Fig. 1b), resulting in only small areas being within the focal plane. This was because the stainless steel had been sectioned from water pipe and hence was curved as well as rough. Using the microscope in the fluorescent mode with acridine orange enabled larger areas of the biofilm to be viewed within the focal plane, even on stainless steel (Fig. 1c), as well as fungal hyphae growing away from the substrata (Fig. 1d). The advantage of the fluorescent mode was that different fluorochromes could be used, including acridine orange, cyanoditolyl tetrazolium chloride (CTC, Sigma UK), propidium iodide and fluorescein. The full microscope setup could be purchased for approximately £30,000, including cameras and computers, and so is within the budget of many research departments. The disadvantage of using this particular set of noncontact lenses was that the specimen could only really be viewed best when dry, a state that would obviously lead to collapse of the biofilm structure. However, this could be overcome by the use of immersion lenses, the most convenient of which is water immersion lenses allowing immediate contact between the lens and the biofilm, thus alleviating the need to use oil and coverslips.

An advancement over the DIC/fluorescence microscope is the scanning confocal laser microscope (SCLM).[11] The advantage of the SCLM, allowing sectioning within the biofilm, enables physiological studies of growth and the relationships of species within the biofilm.[12] SCLM permits optical sectioning of biological materials without optical interference from other focal planes.[11,13–15] SCLM can be used to examine the growth and metabolism of living cells in the biofilm, and with this technique it is possible to analyze and quantify three-dimensional (3-D) biofilm and bioaggregates nondestructively, which is not possible with conventional light microscopy.[16] The study for which the SCLM was used here was to determine whether bacteria could be associated with areas of decreased pH activity and hence increased aggressive biocorrosion of copper plumbing materials.

[11] J. R. Lawrence and T. R. Neu, *Methods Enzymol.* **310**, 131 (1999).

[12] M. Kuehn, M. Hausner, H. J. Bungartz, M. Wagner, P. A. Wilderer, and S. Wuertz, *Appl. Environ. Microbiol.* **64**, 4115 (1998).

[13] D. E. Caldwell, D. R. Korber, and J. R. Lawrence, *J. Microbiol. Meth.* **15**, 249 (1992).

[14] D. E. Caldwell, D. R. Korber, and J. R. Lawrence, *Adv. Microb. Ecol.* **12**, 1 (1992.).

[15] D. De Beer, R. Srinivasan, and P. S. Stewart, *Appl. Environ. Microbiol.* **60**, 4339 (1994).

[16] J. T. Walker, K. Hanson, D. Caldwell, and C. W. Keevil, *Biofouling* **12**, 333 (1998).

a

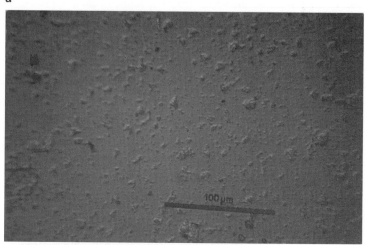

b

FIG. 1. (a) Biofilm generated on glass demonstrating the small microcolonies that initially form on the flat, smooth surface (bar denotes 100 μm). (b) Biofilms formed on stainless steel demonstrating that the biofilm and the surface topography is out of focus, resulting in only small areas being within the focal plane (bar denotes 5 μm). (c) Using the microscope in the fluorescence mode with acridine orange enabled large areas of the biofilm on stainless steel to be viewed within the focal plane (bar denotes 10 μm). (d) Fungi stained with acridine orange showing the fungal hyphae growing away from the substrata (bar denotes 10 μm).

c

d

FIG. 1. (*continued*)

As oil immersion lenses were used, the surfaces had to be flattened before visualization. Carboxyfluorescein was used as the stain and individual bacteria were observed to be surrounded by darkened areas, perhaps indicating reduced pH and increased microbially influenced corrosion (Fig. 2). SCLM is being further

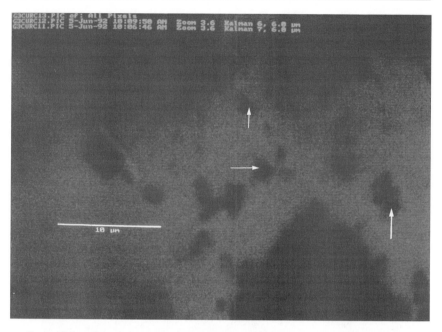

FIG. 2. Carboxyfluorescein was used to stain individual bacteria that were observed to be surrounded by darkened areas (indicated by arrows), perhaps indicating reduced pH and increased microbially influenced corrosion (bar denotes 10 μm).

enhanced with the development of two-photon SCLM, providing greater discrimination with less photobleaching of the fluorescence.[17]

SEM and TEM have for many years been the first choice of biologists for viewing biological specimens including biofilm. We have used SEM (Fig. 3) and TEM (Fig. 4) to visualize biofilm incorporating the human pathogen *Legionella-pneumophila* on polycarbonate EM sample holders. Having been generated using potable water as the nutrient source, the biofilm is observed using the SEM as individual cells on the surface[18] (Fig. 3). With TEM, the advantages are being able to view the specimen in profile and visualizing the interrelationships between the cells demonstrating coaggregation and exopolysaccharides (Fig. 4). Although SEM and TEM provide improved magnification over conventional microscopy, the techniques suffer from the limitation of specimen distortion during preparation. Preparation often involves use of dehydration in graded alcohols (four) and staining with osmium tetroxide. With multiple washing there is a possibility that

[17] J. M. Vroom, K. J. De Grauw, H. C. Gerritsen, D. J. Bradshaw, P. D. Marsh, G. K. Watson, J. J. Birmingham, and C. Allison, *Appl. Environ. Microbiol.* **65,** 3502 (1999).

[18] B. D. Tall, H. N. Williams, K. S. George, R. T. Gray, and M. Walch, *Can. J. Microbiol.* **41,** 647 (1995).

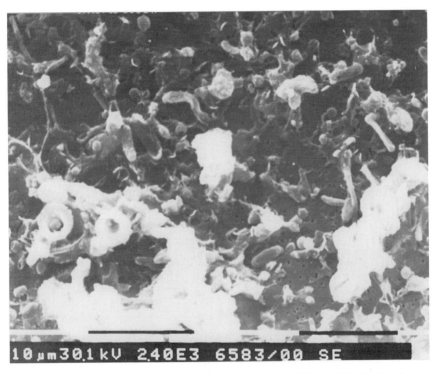

FIG. 3. SEM of *Legionella pneumophila* biofilms on polycarbonate EM sample holders (bar denotes 10 μm).

part of the biofilm structure will be lost from the surface during preparation. The biofilm that remains will have been dehydrated such that the structure may be distorted from that of the original. As such, interpretation of the resulting structure of the biofilm has to be done with great care. Advances in EM to alleviate distortion due to preparation have resulted in the development of the environmental SEM (ESEM).[19] This has the capability to maintain the specimen in a hydrated state, because of the increased pressures attainable within the viewing chamber, and thus maintain the structure of the biofilm, as was observed with a biofilm on copper plumbing materials (Fig. 5).

More recently, advances in AFM microscopy (atomic force microscopy; see Chapter [19]) have provided the biologist with a tool for manipulative studies of the topology of an individual bacterium.[20] Although the AFM may in many cases provide too high a magnification for biofilm structure *per se,* it provides the ability

[19] D. G. Allison, B. Ruiz, C. SanJose, A. Jaspe, and P. Gilbert, *FEMS Microbiol. Lett.* **167,** 179 (1998).
[20] A. Razatos, Y. L. Ong, M. M. Sharma, and G. Georgiou, *Proc. Natl. Acad. Sci. U.S.A.* **95,** 11059 (1998).

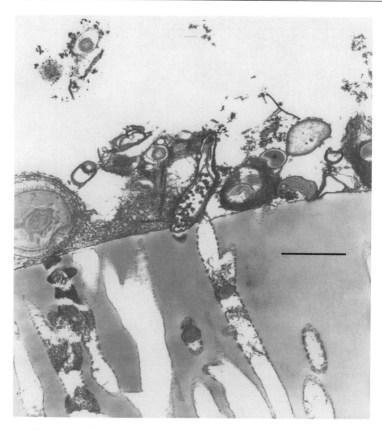

FIG. 4. TEM visualization of biofilm incorporating the human pathogen *Legionella pneumophila* on polycarbonate EM sample holders showing the depth and structure of the biofilm (bar denotes 1 μm).

to study the individual bacteria and their relationships with the surfaces and each other.

Prior to bacterial attachment the mean roughness (Ra) of both the unpolished stainless steel (Ra = 202 ± 30 nm) and abraded stainless steel (Ra = 169 ± 43 nm) was measured using AFM. Images of bacteria attached to the stainless steel substrates are shown in Fig. 6a and 6b. For both substrates bacteria are seen to form "clumps" on the surface. However, individual cells were also apparent, either within the grain boundaries (unpolished) or in the surface defects (roughened). The extent of clumping was significantly reduced for the roughened samples (Fig. 6b), indicating an effect of substratum on cell retention.

From the AFM images it can be seen that surface features on the nanometer scale do affect surface–microbial interactions. These interactions are however,

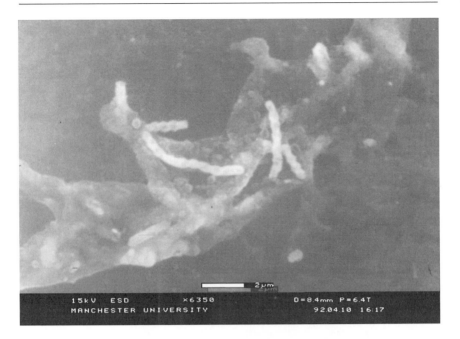

FIG. 5. Biofilm maintained in a hydrated state on copper plumbing tubing as viewed using ESEM (bar denotes 2 μm).

dependent on the size of the surface features. If the features on the surfaces are smaller than the bacteria, the microbes will sit above the features and be unaffected by them. If the features are much larger than the bacteria, the surface area available for bacteria–surface interaction will be similar to that for the smooth surface, and microbial retention will be unaffected. It is only when the size of the features is of the same magnitude as that of the bacteria that significant changes in microbial attachment may occur. In order to quantify the range of sizes of the surface features power spectral density (PSD) analysis was performed on the uncoated substrata (Fig. 6c). PSD gives information on the range of the AFM image with respect to surface feature size (in this case 10 mm down to 0.1 mm) as well as information on the roughness frequency distribution. Unpolished stainless steel has the highest intensity at the low frequency end of the range, corresponding to features of 7–10 mm, i.e., the grain structure, whereas for abraded stainless steel the roughness frequency peaks in the 1 μm range. This corresponds closely to the size of the bacteria. Hence, bacteria should interact more strongly with the abraded surface as the surface features correspond with the size of the bacteria, causing the bacteria to be adsorbed into the surface features and to be retained more successfully. These results, although not conclusive, do suggest a strong correlation between roughness frequency and bacteria–surface interactions. Repeating the work with different

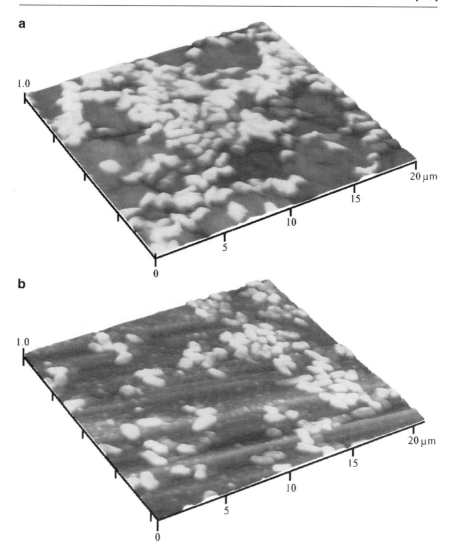

FIG. 6. (a) Clumps of bacteria observed on the unpolished surface using AFM (bar denotes 20 μm). (b) Loss of clumping on roughened surfaces indicating an effect of substratum on cell retention as visualized using AFM (bar denotes 20 μm). (c) Quantification of the size range of the surface features using power spectral density (PSD) analysis on the uncoated substrata.

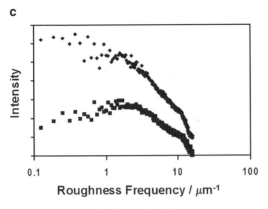

c

FIG. 6. (*continued*)

average roughness as well as different roughness frequency will lead to a more complete model of how roughness influences bacterial retention. However, with this particular instrument the specimens are viewed air dried, which may be a limitation on biofilm surface analysis.

In summary, a range of microscopy techniques have been demonstrated for their usefulness and yield different information about biofilm. In a quest for further information one must utilize as many different techniques as possible in order to increase knowledge of the basics of biofilm structure.

Acknowledgment

The authors thank Bill Keevil for his advice, encouragement, and support; Douglas Caldwell and Keith Hansen for guidance and help with the SCLM; and Andy Skinner for his assistance with the SEM and TEM.

Section IV

Physical–Chemical Characterization of Biofilms

[17] Two-Photon Excitation Microscopy for Analyses of Biofilm Processes

By JAMES D. BRYERS

Introduction

Biofilm Processes

In an aqueous environment, a support, termed a substratum, will be immediately biased by dissolved organic molecules and macromolecules, which adsorb rapidly from the liquid phase. Bacterial cells present in the fluid contact the substratum by a variety of mechanisms. Just prior to or upon arriving at a substratum, bacterial cells alter certain gene expression patterns; enzymes and metabolic pathways are altered (induced or repressed) to create a phenotypically adherent cell. Once at the substratum, the cells can adsorb either reversibly or irreversibly. Provided the cells remain at the surface for sufficient time, they will secrete extracellular polymers that serve to attach the cells tenaciously to the substratum. Attached cells metabolize prevailing energy and carbon sources (either dissolved within the surrounding fluid, adsorbed to the substratum surface, or existing as a constituent of the substratum itself), reduce terminal electron acceptors, grow, replicate, and produce insoluble extracellular polysaccharides, thus accumulating an initial viable biofilm community. Inert particles, bacterial cells of the same or different species, and higher life forms (e.g., algae, amebae, protozoa) continue to be recruited from the fluid and incorporated into the biofilm community. As the biofilm bacterial communities mature, the adherent populations will oxidize electron donors and reduce existing terminal electron acceptors in order of decreasing redox potential. As the biofilm community develops, the penetration depth of one electron acceptor will therefore overlap that of another. Thus, in a natural ecosystem, a succession of different microbial populations can be established: oxygen-guzzling aerobes, facultative denitrifiers using nitrate and nitrite; anaerobic sulfate reducers using sulfate; fermentative microbes partially reducing organic carbon compounds; and finally, anaerobic methane producers. As a result of hydrodynamic forces and stresses exerted by replication, there can be a continual erosion of cells and extracellular material from the biofilm back to the bulk fluid.[1]

Investigation of any one of these many complex processes governing biofilm formation, persistence, and reactivity has historically required destructive sampling of the biofilm to determine cell numbers, polymer compositions, and biofilm

[1] J. D. Bryers, "Biofilms, Principles and Applications," 2nd ed. J. Wiley Interscience, New York, 2000.

METHODS IN ENZYMOLOGY, VOL. 337

composite parameters (thickness, density, viscoelasticity). Although scanning electron microscopy (SEM) or transmission electron microscopy (TEM) provided useful information about biofilm structure in the 1960s–1980s, these are still methods that distort the true structure of the biofilm, because of the required extensive dehydration and gold sputter coatings. SEM and TEM are also inherently unable to quantify species population spatial distributions in a biofilm or quantify metabolic processes taking place in a biofilm. Fluorescence microscopy (epifluorescence) and the many variations in optical image enhancement techniques are sufficient to quantify different species of cells initially attaching to a substratum and their spatial patterns on surfaces, but this method is limited to individual adhering cells and extremely thin films (<10 μm).

The advent of one-photon (1P) laser scanning confocal microscopy (CSLM) and the availability of a myriad of very selective fluorescent probes (cytological stains, immunofluorescent stains, oligonucleotide probes)[2,3] has for the first time made it possible to optically section a biofilm, to observe in real time metabolic processes occurring at different locations with the biofilm structure, and to nondestructively map the biofilm community. Coupled with very powerful image capture instruments and fluorescence detection, CSLMs can provide digital quantification of various kinetic[4–8] and mass transport[9–11] rate processes within a biofilm. So, while some simply use CSLM to just take pictures of real biofilms, others have applied the CSLM to the generation of quantitative information on local dynamic biofilm processes.[12–14]

However, certain inherent properties of CSLM restrict its application to only certain types of biofilms and to the acquisition of only certain types of metabolic process information. The evolution of two-photon or multiple photon excitation

[2] R. P. Haugland, "Handbook of Fluorescent Probes and Research Chemicals," 6th ed. Molecular Probes, Eugene, OR, 1996.

[3] S. E. Cowan, E. Gilbert, A. Khlebnikokov, and J. D. Keasling, *Appl. Environ. Microbiol.*, **66**, 413 (2000).

[4] S. Molin, A. T. Nielsen, B. B. Christensen, J. B. Andersen, T. R. Licht, T. Tolker-Nielsen, C. Sternberg, M. C. Hansen, C. Ramos, and M. Givskov, *in* "Biofilms," 2nd ed. (J. D. Bryers, ed.), p. 89. J. Wiley Interscience, New York, 2000.

[5] R. I. Amann, L. Krumhoiz, and D. A. Stahl, *J. Bacteriol.* **172**, 762 (1990).

[6] J. H. Miller, "Experiments in Genetics," p. 352. Cold Spring Harbor Laboratory, Cold Spring Harbor, NY, 1972.

[7] T. W. J. Gadella, T. M. Jovin, and R. M. Clegg, *Biophys. Chem.* **48**, 221 (1993).

[8] C. G. Morgan, A. C. Mitchell, and J. G. Murray, *Trans. R. Microsc. Soc.* **90**, 463 (1990).

[9] J. D. Bryers and F. E. Drummond, *Biotechnol. Bioeng.* **60**, 462 (1998).

[10] S. Singleton, R. Treloar, P. Warren, G. K. Watson, R. Hodgson, and C. Allison, *Adv. Dent. Res.* **11**, 133 (1997).

[11] J. J. Birmingham, N. P. Hughes, and R. Treloar, *Phil. Trans. Royal Soc. London B* **350**, 312 (1995).

[12] D. E. Caldwell, D. R. Korber, and J. R. Lawrence, *Microbiol. Methods* **15**, 249 (1992).

[13] D. E. Caldwell, D. R. Korber, and J. R. Lawrence, *J. Appl. Bacteriol.* **74**, 52 (1993).

[14] J. R. Lawrence, G. M. Wolfgaardt, and D. R. Korber, *Appl. Environ. Microbiol.* **60**, 1166 (1994).

microscopy (2PE or MPE microscopy) may alleviate a number of the restrictions placed on CSLM in the biofilm diagnostics. This chapter discusses the basics of 2PE microscopy and its potential for biofilm analyses. For greater detail on two- or multiple-photon microscopy, the reader is directed to the following basic literature reviews[15–17] and Internet Web sites (http://www):

> cbit.uchc.edu/microscopy/two_photon.html
> microcosm.com/tutorial/tutorial3.html
> broccoli.caltech.edu/~pinelab/2photon.html.

One-Photon Confocal Microscopy

Fundamentals

With normal fluorescence microscopy, one cannot resolve deep structures within a specimen because of light emitted and scattered by the out-of-focus specimen. By placing small apertures in the light path at points confocal to the focal point within the specimen, almost all of the out-of-focus fluorescence is blocked, allowing detection of just the point of interest. By scanning the light source, a laser beam, in the X and Y directions, a high-resolution image of a thin slice of the sample can be digitally constructed within seconds. By making a series of such optical "slices" through the thickness (Z-direction) of the sample, a three-dimensional representation can be generated and manipulated with image-processing software. The number of papers regarding the visualization of biofilm architecture and the interactions of this architecture with the surrounding bulk fluid phase has increased dramatically over the past 10 years.[18–21]

Numerous CSLM systems are available commercially from most of the major microscopy companies. They are available in a wide range of options (software and hardware) and auxiliary devices. CSLM is essentially a standard epifluorescence microscope with an additional laser light source, specialized scanning equipment, and computerized digital imaging. A general schematic of 1P confocal microscopy is shown in Fig. 1. Lasers used include helium–neon, argon ion,

[15] O. Nakamura, *Microsc. Res. Technique* **47**, 165 (1999).

[16] A. Jenei, A. K. Kirsch, V. Subramanian, D. J. Arndt-Jovin, and T. M. Jovin, *Biophys. J.* **76**, 1092 (1999).

[17] W. Denk, D. W. Piston, and W. W. Webb, *in* "Handbook of Biological Confocal Microscopy" (J. B. Pawley, ed.). Plenum Press, New York, 1995.

[18] B. A. Sanford, A. W. de Feijter, M. H. Wade, and V. L. Thomas, *J. Industr. Microbiol.* **16**, 48 (1996).

[19] S. R. Wood, J. Kirkham, P. D. Marsh, R. C. Shore, B. Nattress, and C. Robinson, *J. Dent. Res.* **79**, 21 (2000).

[20] D. Cummins, M. C. Moss, C. L. Jones, C. V. Howard, and P. G. Cummins, *Binary* **4**, 86 (1992).

[21] J. R. Lawrence, D. R. Korber, B. D. Hoyle, J. W. Costerton, and D. E. Caldwell, *J. Bacteriol.* **173**, 6558 (1991).

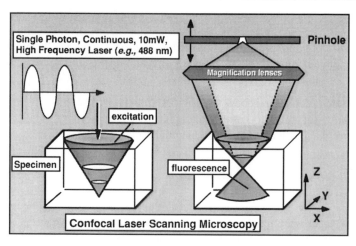

FIG. 1. Principles of one-photon confocal laser scanning microscopy (1P-CSLM), illustrating one-photon fluorescence of entire sample with confocal pinhole exclusion of undesired image planes.

helium–cadmium, krypton–argon, and UV lasers. Typically, CSLM systems are equipped with helium–neon (543 or 633 nm) and krypton–argon lasers (488-nm blue, 568-nm yellow, and 647-nm red lines). UV/Vis lasers (157–351 nm) may also be obtained with commercially available CSLM systems. Lasers are connected to the scanning head, which is a unit with galvanometric mirrors that scan the beam onto the specimen and a system of mirrors and beam splitters that direct the return signal to specific photomultiplier. The scan rate can be varied; however, resolution decreases with increasing scan rate. When the scan rate is too low, the potential for photobleaching is increased. Pinhole(s) in the light path allow only those fluorescence signals that arise from a focused XY plane to be detected by the photomultiplier. The system is said to be "confocal" since the pinholes exclude the emitted fluorescent signals originating from above, below, or beside the point of focus from reaching the photodetector. Different wavelength-specific filters are used to supply specific excitation and emission wavelengths for the fluorescent probes used.

One problem with CSLMs is that the fluorophores release toxic products when they are excited.[22] Additionally, the fluorochromes are rapidly photobleached by the excitation beam. To avoid such a situation, CSLMs must limit the amount of excitation that can be used, which results in low signal levels. This toxicity problem limits the amount of time during which living cells and tissues can be observed in fluorescence. However, the major problem with CSLMs is that fluorophore excitation occurs throughout the sample, since the entire field of view is being

[22] R. M. Williams, D. W. Piston, and W. W. Webb, *FASEB J.* **8,** 804 (1994).

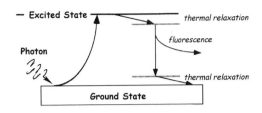

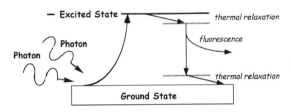

FIG. 2. Energy transition diagrams for one photon- and two-photon excitation:fluorescence.

irradiated at any one time, although only one focal section of $\sim$0.5 μm thickness is being observed.

Multiple-Photon Excitation Microscopy

Fundamentals

Figure 2 shows the energy transfer diagrams for 1PE and 2PE absorbed fluorescence. Two photon excitation refers to the simultaneous adsorption of two photons of longer, not necessarily identical wavelengths ω_1, ω_2, that combine their energies to effect a molecular excitation that otherwise would require a single photon of shorter wavelength $(1/\omega_1 + 1/\omega_2)^{-1}$. This is not the same situation as a molecule being excited to an intermediate level first by one photon and then to the final state by a second photon.

The transition probability for simultaneous two photon absorption depends on the square of the instantaneous light intensity, which is why very brief intense pulses increase the average 2PE absorption probability for a constant average power input. One wants to minimize the average excitation power to avoid one-photon adsorption, which is responsible for most of the heating and photodamage. The use of short pulses, $\sim$100 fsec (1 fsec $= 10^{-15}$ sec), at 100 MHz and small duty cycle is critical to image acquisition in reasonable collection times while maintaining "biologically pristine" power levels.

The probability, N_a, that a given fluorophore at the center of the focus absorbs a photon-pair during a single burst is derived by Denk *et al.*[23] as Eq. (1),

$$N_a \approx \delta \langle P \rangle^2 (F_p)^{-1} \left\{ \frac{\pi A_p^2}{hc\omega} \right\}^2 \zeta \tag{1}$$

where δ is the two-photon cross section, P is average power, F_p is repetition frequency, A_p is numerical aperture, ω is wavelength, c is the speed of light in a vacuum, h is Planck's constant, and ζ is a two-photon "advantage factor" $\cong 0.66(F_p \tau_p)^{-1}$, with τ_p the time between the half-power points of the burst.

Equation (1) is only correct as long as the probability for each fluorophore to be excited in a single pulse is less than 1. During a pulse ($\sim$100 fsec and an excitation state lifetime, τ_f, of nanoseconds), the prerequisite for absorbing a second photon is that the molecule has insufficient time to relax to its ground state. Thus, as N_a approaches 1, saturation effects are seen. For maximum fluorescence generation, the desirable time between pulses is on the order of the τ_f of the fluorophore, which happens to be a few nanoseconds for most biological fluorescent stains. Repetition rates of $\sim$100 MHz (one pulse every 10 nsec) are therefore ideal and also provided by available mode-locked lasers.

The setup and operation of a 2PE or MPEM are very similar to those of 1P CSLM with the main differences being the type of laser used and the increased number of detection options. At the time of this publication, the three major CSLM manufacturers (Zeiss, Bio-Rad, and Leica) were offering commercially available integrated 1P and 2PE CSLM systems and accompanying image analysis systems. Currently, whether an integrated package or whether the 2PE excitation laser and optics are bought as separate components, the shift to 2PE is expensive ($\sim$\$80,000–100,000), increasing the cost of a 1P CSLM system by at least 50%.

For most applications, the 2PE light source is the self-mode-locked Ti:sapphire laser pumped by a continuous wave argon-ion laser, which can be tuned from just below 700 nm to slightly above 1050 nm with pulses $\approx$100 fsec. To cover this entire wavelength range requires three different mirror optic combinations, with the changeover between sets being tedious, requiring several hours. These lasers have significant power (20–30 kW) and cooling water ($\sim$40–60 liter/min) requirements and, whether in-house fabricated or purchased as an integrated system, require dedicated space of $\sim$250–400 ft^2 and a stable optical table.

Selection criteria for fluorophores in 2PE microscopy are the same as in any type of fluorescence microscopy: large absorption cross section at desired wavelengths, high quantum yield, low rate of photobleaching, and minimal chemical or photochemical toxicity to the stained cells. From an empirical approach, most stains used in 1P CSLM have been tried in 2PE microscopy. In general, 2PE will excite a fluorophore whenever there is 1P absorption at a wavelength about twice

[23] W. Denk, J. H. Strickler, and W. W. Webb, *Science* **248**, 73 (1990).

the energy of the excitation photons. Although photodamage to a cell is one suggested advantage of 2PE over 1P, one must realize that in the focal plane itself, the damage could be as large as for 1PE. Any reactions brought on by the excited state of the fluorophore are still possible. Also, it is possible that indigenous biological molecules may have unusually high 2PE absorption cross sections (e.g., bacteriorhodopsin) and be susceptible to photodamage.

Application of 2PE to Biofilms Diagnosis

Since its inception, 2PE (and MPEM) microscopy has been applied to a number of biological specimens in direct comparisons to 1PE CSLM and conventional digital deconvolution fluorescence microscopy: mostly intact mammalian cells or tissue sections, including rat cardiac myocytes, sea urchin embryo, frog embryo, monkey kidney tissue sections, neonatal hamster brain sections, and living bovine embryo.[17,24,25]

However, application of 2PE microscopy to bacterial biofilms is only just beginning, with a paucity of publications evaluating its potential.[26] Vroom *et al.*[26] quantitatively compare 1PE-CSLM to 2PE microscopy for their respective abilities to noninvasively dissect deep, *in vitro* bacterial biofilms. In addition, pH gradients were determined by fluorescence lifetime imaging and the fluorescent pH-sensitive stain carboxyfluorescein. A defined mixed culture bacterial biofilm was cultivated in a constant depth biofilm system[27] at a fixed depth of ∼100 μm. 2PE was able to collect clear images at greater resolution at depths four times greater (140 μm vs 23 μm) than CSLM. Periasamy *et al.*[25] also reported the same depth of penetration advantage of 2PE over both CSLM and digital deconvolution fluorescence microscopy, imaging *Xenopus* (frog) gastrula sections. Vroom *et al.*[26] also directly measured pH within the biofilms in both the lateral and the axial directions. When biofilms were covered with 14 mM sucrose for 1 hr, the development of distinct pH gradients was imaged in real time over a time period of 1 hr without any appreciable photobleaching.

Autofluorescence of the biofilm can be an interference in any fluorescence microscopy. To minimize the autofluorescence while avoiding excessive stain concentration, Vroom *et al.* selected a probe concentration of 100 μm and an interference bandpass filter with a transmission band from 500 to 530 nm that transmits 50% of the fluorescein fluorescence, but only 20% of the autofluorescence. In addition, this group employed delayed opening of the time gates to further minimize the contribution of autofluorescence. Autofluorescence decay times are approximately 1 nsec. By employing a delay of 1 nsec between excitation pulse and the opening

[24] V. E. Centonze and J. White, *Biophys. J.* **75,** 2015 (1998).

[25] A. Periasamy, P. Skoglund, C. Noakes, and R. Miller, *Microsc. Res. Technique* **47,** 172 (1999).

[26] J. M. Vroom, K. J. de Grauw, H. C. Gerritsen, D. J. Bradshaw, P. D. Marsh, G. K. Watson, J. J. Birmingham, and C. Allison, *Appl. Environ. Microbiol.* **65,** 3502 (1999).

[27] S. L. Kinniment, J. W. T. Wimpenny, D. Adams, and P. D. Marsh, *Microbiology* **142,** 631 (1996).

of the first gate, autofluorescence was reduced by 80% with little impact on probe fluorescence.

Another practical point is to match the index of refraction of the immersion liquid of the objective to the index of refraction of the sample. A discontinuity leads to spherical aberration, which increases with increasing depth. Thus, with bacterial biofilms, it is suggested to employ a water immersion objective rather than oil immersion.

In 1998, we reported[9,28,29] on the use of conventional microscopic FRAP to determine local mass transfer coefficients in thin biofilms of heterogeneous architecture. We are presently employing 2PE microscopy to repeat those experiments in much thicker biofilms. Our two-photon microscope (http://www.cbit.uchc.edu/microscopy/two_photon.html) comprises a modified Bio-Rad MRC600 laser scanning confocal microscope and a Coherent 900-F femtosecond titanium sapphire laser system. The microscope has been modified to permit excitation throughout the visible and near infrared. Data can be collected by the system confocal software using two different optical detection schemes. In the simplest, i.e., descanned detection, the confocal pinhole aperture is fully opened, increasing collection efficiency over confocal while utilizing the intrinsic multiphoton excitation sectioning. Even greater sensitivity can be obtained by using a nondescanned, otherwise known as an all-area, detection scheme. Here, a dichroic mirror is placed between the objective and scanning mirrors and fluorescence is detected with an external photomultiplier. This scheme increases collection efficiency not only by eliminating many optical components, but in addition by using the whole area of the photomultiplier tube, which is made possible by eliminating the confocality requirement of using one-photon excitation. Signal processing electronics have been developed in house to allow an external photomultiplier tube to be read by the original integration electronics. Nonscanned detection increases sensitivity over simple descanned detection by factors of 3–5, depending on the numerical aperture of the objective. This sensitivity will prove essential in probing deep (100–180 μm) into the highly turbid environment of bacterial biofilms.

The laser system is tunable from 700 to 1000 nm with average power of 500–1000 mW at 76 MHz repetition rate and pulse width of 100 fsec. The one and two-photon absorption spectra of fluorescein[30,31] differ significantly in that the 2P maxima are at 790 and 920 nm. The corresponding band in the former peak in 1P excitation, at half the wavelength, is symmetry forbidden and consequently very weak. Greater depth of penetration is possible at 920 nm because of the Rayleigh scattering wavelength dependence; however, greater laser power

[28] F. E. Drummond, "Macromolecule Transport Mechanisms in Living Bacterial Biofilms of *Pseudomonas putida*." M.S. Thesis, Duke University, Durham, NC, 1993.

[29] J. D. Bryers and F. E. Drummond, *in* "Progress in Biotechnology 11: Immobilized Cells" (R. H. Wijffels *et al.*, eds.), p. 196. Elsevier Science BV, Amsterdam, 1996.

[30] C. Xu, R. M. Williams, W. Zipfel, and W. W. Webb, *Bioimaging* **4**, 198 (1996).

[31] C. Xu and W. W. Webb, *J. Opt. Soc. Am. B* **13**, 481 (1996).

and optical throughput of the microscope are possible at 790 nm. The power necessary for saturation of a given fluorophore per laser pulse can be approximated by Eq. (2):

$$\text{abs} \approx \Phi F^2 \tau \tag{2}$$

where Φ is the two-photon absorption cross section, F is the laser flux in photons per unit time per unit area, and τ is the laser pulse width. The 2P absorption cross section for fluorescein is 10^{-49} cm^4 s, and with 100 femtosecond pulse width and 1.4 N.A. lens, saturation will occur with approximately 10 mW at the sample.

The 2PE microscope has been adapted for 2P fluorescence recovery after photobleaching (FRP) experiments. An electro-optic modulator (EOM) is used to control the intensity of light entering the microscope. A bias voltage on the EOM sets the static light level and then the modulator is actively switched to obtain maximum throughput. The ratio of these conditions is set to be approximately 10 to 1. Significant pulse broadening, approximately twofold, occurs through the EOM; hence, dispersion compensation with a prism compressor is used to maintain the highest peak power possible. Additional pulse broadening occurs through the microscope and objective lens, and to obtain maximum bleaching power, this dispersion can be precompensated with the use of a second prism pair. The most straightforward FRAP geometry with our microscope is line scanning, where data is obtained by the following method. A normal 2P image is obtained for reference. Next, a line of interest is selected and that line is scanned repetitively to obtain a steady-state fluorescence level and then the EOM is activated for a duration sufficient to obtain a bleach, typically 20–200 msec, i.e., 20–200 lines, triggered off the microscope clock. The line scanning continues and the fluorescence recovery is again monitored at low power. This procedure has been successfully used for studies of diffusion in sea urchin oocytes.[32] Other geometries can be utilized as well, as mitigated by the particular application, such as either spot bleaching followed by line probing, or bleaching a larger spot (1 μm) from either below or an oblique angle followed by normal line probing. In addition, the system software has been modified to allow for rectangular bleaching and probing as well as accommodate user-defined pixel dwell times. We have employed 2PE FRAP to verify earlier diffusion coefficients for several solutes (fluorescein, dextran, and bovine serum albumin) but now in much thicker biofilms (total average thickness ≈390 μm) where the deepest plane of observation is ~120–140 μm. Table I summarizes the ratio of solute tracer diffusion coefficient relative to its diffusion in water as a function of depth into a biofilm cluster. As one can see for the three solutes tested, there is a spatial dependency on mass transport coefficient, which we attribute to local density variations in the biofilm cluster.

---

[32] M. Terasaki and P. J. Campagnola, in preparation.

TABLE I
RATIOS OF DIFFUSION COEFFICIENTS FOR FLUORESCENTLY LABELED
MOLECULES IN WATER AND IN BACTERIAL BIOFILM USING
2PE-FRAP METHOD[a]

	$D_{\text{i-biofilm}}/D_{\text{i-water}}$[b]		
Depth in biofilm	Fluorescein	Dextran (10,000 MW)	BSA (68,000 MW)
10.	0.94	0.34	0.60
25.	0.92	0.32	0.57
50.	0.92	0.30	0.55
100.	0.89	0.28	0.52
120.	0.90	0.28	0.54

[a] Biofilm was 390 μm in overall thickness; results in this table were averaged from FRAP scans taken at an overall depth of 120 μm.

[b] Calculated from Einstein's equation: $D_{\text{i-water}} = k_b T/6 \pi \mu r_s$, where $k_b =$ Boltzmann's constant; $T =$ temperature, K; $\mu =$ viscosity of solvent : solute mixture at temperature, T; and $r_s =$ Stokes molecular radius. BSA, bovine serum albumin.

Summary: 1PE Confocal vs 2PE Microscopy for Biofilm Diagnostics

The confocal laser scanning microscope (CSLM) has dramatically improved the ability to noninvasively dissect a bacterial biofilm. Coupled with the development of various fluorescent stains, the CSLM can provide observations of various cellular and biofilm community processes (e.g., multiple species enumeration, plasmid presence and transfer,[33,34] bacterial cell viability, and specific gene expression via *in situ* hybridization.[35]

The aspect of using CSLM together with a myriad of viable stains,[36] oligonucleotide probes, and metabolically specific fluorescent markers[37] to view, noninvasively, dynamic process within a biofilm is all too tempting. However, use of CSLM should be avoided for dissecting thick biofilm samples and for quantifying rate processes that would require lengthy scan times of the same sample (e.g., determination of mass transfer parameters, *in situ* hybridization of RNA, *in situ* determination of antibiotic efficacy, real-time observation of plasmid-DNA transfer between cells) because of the inherent limitations of CSLMs listed above. Most researchers resort to using CSLM to monitor these dynamic processes by

[33] C. L. Bender and D. A. Cooksey, *J. Bacteriol.* **165,** 535 (1986).

[34] L. Diels, A. Sadouk, and M. Mergeay, *Toxicol. Environ. Chem.* **23,** 79 (1989).

[35] D. D. Focht, D. B. Searles, and S.-C. Koh, *Appl. Environ. Microbiol.* **62,** 3910 (1996).

[36] B. L. Roth, M. Poot, S. T. Yue, and P. J. Millard, *Appl. Environ. Microbiol.* **63,** 2421 (1997).

[37] J. C. Fry and M. J. Day, *in* "Bacterial Genetics in Natural Environments" (J. C. Fry and M. J. Day, eds.). Chapman & Hall, London, 1990.

collecting a series of "snapshots" taken from different locations in a sample or from different samples.

CSLM problems can be minimized with two-photon (2P) excitation microscopy. In 2P microscopy, the sample is illuminated with a light of a wavelength of about twice the wavelength of the absorption peak of the fluorochrome being used. For example, in the case of fluorescein, which has an absorption peak around 500 nm, 1000 nm excitation could be used. If a normal continuous laser is used, no excitation of the fluorochrome will occur at this wavelength and hence no bleaching will occur in the bulk of the sample. However, if a high-powered pulsed laser is used that has a peak power of >2 kW and is contained in pulses shorter than a picosecond (so that the mean power levels are moderate and do not damage the specimen), two-photon events will occur at the point of focus.

In summary, advantages of two-photon excitation include the following:

- Optical sections may be obtained from deeper within a sample than can be achieved by confocal or wide-field imaging. There are three main reasons for this: the excitation source is not attenuated by absorption by fluorochrome above the plane of focus; there is little electron transition and molecular vibration absorption in the near infrared; the longer excitation wavelengths used suffer less Rayleigh scattering; and the fluorescence signal is not degraded by scattering from within the sample as it is not imaged.

- Microscope optics are much better corrected for use with visible wavelength light than for UV. Therefore, the 2P system gives better optical performance than a UV CSLM system and does not require the additional complicated optics for correction of lateral and axial chromatic aberration of the UV fluorescence system.

- All the emitted photons from multiphoton excitation can be used for imaging (in principle); therefore, no confocal blocking apertures have to be used.

- It is possible to excite UV fluorophores using a lens that is not corrected for UV as these wavelengths never have to pass through the lens.

- The major advantages is that only the focal plane being imaged is excited, compared to the whole sample in the case of CSLM. This considerably reduces total bleaching and the consequent production of toxic products during live-cell imaging.[22,38]

- The absorbing, fluorescing, or photochemically reacting area is localized because the probability of two-photon absorption is proportional to the square of the intensity of the excitation field.

- In 2PE microscopy, the pulsed laser's oscillating wavelength is twice as long as that for single-photon CSLM, resulting in a $1/16$ reduction in Rayleigh scattering with a sample.

[38] H. C. Gerritsen, J. M. Vroom, and C. J. de Grauw, *Microsc. Microanal* 4 (supp) 2, 420 (1998).

[18] Measurements of Softness of Microbial Cell Surfaces

By HENNY C. VAN DER MEI, PASCAL KIERS, JOOP DE VRIES, and HENK J. BUSSCHER

Introduction

Microbial adhesion to surfaces is in essence the approach of two charged surfaces and compression of their diffuse double layer charges, which leads to the traditional electrostatic repulsion as accounted for in the DLVO theory.[1,2] In order for microbial adhesion to occur, the electrostatic double layer repulsion has to be overcome by electrodynamic Lifshitz–van der Waals attraction. When the interacting surfaces are soft and ion penetrable,[3,4] however, double layer charges are driven into the ion-penetrable layers on approach, causing an effective decrease in surface potential and electrostatic repulsion. Until now, this has not been accounted for in the DLVO theory of colloidal stability as applied toward microbial adhesion, because microorganisms were generally considered as ion-impenetrable particles.[1,5,6]

We have compared the dependence of the electrophoretic mobility on the ionic strength of a fibrillated and a nonfibrillated oral streptococcal strain and reported that a soft, ion-penetrable strain was fibrillated, while an electrophoretically harder strain was devoid of a proteinaceous fibrillar surface layer,[7] demonstrable by electron microscopy after negative staining.[8] Interestingly, the bald streptococcal strain had a diffusion coefficient, as measured by dynamic light scattering, that was independent of ionic strength and composition of the suspending fluid.[9] The fibrillar strain showed an almost twofold decrease in diffusion coefficient on increasing the suspension pH from 2 to 7. Direct probing of the softness of microbial cell surfaces has been described using atomic force microscopy[10] and

---

[1] P. R. Rutter and B. Vincent, *in* "Physiological Models in Microbiology" (M. J. Bazin and J. I. Prosser, eds.), p. 87. CRC Press, Boca Raton, FL, 1988.

[2] C. J. van Oss, *Colloids Surf. B: Biointerfaces* **5**, 91 (1995).

[3] H. Oshima and T. Kondo, *Biophys. Chem.* **39**, 191 (1991).

[4] H. Oshima, *Colloids Surf. A: Physicochem. Eng. Aspects* **103**, 249 (1995).

[5] R. Bos, H. C. van der Mei, and H. J. Busscher, *in* "Oral Biofilms and Plaque Control" (H. J. Busscher and L. V. Evans, eds.), p. 163. Harwood Academic Publishers, India, 1998.

[6] H. J. Busscher, P. S. Handley, R. Bos, and H. C. van der Mei, *in* "Physical Chemistry of Biological Interfaces" (A. Baszkin and W. Norde, eds.), p. 431. Marcel Dekker, Inc., New York, 2000.

[7] R. Bos, H. C. van der Mei, and H. J. Busscher, *Biophys. Chem.* **74**, 251 (1998).

[8] P. S. Handley, *Biofouling* **2**, 239 (1990).

[9] H. C. van der Mei, J. M. Meinders, and H. J. Busscher, *Microbiology* **140**, 3413 (1994).

[10] G. Binning, C. F. Quate, and Ch. Gerber, *Phys. Rev. Lett.* **56**, 930 (1986).

suggested that the mechanical softness of microbial cell surfaces related with the electrophoretic one and the dynamic nature of microbial cell surfaces upon changes of their ionic environment.[11] Thus, particulate microelectrophoresis, dynamic light scattering, and atomic force microscopy have become valuable, partly complementary methods to assess the softness and related ion penetrability of microbial cell surfaces. Considering the implications of these dynamic properties of microbial cell surfaces for their adhesion to surfaces, we will describe the application of these three methods as geared toward the analysis of microbial cell surfaces.

Atomic Force Microscopy: Mechanical Softness

Two *Serratia marcescens* strains with (RZ30) and without (RZ37) fibrils as demonstrated by electron microscopy were suspended in water to a density of 10^5 per ml, after which 10 ml of this suspension was filtered through an Isopore polycarbonate membrane (Millipore) with pore size of 0.8 μm, i.e., slightly smaller than the bacterial dimensions, to immobilize the bacteria through mechanical trapping as schematically indicated in Fig. 1.[12] After filtering, the filter was carefully fixed with double-sided sticky tape on a small sample disk and transferred to the AFM while avoiding dewetting. Force–distance measurements were made under water at room temperature, either in deionized water or in 0.1 *M* KCl, using an optical lever microscope (Nanoscope III, Digital Instruments, Santa Barbara, CA). Oxide-sharpened microfabricated Si_3N_4 cantilevers from Park Scientific Instruments (Mountain View, CA) with a spring constant of 0.03 N m^{-1} and a probe curvature radius of 20 nm (according to manufacturer specifications) were used. The approach curves in Fig. 1 can be fitted with an exponential function:

$$F = F_0 e^{-d/\lambda'} \qquad (1)$$

in which F is the measured force, F_0 is the force at zero separation distance, d is the separation distance, and λ' is a decay length. From Fig. 1 it can be seen that the fibrillated strain RZ30 in water showed a long-range repulsion, starting at a separation of 100 nm, and on retraction multiple adhesion forces were found. In 0.1 *M* KCl, however, the tip experienced a repulsion upon approach at a much shorter distance. For the nonfibrillated *S. marcescens* RZ37, identical force–distance curves were observed in water and in 0.1 *M* KCl, which were similar to the force–distance curves of *S. marcescens* RZ30 in 0.1 *M* KCl.

[11] H. C. van der Mei, H. J. Busscher, R. Bos, J. de Vries, C. J. P. Boonaert, and Y. F. Dufrêne, *Biophys. J.* **78**, 2668 (2000).
[12] S. Kasas and A. Ikai, *Biophys. J.* **68**, 1678 (1995).

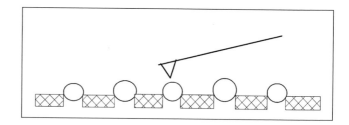

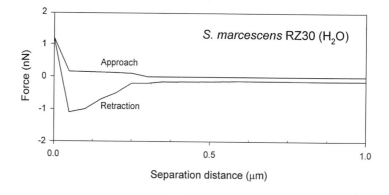

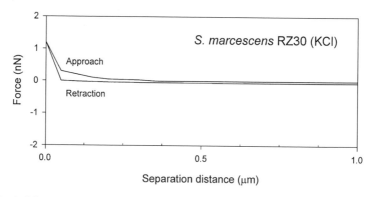

FIG. 1. Schematic presentation of atomic force microscopy on microbial cells immobilized in a membrane filter (insert) and force-distance curves for *S. marcescens* RZ30 immersed in water and in 0.1 *M* KCl.

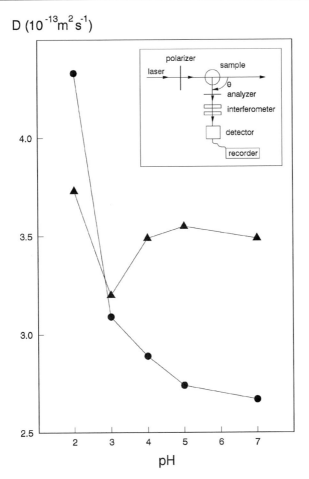

FIG. 2. Dynamic light scattering (insert) and the pH dependence of the diffusion coefficients as measured by dynamic light scattering of *S. salivarius* HB (●) and *S. salivarius* HBC12 (▲).

Dynamic Light Scattering: Dynamic Nature of Soft Cell Surface Layers

A fibrillated and a nonfibrillated oral streptococcal strain were suspended in potassium phosphate solutions of different pH to a density of approximately 5×10^6 bacteria ml^{-1}. Bacterial suspensions were put in a cuvette and the intensity of the light (wavelength 632.4 nm) was measured as a function of time at a detection angle of 90°, as schematically shown in Fig. 2. Intensity fluctuations due to Brownian motion are subsequently expressed in a so-called correlation function from which the diffusion coefficient of the suspended bacteria can be calculated[13,14]:

$$D_\infty = \frac{kT}{6\pi\mu r_h} \tag{2}$$

in which k is the Boltzmann constant, T is the absolute temperature, μ is the dynamic viscosity of the medium, and r_h is the hydrodynamic radius of the bacterium. In Fig. 2 it can be seen that the nonfibrillated bald *Streptococcus salivarius* HBC12 has a diffusion coefficient that is nearly independent of pH, whereas the fibrillated *S. salivarius* HB shows a diffusion coefficient that decreases with increasing ionic strength.[9] Likely, stabilizing electrostatic repulsion between fibrillar structures at higher pH values causes extension of the surface layer, yielding a larger hydrodynamic radius and consequently a smaller diffusion coefficient. Clearly, such a dynamic behavior of the cell surface of a nonfibrillar strain upon changing the ionic environment of the organism does not exist.

Particulate Microelectrophoresis: Electrophoretic Softness or Ion Penetrability

Oshima and Kondo[3,4] have developed a model that describes the electrophoretic mobility of particles, including bacteria,[7] for which the surface charge is distributed through an ion-penetrable surface layer.[15] In the Oshima model, the ion-penetrable layer is characterized by its charge density ρ and a parameter λ^{-1}, which is referred to as the "softness" of the ion-penetrable layer and which depends on the frictional force exerted on water when it flows through the ion-penetrable layer. For ion-impenetrable and ion-penetrable hard (i.e., $\lambda^{-1} \to 0$) bacterial cell surfaces, electrophoretic liquid flow is zero at the outermost cell surface and increasing exponentially with distance from the surface, whereas for the soft, ion-penetrable cell surface a substantial electrophoretic flow is assumed to have developed already in the ion-penetrable layer. In consequence, despite having a similar charge distribution and electric potential, the hard and soft ion-penetrable bacteria demonstrate distinctly different electrophoretic mobilities, as can be measured by particulate microelectrophoresis (see Fig. 3).

Electrophoretic mobilities of two staphylococcal strains with encapsulation, as demonstrated by India ink staining, suspended in KCl solutions (approximately 10^8 cells ml^{-1}) of different ionic strength, were measured as shown in Fig. 3 at an applied voltage of 150 V. Subsequently, the electrophoretic mobilities were

[13] Z. Xia, L. Woo, and T. G. M. Van de Ven, *Biorheology* **23**, 359 (1989).

[14] J. Lyklema, *in* "Fundamentals of Interface and Colloid Science," p. A11.1. Academic Press, London, 1991.

[15] A. van der Wal, M. Minor, W. Norde, A. J. B. Zehnder, and J. Lyklema, *Langmuir* **13**, 165 (1997).

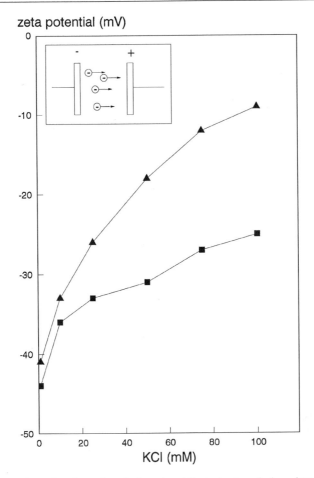

FIG. 3. Particulate microelectrophoresis (insert) and the ionic strength dependence of the electrophoretic mobility of *S. epidermidis* ATCC 35984 (▲) and *S. epidermidis* 3399 (■) in KCl.

fitted to

$$\mu = (\varepsilon_r \varepsilon_0 / \eta)[(\Psi_0 / \kappa_m + \Psi_{DON} / \lambda) / (1/\kappa_m + 1/\lambda)] + (zeN/\eta\lambda^2) \qquad (3)$$

in which μ is the electrophoretic mobility, ε_r the relative permittivity, ε_0 the permittivity of vacuum, η the viscosity of the solution, κ_m the Debye–Hückel parameter of the polymer layer, $1/\lambda$ a measure of the softness of the polyelectrolyte layer, z the valence of charged groups in the soft layer, e the electrical unit charge, N the density of charged groups in the soft layer, Ψ_0 the potential at the boundary between the polyelectrolyte layer and the surrounding solution, and Ψ_{DON} the

Donnan potential within the polyelectrolyte layer. Interestingly, from the dependences upon ionic strength of the two staphylococcal strains in Fig. 3, it is clear that *S. epidermidis* ATCC 35984 is more sensitive to ionic strength than strain 3399, corresponding to a softness of 1.7 nm and 5.0 nm, respectively. The softness of strain 3399 is likely due to charged polymers around the cell surface that are probably lacking around the cell surface of strain ATCC 35984. In addition, the more rigid strain ATCC 35984 had a slightly higher volumetric charge density (zN is $-0.026\ M$) than the softer strain 3399 (zN is $-0.021\ M$).

Synthesis

The mechanical cell surface properties of microorganisms as probed directly by the AFM can be interpreted in line with conclusions from dynamic light scattering and particulate microelectrophoresis. Mechanically soft cell surfaces respond more strongly to changes of their ionic environment than mechanically hard surfaces. This dynamic behavior of the outermost bacterial cell surface coincides with an electrophoretically defined softness of the outermost layer allowing electrophoretic liquid flow.

The dynamic behavior of microbial cell surfaces on changes in their ionic environment is of great importance for proper understanding of their interaction with surfaces and likely bears relation with older observations by Fletcher[16] showing alterations in contact area between adhering microorganisms and a substratum surface on changes of the ionic environment determined by interference reflectance microscopy. In addition, Morisaki *et al.*[17] reported that electrophoretically soft cell surfaces experience far less electrostatic repulsion than expected for a corresponding electrophoretically hard cell surface.

[16] M. Fletcher, *J. Bacteriol.* **170**, 2027 (1988).
[17] H. Morisaki, S. Nagai, H. Oshima, E. Ikemoto, and K. Kogure, *Microbiology* **145**, 2797 (1999).

[19] Application of Atomic Force Microscopy to Study Initial Events of Bacterial Adhesion

By Anneta Razatos

Introduction

The first and arguably most important step in the establishment of biofilm communities and surface-associated infections is bacterial adhesion. Bacterial adhesion consists of two kinetically distinct events: the initial, nonspecific, reversible, long-range interaction of bacteria with a surface, followed by specific, irreversible, short-range interactions such as receptor–ligand binding events.[1-4] Whether a bacterial cell is attracted to or repelled away from a surface depends on the balance between the long-range physiochemical forces of interaction, which include van der Waals, electrostatic interactions, short-range hydration and/or hydrophobic interactions, and steric effects due to the overlap of surface bound polymers.[5,6] Only when bacteria are in close proximity to a surface do specific, short-range interactions become significant. In other words, microorganisms attempting to colonize an unknown substrate must do so by reversible physicochemical interactions before any specific, irreversible binding events can occur.[7] The physicochemical forces of interaction that initially drive bacterial adhesion depend on properties and characteristics of both the bacterial cell and the substrate surfaces, as well as the nature of the intervening medium. Accordingly, control and prevention of bacterial adhesion requires an in-depth understanding of the surface properties and physicochemical forces of interaction involved in the initial events of bacterial adhesion.

Until recently, traditional bacterial adhesion studies have consisted of incubating a substrate in a cell suspension for a specified amount of time, rinsing the substrate, and counting the number of bacteria that remain attached to that substrate.[4,8,9] There are several problems associated with this experimental approach. First, the number of bacteria that remain associated with a substrate depends on both long-range, nonspecific physicochemical interactions and short-range biospecific interactions. Second, the data obtained from such studies are difficult to reproduce because they depend strongly on experimental protocol, especially rinsing conditions. Consequently, results from such studies are insensitive, inexact, and at best qualitative. Third, this experimental approach provides no quantitative information on the magnitude or nature of the forces of interaction between cells and surfaces. As a result, progress in bacterial adhesion entailed the development of new technologies capable of providing more quantitative data on the forces of interaction between bacteria and substrates. For this purpose, the

[1] G. D. Christensen, L. Baldassarri, and W. A. Simpson, *Methods Ezymol.* **253**, 477 (1995).
[2] J. W. Costerton, Z. Lewandowski, D. E. Caldwell, D. R. Korber, and H. M. Lappin-Scott, *Annu. Rev. Microbiol.* **49**, 711 (1995).
[3] A. G. Gristina, *Science* **237**, 1588 (1987).
[4] I. Ofek and R. J. Doyle, *in* "Bacterial Adhesion to Cells and Tissues." Chapman Hall, Inc., New York, 1994.
[5] Y. L. Ong, A. Razatos, G. Georgiou, and M. M. Sharma, *Langmuir* **15**, 2719 (1999).
[6] J. Israelachvili, *in* "Intermolecular and Surface Forces." Academic Press, San Diego, 1992.
[7] H. J. Busscher and H. C. van der Mei, *Adv. Dent. Res.* **11**, 24 (1997).
[8] M. Fletcher and J. H. Pringle, *J. Colloid Interface Sci.* **140**, 5 (1985).
[9] P. Gilbert, D. J. Evans, I. G. Duguid, and M. R. W. Brown, *J. Appl. Bacteriol.* **71**, 72 (1991).

tool best suited to directly measure the forces of interaction between colloid particles and planar substrates is the atomic force microscope.[10,11] Razatos *et al.*[12] developed an experimental technique using the atomic force microscope (AFM) to directly measure the forces of interaction between bacteria and planar substrates as bacteria initially approach a substrate. This chapter describes in detail the AFM-based methodology, proved to be a highly sensitive, reproducible, and quantitative means of investigating bacterial adhesion.[12–15]

The Atomic Force Microscope

The atomic force microscope (AFM) is a scanning probe microscope invented by Binnig, Quate, and Gerber in 1986 to image rigid surfaces with atomic resolution.[16–18] Atomic force microscopy is ideally suited for biological applications because it does not require a conductive surface. As a result, pretreatment of biological samples is minimal in comparison to other microscopic techniques and all experiments can be conducted under physiological conditions. In addition to imaging, the AFM can also be used to measure the forces of interaction between cantilever tips and planar substrates. In force mode, the surface of interest is advanced toward a stationary AFM cantilever with a sharp, pyramidal, silicon nitride tip (Fig. 1). Physicochemical forces of interaction are measured between the silicon nitride tip and the substrate of interest. Initially, the distance of separation between the cantilever and substrate is sufficiently large that no force of interaction is exerted on the cantilever by the substrate (Fig. 1A). However, as the surface approaches the cantilever, at a critical distance of separation, the cantilever tip will either be attracted to and deflected down toward the surface (Fig. 1C) or repelled by and deflected up away from the surface (Fig. 1B). During force measurements, tip deflection is monitored as a function of distance of separation. Tip deflection (nm) data can then be converted to force (nN) according

[10] W. A. Ducker, Z. Xu, and J. N. Israelachvili, *Langmuir* **10,** 3279 (1994).
[11] R. H. Yoon, D. H. Flinn, and Y. I. Rabinovich, *J. Colloid Interface Sci.* **185,** 363 (1997).
[12] A. Razatos, Y. L. Ong, M. M. Sharma, and G. Georgiou, *Proc. Natl. Acad. Sci. U.S.A.* **95,** 11059 (1998).
[13] A. Razatos, Y. L. Ong, M. M. Sharma, and G. Georgiou, *J. Biomaterials Sci., Polymer Ed.* **9,** 1361 (1998).
[14] A. Razatos, "Factors and Forces Involved in the Initial Events of Bacterial Adhesion as Monitored by Atomic Force Microscopy." Dissertation, the University of Texas at Austin, Austin, TX, 1999.
[15] A. Razatos and G. Georgiou, *in* "Handbook of Bacterial Adhesion" (Y. H. An and R. J. Friedman, eds.), p. 285. Humana Press. Totowa, NJ, 2000.
[16] G. Binnig, C. F. Quate, and C. Gerber, *Phys. Rev. Lett.* **56,** 930 (1986).
[17] S. A. C. Gould, B. Drake, C. B. Prater, A. L. Weisenhorn, S. Manne, G. L. Kelderman, H. J. Butt, H. Hansma, and P. K. Hansma, *Ultramicroscopy* **33,** 93 (1990).
[18] P. K. Hansma, V. B. Elings, O. Marti, and C. E. Bracker, *Science* **242,** 209 (1988).

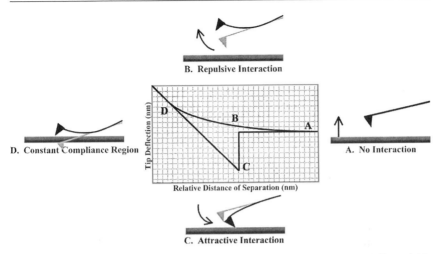

FIG. 1. Operation of AFM in force mode. The AFM cantilever (with a sharp pyramidal silicon nitride tip) is held stationary while the planar substrate is advanced toward the cantilever. (A) Cantilever and substrate are sufficiently separated so there is no interaction and hence no tip deflection. (B) Silicon nitride tip is repelled by substrate so cantilever is deflected up away from substrate. (C) Silicon nitride tip is attracted to substrate so cantilever is deflected down toward substrate. (D) Cantilever and substrate are in contact and move together. This portion of the tip deflection curve is known as the constant compliance region.

to Hooke's law by treating the cantilever as a spring with a characteristic spring constant.[19]

Forces of interaction between bacteria and substrates can be measured via two possible configurations: (1) bacteria can be irreversibly immobilized onto planar glass and probed by standard AFM cantilevers and (2) bacteria can be irreversibly immobilized onto AFM cantilever tips and used to probe planar substrates (Fig. 2). The immobilization protocol, which is described in the following section, results in uniform, stable, confluent bacterial layers on both planar glass and AFM cantilever tips. The presence of confluent bacterial cell layers on glass or cantilever tips is required to ensure that force measurements reflect interactions between substrates and bacterial cells. For example, if planar glass is completely coated with bacteria, then the AFM cantilever is certain to interact with bacterial cells and not the underlying support surface. Moreover, the immobilization protocol described herein produces stable bacterial layers such that individual bacterial cells are neither suspended in buffer solutions, nor dislodged under the force of the AFM probe.[12]

[19] W. A. Ducker, T. J. Senden, and R. M. Pashley, *Langmuir* **8,** 1831 (1992).

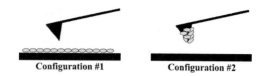

Configuration #1   Configuration #2

FIG. 2. Configuration #1: Silicon nitride tips of AFM cantilevers probing immobilized bacterial lawns. Configuration #2: Bacteria-coated tips of AFM cantilevers probing planar substrates.

Immobilization Protocol

Bacterial Cell Preparation

Bacteria grown to the desired cell density are collected by centrifugation and washed with phosphate buffered saline (PBS, pH 7.2). Cells resuspended in PBS are stirred into a 2.5% v/v glutaraldehyde solution for 2 hr at 4° at a density of 0.6–0.8 mg dry cell weight/ml.[20] The glutaraldehyde solution is prepared from a 25% v/v stock solution diluted to 2.5% v/v in PBS and purified by stirring with 50 mg/ml charcoal at 4° for 24 hr.[20] The pH of the glutaraldehyde solution is adjusted to 7.5 prior to the addition of charcoal. After treatment with glutaraldehyde, the bacterial cells are rinsed repeatedly in PBS and can be stored overnight at 4°.

Substrate Preparation

Prior to immobilization of bacteria, planar glass substrates and AFM cantilevers are coated with polyethyleneimine (PEI). PEI stock solution (100%, MW 1200) is diluted to 1% v/v in distilled, deionized water (ddH$_2$O); pH is adjusted to 8 with HCl. Glass to be coated with PEI is cleaned by soaking in 1 M HNO$_3$ overnight, rinsed with ddH$_2$O followed by methanol, and finally dried with sterile air. A drop of 1% v/v PEI is placed on one side of the glass and allowed to adsorb for 3 hr. The PEI solution is decanted; the glass slides are rinsed in ddH$_2$O and stored at 4°. Standard AFM cantilevers (Digital Instruments, Santa Barbara, CA) are immersed in 1% v/v PEI for 3 hr, rinsed in ddH$_2$O, and stored at 4°. In order to immobilize bacteria onto planar glass substrates, a concentrated drop of the glutaraldehyde-treated cells in 1 mM Tris buffer [tris(hydroxymethyl)aminomethane, pH 7.5] is placed on PEI-coated glass slips that are then placed in a vacuum desiccator at room temperature until excess water evaporates (2–3 hr); the cells themselves are not desiccated. To immobilize bacteria onto AFM cantilevers, a pellet of glutaraldehyde-treated cells is manually transferred onto the PEI coated tips. The pellet is further treated with a drop of 2.5% v/v glutaraldehyde and incubated at 4° for 1–2 hr. Bacteria-coated cantilevers are rinsed in ddH$_2$O and excess water is allowed to evaporate at room temperature.

[20] A. Freeman, S. Abramov, and G. Georgiou, *Biotechnol. Bioeng.* **52,** 625 (1996).

Force Measurements

AFM Operation

The AFM-based methodology was developed using a Nanoscope III Contact Mode AFM and Nanoprobe cantilevers with silicon nitride tips (Digital Instruments, Santa Barbara, CA). Experiments can be conducted in an AFM fluid cell (Digital Instruments, Santa Barbara, CA) in which the AFM cantilever and planar substrate are completely immersed in solution. Force measurements are carried out by first engaging the AFM without touching the tip to the substrate surface. The substrate, mounted on a piezo motor, is then approached toward the AFM cantilever in 100-nm increments with a specified Z scan size of 300 nm at a frequency of 1 Hz until a tip deflection curve is recorded as a function of distance of separation. The bacterial coated substrates are imaged with the AFM after every force measurement to confirm the presence of confluent bacterial layers. Bacterial coated AFM cantilevers are imaged under SEM after every force measurement to confirm the presence of bacterial cells at the apex of the tip. Fresh, clean, cantilevers and new substrates should be used for every force measurement.

Data Analysis

Data are initially acquired in terms of tip deflection (nm) vs relative distance of separation (nm). All deflection curves are normalized so that tip deflection is zero where there is no interaction, and the slope of the constant compliance region (portion of deflection curve where the cantilever moves with the surface; Fig. 1D) is equal to the rate of piezo displacement.[19] Multiple curves can be plotted together by aligning the zero deflection and constant compliance regions of the data curves resulting in tip deflection (nm) vs relative distance of separation (nm) curves. In order to convert data from tip deflection (nm) vs relative distance of separation (nm) to force (nN) vs distance of separation (nm), the point of contact between the AFM tip and the sample has to be established. The point of contact between the surface and the tip, zero distance of separation, is defined as the onset of the constant compliance region. Distance of separation (nm) is calculated as the sum of tip deflection and piezo position relative to zero distance of separation.[19] Force is calculated by treating the cantilever as a spring with a characteristic spring constant (k in nN/nm) according to Hooke's law: $F = k \Delta Y$ where ΔY is tip deflection.[19] Values for the spring constants of AFM cantilevers used to convert tip deflection data can be obtained from the manufacturer or can be determined experimentally according to Senden and Ducker.[21] For example, the spring constant reported for the long thin cantilever manufactured by Digital Instruments (Santa Barbara, CA) is $k = 0.06$ nN/nm.

[21] T. J. Senden and W. A. Ducker, *Langmuir* **10**, 1003 (1994).

Discussion

Control Studies

Extensive control studies were carried out to ensure that the immobilization protocol itself did not alter the physicochemical properties of the bacterial cell surface and did not introduce artifacts into AFM force measurements.[5,12,14] Physicochemical properties of surface free energy and surface charge density were evaluated by contact angle and zeta potential measurements, respectively. Contact angle measurements, and hence surface free energies, were identical for bacterial cells before and after glutaraldehyde treatment.[5,12] Similarly, zeta potential measurements confirmed that glutaraldehyde treatment did not alter the surface charge density of bacterial cells.[5,12] Moreover, traditional bacterial adhesion studies consisting of enumerating cells remaining attached to substrates confirmed that bacterial adhesion was not affected by glutaraldehyde treatment.[14] Finally, control studies using the AFM verified that PEI did not alter the AFM cantilever tip in such a way as to affect AFM force measurements.[5]

In addition to the aforementioned control studies, standard deviations calculated for multiple tip deflection curves were small in comparison to other bacterial adhesion techniques illustrating the reproducibility of this experimental technique.[15] Moreover, analysis of tip deflection curves confirmed that the bacterial lawns do not become deformed under the force of the AFM probe.[15] In other words, bacterial cells are not elasticity deformed during force measurements, but rather behave as rigid interfaces. Therefore, tip deflection curves do in fact depict the physicochemical forces of interaction between bacteria and substrates, and not artifacts due to sample deformation or introduced by the immobilization protocol.

Agreement of Two Configurations

Silicon nitride tips of AFM cantilevers are hydrophilic and negatively charged at pH 7.5.[22,23] The surface properties of silicon nitride are known to be similar to those of glass.[22,23] As expected, force curves measured between silicon nitride tips and immobilized bacterial lawns (configuration #1 of Fig. 2) were in excellent agreement with force curves measured between bacteria-coated cantilevers and clean glass (configuration #2 of Fig. 2).[12,13] In both cases, the forces measured by the atomic force microscope reflect the interaction between bacteria and anionic, hydrophilic substrates.

[22] T. J. Senden and C. R. Drummond, *Colloids Surf. A* **94**, 29 (1995).

[23] C. B. Prater, P. G. Maivald, K. J. Kjoller, and M. G. Heaton, *in* "Probing Nano-scale Forces with the Atomic Force Microscope." Digital Instruments, Santa Barbara, CA, 1995.

Interpretation of Results

Figure 3 has been included in this chapter as an example of typical AFM data. These tip deflection curves are representative of silicon nitride tips probing bacterial lawns or bacteria-coated tips probing planar substrates. Figure 3A is a plot of tip deflection vs relative distance of separation depicting both repulsive (upper curve) and attractive (lower curve) interactions. The tip deflection data in Fig. 3A is converted to force vs distance of separation (Fig. 3B) using a spring constant of $k = 0.06$ nN/nm. In Fig. 3B, the attractive force of interaction measured by the atomic force microscope is roughly -1 nN (Fig. 3). In previous bacterial adhesion

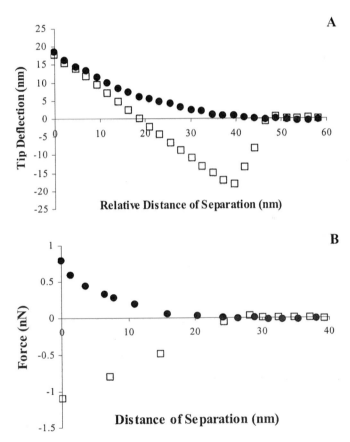

FIG. 3. Typical results from AFM force measurements. (A) Tip deflection (nm) vs relative distance of separation (nm) depicting repulsive (upper curve) and attractive (lower curve) interactions. (B) Tip deflection data from (A) is converted to force (nN) vs distance of separation (nm) using the spring constant $k = 0.06$ nN/nm.

studies, attractive forces of interaction measured by the AFM ranged from −0.25 to −50 nN.[5,14]

Data from AFM measurements are presented as a function of distance of separation depicting the behavior of bacteria as cells initially approach a surface. Therefore, results from AFM studies are quantitative and much more informative in comparison to traditional bacterial adhesion studies in which data are presented as relative numbers of cells. Moreover, AFM results in terms of force vs distance can be compared to theoretical predictions based on colloid theory.[5,6]

In previous experiments, the AFM-based methodology was successful in determining the contribution of specific physicochemical forces and biological factors involved in the initial events of bacterial adhesion. For example, the contribution of electrostatic interactions to bacterial adhesion can be determined by changing the salinity or pH of the intervening medium.[12,23] Similarly, Razatos *et al.*[12] used the AFM to determine the role of lipopolysaccharide molecules coating the bacterial cell surface in adhesion. In this study, the AFM methodology was sensitive enough to detect differences in bacterial adhesion due to subtle changes in lipopolysaccharide composition that were not detected by conventional bacterial adhesion techniques.[12,14]

Conclusions

Advantages and Disadvantages

In summary, results from AFM-based bacterial adhesion studies are quantitative, sensitive, and reproducible. The AFM provides an excellent dynamic representation of how bacteria interact with planar substrates as a function of distance of separation. Moreover, the AFM technique allows for the dissection of specific physicochemical forces and biological factors involved in the initial events of bacterial adhesion. The down side is that the AFM-based methodology is more labor intensive than traditional bacterial adhesion techniques. Bacteria-coated cantilevers and bacteria-coated planar glass can only be used once, necessitating the preparation of numerous samples. Once samples are prepared, the actual AFM measurements take 0.5–1.0 hr. After every force measurement, bacteria-coated tips must be imaged under SEM to verify the presence of bacterial cells. Traditional bacterial adhesion assays are appropriate when the goal is to simply determine if a bacterial strain "sticks" to a surface.

Limitations

The following considerations need to be kept in mind when determining the forces of interaction between bacteria and substrates using atomic force microscopy: (1) In the conversion of tip deflection vs relative distance of separation curves to force vs distance of separation curves, the point of contact between the

cantilever and the substrate is chosen as the onset of the constant compliance region. The point of contact, however, is not known with certainty, and therefore
the calculated distance of separation is not absolute.[6] (2) The exact tip geometry
and radius of curvature of the contact area are unknown, not only for AFM tips
modified with bacterial cells, but for silicon nitride tips of AFM cantilevers in general, thus complicating direct comparison of AFM results [in terms of force (nN)]
to theoretical predictions [in terms of force over radius of curvature (nN/nm)].[6]
Despite these limitations, the AFM-based methodology is a significant improvement over traditional bacterial adhesion assays facilitating quantitative analysis of
the dynamic behavior of bacteria at solid/liquid interfaces.

[20] Study of Biofilm within a Packed-Bed Reactor by Three-Dimensional Magnetic Resonance Imaging

By Marion Paterson-Beedle, Kevin P. Nott, Lynne E. Macaskie, and
Laurance D. Hall

Introduction

Magnetic resonance imaging (MRI) has been predominantly associated with
clinical medicine; it is a powerful technique because it can provide, *in vivo,* noninvasively, and nondestructively, spatially resolved information about living tissues.
However, the potential application of MRI to problems of industrial interest is now
an area of increasing activity. Thus, it is well recognized that the MRI technique
can be used to help optimize product properties and to study processing operations
in situ, e.g., adsorption, transport, and reaction processes.

It can be used to study chemical and physical properties of small samples and
for spatial mapping, or imaging, of larger systems. For example, Fig. 1a shows a
static three-dimensional (3D) MR image of the distribution of water that fills the
pore space of a packed-bed column, containing ceramic raschig rings, a support
typically used in the chemical industry. Figure 1b shows a single raschig ring. MRI
is capable of imaging not only the packing distribution but also the flow processes
occurring within the reactor. This prompts the question: What is the potential of the
three-dimensional MRI technique to study the behavior of a bioreactor containing
a biofilm immobilized onto the support?

The importance of biofilms to industrial, medical, and environmental systems
has greatly motivated research into the understanding of their development, structures, and physical properties, and of the transport processes occurring within
biofilm reactors. However, in the literature there are few citations available that

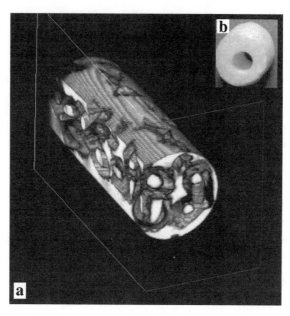

FIG. 1. (a) A 3D MR image (obtained using a 31 cm bore magnet) of the distribution of Cu^{2+} (10 mM) doped water occupying the pore space of a packed-bed column (length 9 cm, internal diameter 1.5 cm), containing ceramic raschig rings. The image has been "volume rendered" to produce a 3D image; areas shown in white are from water (interstitial between the rings and occupying the ring's lumina) and those in mottled gray show water within the porous matrix of the rings (pore size ca. 100 Å). (b) Photograph of a single raschig ring (length 7 mm, external diameter 6.6 mm, and internal diameter 2 mm).

describe the applications of the MRI technique to the study of biofilm, and most of those studies have utilized two-dimensional imaging. Gladden and Alexander[1] and Lens and Hemminga[2] have reviewed the application of nuclear magnetic resonance (NMR) imaging in process and environmental engineering, respectively. Lewandowski et al.[3] used MRI to study flow velocity distributions in a flat plate reactor as a function of biofilm formation. Hoskins et al.[4] used MRI relaxation techniques for the imaging of biofilms in aqueous (nonporous and porous) systems.

An enzymatically mediated metal-accumulation process, which uses a strain of *Citrobacter* sp., has been extensively studied for the potential application in

[1] L. F. Gladden and P. Alexander, *Meas. Sci. Technol.* **7**, 423 (1996).

[2] P. N. L. Lens and M. A. Hemminga, *Biodegradation* **9**, 393 (1998).

[3] Z. Lewandowski, P. Stoodley, and S. Altobelli, *Wat. Sci. Technol.* **31**, 153 (1995).

[4] B. C. Hoskins, L. Fevang, P. D. Majors, M. M. Sharma, and G. Georgiou, *J. Magn. Reson.* **139**, 67 (1999).

the rehabilitation of metal-bearing wastewaters.[5,6] *Citrobacter* sp. produces an atypical heavy metal–resistant acid-type phosphatase that catalyzes the hydrolysis of an organic "donor" molecule (e.g., glycerol 2-phosphate), incorporated to the metal solution, releasing inorganic phosphate (HPO_4^{2-}), thus allowing the metal cations (M) to precipitate, stoichiometrically, as $MHPO_4$. That system will be used here as an illustrative model, since *Citrobacter* biofilm reactors have been well described and mathematically modeled to obtain a description of the biochemical and chemical processes within.[6-9] However, such "black box" approaches cannot provide the dynamic and spatial information that can be yielded by MRI. Figure 2a shows a 3D MR image of the static fluid distribution around and within the pore spaces of a packed-bed reactor containing *Citrobacter* biofilm immobilized onto cubes of polyurethane reticulated foam[8] and challenged with a sodium glycerol 2-phosphate (5 m*M*) solution in sodium citrate buffer (2 m*M*), pH 6.0. Figure 2b shows a two-dimensional (2D) slice through a 3D image data set of a similar packed-bed reactor challenged with a lanthanum nitrate solution (1 m*M*) in sodium citrate buffer (2 m*M*), pH 6.0. Figure 2c shows a single cube of polyurethane reticulated foam.

This chapter describes the 3D MRI technique used to acquire the images illustrated in Figs. 1 and 2. It also provides an overview of the underlying principles of MRI and of some of the limitations of the technique with respect to imaging of biofilms. Finally, other applications of MRI, which are relevant to the study of bioreactors, are described. It is important to emphasize that no attempt has been made to provide a comprehensive array of the literature; however, references to publications from the laboratories of the authors have been included to provide at least some direct access to other areas.

Reactors

Supports

Ceramic raschig rings based on aluminium oxide (Product number SA6575, length 7 mm, external diameter 6.6 mm, internal diameter 2 mm, and pore size ca. 100 Å) and Filtren TM30, a polyurethane reticulated foam (1 cm^3 cubes), supplied, respectively, by Norton Chemical Process Products Corporation (USA)

[5] L. E. Macaskie, B. C. Jeong, and M. R. Tolley, *FEMS Microbiol. Rev.* **14**, 351 (1994).

[6] L. E. Macaskie, P. Yong, T. C. Doyle, M. G. Roig, M. Diaz, and T. Manzano, *Biotechnol. Bioeng.* **53**, 100 (1997).

[7] L. E. Macaskie, R. M. Empson, F. Lin, and M. R. Tolley, *J. Chem. Technol. Biotechnol.* **63**, 1 (1995).

[8] J. A. Finlay, V. J. M. Allan, A. Conner, M. E. Callow, G. Basnakova, and L. E. Macaskie, *Biotechnol. Bioeng.* **63**, 87 (1999).

[9] P. Yong and L. E. Macaskie, *J. Chem. Technol. Biotechnol.* **74**, 1149 (1999).

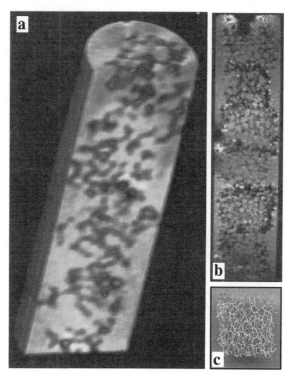

FIG. 2. (a) A 3D MR image (volume rendered) of the distribution of fluid occupying the pore space of a packed-bed column (length 9 cm and internal diameter 1.5 cm), containing *Citrobacter* biofilm immobilized onto cubes of polyurethane foam, challenged with a sodium glycerol 2-phosphate solution (feedstock) (5 mM) in sodium citrate buffer (2 mM), pH 6.0. The image was obtained using a 31 cm bore magnet. (b) A 2D slice through a 3D image data set of a packed-bed reactor, containing *Citrobacter* biofilm immobilized onto cubes of polyurethane foam, and challenged with a solution of lanthanum nitrate (1 mM) in sodium citrate buffer (2 mM), pH 6.0. The image was obtained using a 31 cm bore magnet. Note that extensive metal bioprecipitation (which would block the pores) has been avoided by the omission of feedstock, thereby maintaining the biofilm in a "resting" state. (c) Photograph of a single polyurethane reticulated foam (1 cm^3) prior to biomass colonization (invisible by MRI because it contains no water).

and Recticel (Belgium), are used as example supports for biofilm colonization and growth.

Biofilm Production

The conditions for the controlled growth of *Citrobacter* sp. biofilm have been established.[8] *Citrobacter* sp. cells are grown in a carbon-limiting continuous culture in an air-lift fermenter containing an appropriate support (e.g., cubes of polyurethane foam or raschig rings) for biofilm development. Under these conditions, *Citrobacter* sp. cells adhere to the surface of the support, aggregate, and grow in a hydrated polymeric matrix of their own synthesis to form a biofilm of heterogeneous "stack and channel" format[10] as described in the literature.[11]

Reactor Packed-Bed Systems

1. For Fig. 1, ceramic raschig rings were packed in a cylindrical glass column, working volume ca. 13 ml, and the column filled with water that had been doped with copper sulfate (10 mM) to provide greater contrast (see later).

2. For Fig. 2, cubes of polyurethane foam containing *Citrobacter* biofilm were packed into similar glass columns as described in (1). The columns were challenged for 60 hr with a solution of sodium glycerol 2- phosphate (5 mM) or, here, lanthanum nitrate (1 mM) in sodium citrate buffer (2 mM), pH 6.0. The flow rates (ca. 10 ml/hr) were controlled by an external peristaltic pump (Watson-Marlow, 505U).

3. For Fig. 10a, a similar reactor was prepared as described in step 2. The column was challenged with a solution of sodium glycerol 2-phosphate (5 mM) in sodium citrate buffer (2 mM), pH 6.0.

Principles of MRI

A brief overview of the principles of the MRI technique, which are relevant to this work, is given here. More detailed discussions on the subject can be found in Foster and Hutchinson,[12] Partain *et al.*,[13] Callaghan,[14] Westbrook and Kaut,[15] and Hashemi and Bradley.[16] The physical principles of nuclear magnetic resonance (NMR) are common to MRI and knowledge of magnetic fields, their origin, their interactions, and their measurement is required to understand the basis of the NMR phenomenon and its limitations.

[10] V. J. M. Allan, "The structure, formation, and activity of a *Citrobacter* N14 biofilm." Ph.D. thesis, University of Birmingham, U.K. 1999.
[11] J. W. Costerton, Z. Lewandowski, D. DeBeer, D. Caldwell, D. Korber, and G. James, *J. Bacteriol.* **176,** 2137 (1994).
[12] M. A. Foster and J. M. S. Hutchinson, "Practical NMR Imaging." IRL Press, Oxford, 1987.
[13] C. L. Partain, R. R. Price, J. A. Patton, M. V. Kulkarni, and A. E. James, Jr., "Magnetic Resonance Imaging." 2nd ed., Vol. 2. W. B. Saunders Company, Philadelphia, 1988.
[14] P. T. Callaghan, "Principles of Nuclear Magnetic Resonance Microscopy." Clarendon Press, Oxford, 1991.
[15] C. Westbrook and C. Kaut, "MRI in Practice." Blackwell Scientific Publications, Oxford, 1993.
[16] R. H. Hashemi and W. G. Bradley, "MRI: The Basics." Williams and Wilkins, Baltimore, 1997.

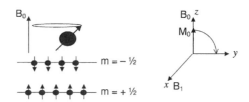

FIG. 3. The application of a static magnetic field (B_0) aligns the nuclear spins either parallel ($m = +1/2$) or antiparallel ($m = -1/2$) to the field; the energy difference between the two states results in a population difference producing a net magnetization in one direction. Circle denotes precessional motion of the net magnetization vector. The application of a time-varying magnetic field (B_1) with a frequency equal to that of the B_0 field will produce magnetization on the transverse plane; this is detected as a voltage induced in an RF-receiver coil near the sample and constitutes the magnetic resonance signal.

Nuclear Magnetic Resonance

Nuclei that contain an odd number of protons and neutrons, such as the proton (^{1}H), possess a property known as nuclear spin. When such nuclei are subjected to a static external magnetic field, B_0, the individual magnetic moments of each spin will tend to align either parallel or anti-parallel with the applied field (represented by quantum numbers $+1/2$ and $-1/2$, respectively; Fig. 3). The energy difference between the two states creates a population difference that gives rise to an overall magnetization (the net magnetization vector, $\mathbf{M}$). For $\mathbf{M}$ to undergo a precessional motion, a component of the magnetic moment needs to be perpendicular to B_0 (M_t). The frequency of the precessional motion is called the Larmor frequency (ω_0) and is defined by

$$\omega_0 = \gamma B_0 \tag{1}$$

where γ is the gyromagnetic ratio of the nucleus under study and B_0 is the strength of the magnetic field. Every nucleus possesses a unique gyromagnetic ratio and hence a unique Larmor frequency, thereby explaining the element-specific nature of the NMR technique. The hydrogen nucleus contains a single proton and is the most useful nucleus used in NMR because of its abundance in nature, e.g., water in biological tissues, and because it gives the greatest NMR signal of all the nuclei susceptible to the NMR phenomenon.

The magnetization $\mathbf{M}$ can be displaced away from the B_0 direction by applying an additional time-varying magnetic field (B_1) perpendicular to B_0. This is accomplished by applying short pulses (microsecond to millisecond) of radiofrequency (RF) radiation from a transmitter coil; the angle (θ) through which the magnetization vector is tipped from the direction of B_0 is proportional to the total amount of power applied. When the frequency of the B_1 field is equal to the Larmor frequency of the B_0 field, a condition called resonance is produced in which the RF radiation can interact with the magnetic nuclei (spins). Two independent processes will take place: (1) a change in the populations of the two spin states and, as a consequence, in the net longitudinal magnetization (z direction in Fig. 3); (2) phase coherence in the precessional motion of the spins, and therefore a net gain in magnetization within the transverse plane (x,y-plane in Fig. 3). The magnetization in that transverse plane is detected as a voltage induced in an RF receiver coil near the sample; this constitutes the magnetic resonance signal. Typically, complete rephasing of the spins in the transverse plane caused by a "90° pulse" gives a maximum NMR signal, whereas complete inversion of magnetization in the longitudinal direction (inversion of spin populations) caused by a "180° pulse" gives a minimum, or no, signal. The magnitude of a signal is directly proportional to the number of resonant spins in the system under study and the strength of the external magnetic field; hence, NMR is an inherently a quantitative technique.

Relaxation Processes

As described above, the magnetization of the spin system is perturbed from its equilibrium by the application of RF energy. Two relaxation processes are involved in the return of the magnetization of the spin system to equilibrium, each characterized by its own time constant:

1. Longitudinal relaxation, with a time constant T_1. In this process, the spin system loses energy to the surrounding lattice. The longitudinal relaxation is responsible for the return of the longitudinal component of magnetization (relative to the direction of the magnetic field) to its equilibrium value. T_1 is also known as the spin lattice relaxation time.
2. Transverse relaxation, with a time constant T_2. In this process the spin system loses no energy as a whole, but the individual spins exchange energy among themselves. The net effect is a gradual loss of phase coherence in the precessional motion of the spins, which results in a loss of transverse magnetization. T_2 is also known as spin–spin relaxation.

During the spin–lattice relaxation (T_1), the loss of energy from the spin system as a whole also causes a decay of transverse magnetization; consequently, the observed T_2 depends on both processes and, as a result, T_2 can never exceed T_1. The T_1 and T_2 properties of the nuclei in a liquid are descriptive of energy exchange and are dependent on the extent of interaction between the spins and their surrounding thermal reservoir or lattice. As a result, measurement of T_1 and T_2 of a liquid in a porous medium can be used to study the physical and chemical properties of those materials.

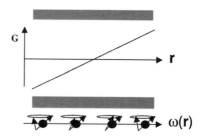

Fig. 4. The magnetic field gradient (**G**) across a sample space (**r**) acts to produce a similar spatial dependence in the Larmor frequency of the spins $\omega(\mathbf{r})$.

Spatial Localization of the MR Signal

To observe magnetic resonance phenomena of nuclei as a function of their position in real space (i.e., to produce an image) a small magnetic field gradient, **G**, must be applied to the uniform polarizing field $\mathbf{B}_0$. As a result, the resonant frequency of the nuclear spins is then dependent on the spatial position of the molecules within the sample and is given by Eq. (2):

$$\omega(\mathbf{r}) = \gamma B_0 + \gamma \mathbf{G} \cdot \mathbf{r} \tag{2}$$

where **r** is the position vector of the nuclear spin (Fig. 4). MRI uses three orthogonal magnetic field gradients for spatial localization of an object in three dimensions.

MRI is essentially a line scanning technique, which can be demonstrated using the original experiment carried out by Lauterbur[17] on two tubes of water (see Fig. 5); the caption summarizes the basic mechanism. In practice, a number of one-dimensional projections is required in order to define the position of the tubes in two-dimensional space; usually that 2D is for a defined slice. Many more projections and hence much greater scan times are required to produce a high-resolution 3D image.

Although Lauterbur's "projection–reconstruction" methodology has been largely replaced by Fourier methods of image acquisition and reconstruction, the basic principles remain the same. This is a line scan method and each additional scan induces higher resolution in the final image. The image can be produced by a variety of pulse sequences, which consist of a number of carefully timed RF and magnetic field gradient pulses (see Hashemi and Bradley[16]). The most commonly used scan methods are the spin and gradient echo sequences described below.

Spin Echo Imaging

Inhomogeneities of the $\mathbf{B}_0$ magnetic field will cause spins to precess at different Larmor frequencies in different parts of the sample; those inhomogeneities arise

[17] P. C. Lauterbur, *Nature* **242**, 190 (1973).

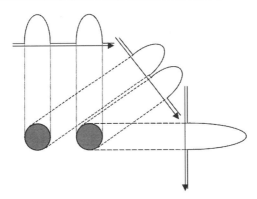

FIG. 5. Lauterbur's "projection–reconstruction" experiment using two tubes of water. Each arrow represents the direction of the magnetic field gradients used to encode the NMR signal with spatial information, and the associated profiles represent the one-dimensional projections of the tubes onto the direction of the gradient. Projecting those profiles back along the gradients (dotted lines) creates the two-dimensional MR image of the tubes.

from imperfections in the main magnetic field as well as variations in magnetic field susceptibility within the sample. This induces dephasing of transverse magnetization after the application of a 90° pulse (θ), in addition to the T_2 relaxation process (Fig. 6). However, the two decay processes differ in that the loss of phase coherence due to magnetic field inhomogeneities is inherently reversible. Thus, application of a 180° pulse can reverse the dephasing process, resulting in a phase

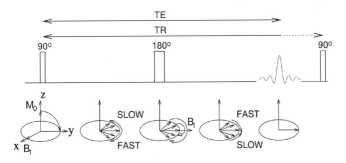

FIG. 6. The spin echo pulse sequence. The 90° pulse produces magnetization in the transverse (x,y) plane, which dephases because of T_2 relaxation and magnetic field inhomogeneities. The loss of phase coherence through field inhomogeneities is reversed by application of the 180° pulse; this produces a phase coherence known as a spin echo, whose magnitude is attenuated by T_2 relaxation. A period (the repetition time, TR) must be left before the next acquisition to enable the longitudinal magnetization to recover (by T_1 relaxation). The total time between the 90° pulse and the echo is called the echo time (TE).

coherence known as a "spin echo," solely due to T_2 relaxation. The time between the 90° and 180° pulses is the same as that between the 180° pulse and the spin echo, and the total time between the 90° pulse and echo is called the echo time, TE. Acquisition of a series of scans for different values of TE produces a series of echoes whose amplitudes are modulated by transverse relaxation and have an associated decay constant, T_2.

A simple one-dimensional (1D) spin echo projection is acquired by the application of a "frequency encoded" gradient in one direction (described above) as the spin echo information is acquired. The image, or spatial distribution of frequency, is formed by Fourier transformation of the echo, or time domain signal.

Gradient Echo Imaging

A gradient echo pulse sequence utilizes a variable RF excitation pulse and, therefore, can flip the net magnetization vector through an angle (θ) less than 90° (Fig. 7). A transverse component of magnetization is created, the magnitude of which is less than in spin echo, and the longitudinal magnetization is only partially converted to the transverse plane. Application of a 180° pulse would produce a negative magnetization that has a long T_1 recovery; in contrast, for a spin echo experiment this component is very small. This is not appropriate for fast imaging (see later), so instead a frequency encoding (or read) gradient is initially applied negatively to speed up the dephasing of the signal, and then its polarity is reversed in order to rephase the magnetization to produce a "gradient echo" (Fig. 7). As before, the signal is Fourier transformed to form a 1D projection.

The gradient echo sequence differs from the spin echo sequence in that dephasing due to magnetic field inhomogeneities is not reversed by a gradient refocusing pulse; consequently, gradient echo sequences have a greater sensitivity toward magnetic susceptibility differences within the sample. Thus, transverse magnetization decays faster, with a new time constant T_2^* that is shorter than T_2.

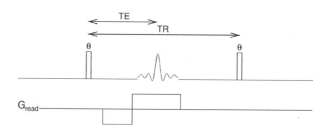

Fig. 7. The gradient echo sequence differs from the spin echo sequence in that it uses inversion of the direction of a magnetic field gradient to refocus the magnetization rather than the 180° refocusing pulse. The frequency encoding gradient is initially applied negatively to speed up the dephasing of the signal, and then its polarity is reversed in order to rephase the magnetization to produce a gradient echo.

MRI Equipment

Since the equipment used for MRI has been described elsewhere (for instance, Morris,[18] Attard and Hall,[19] or for MR microscopy, Callaghan[14]), only a brief description is given here. Bushong[20] has detailed the practicalities of installing a MRI scanner.

The majority of MR scanners use a superconducting magnet (Fig. 8), the coils of which operate near absolute zero temperatures and, therefore, require liquid helium and nitrogen. The strong field external to the magnet dictates that the building and surrounding objects be largely free of iron.

Within the cylindrical magnet bore are other cylindrical formers, which house the shim and gradient coils. The shim coils allow the B_0 magnetic field inhomogeneity to be optimized. The gradient coils provide the magnetic field gradients in the x, y and z, directions used for spatial encoding of the MR information. Within the former of the gradient coils is the RF probe that is used both to transmit pulses of RF radiation to manipulate the magnetization of the nuclei and to receive the signal from the sample. The shape of the coils in the probe can be matched to the dimensions of the object and can vary from a basic RF antenna or loop of wire to a more complex cylindrical cage.

The magnet system is generally housed in a screened enclosure (Faraday cage), usually made of copper mesh, which shields the radiofrequency receiver coil from outside RF interference. The rest of the MR system must be placed at a reasonable distance outside the cage since the external magnetic field can, for example, corrupt computer hard disks and distort computer screens. All power and control lines entering the Faraday cage must be filtered to reject noise.

Magnet Size and Imaging Volume

The region of acceptable field homogeneity within a magnet limits the maximum volume of object that can be scanned. Most of the images in this article were acquired using a 2.35 Tesla, 31 cm bore superconducting magnet (Oxford Instruments, Oxford, U.K.) (Fig. 8, top), which has a useful imaging volume equivalent to that of a 6 cm diameter sphere. In contrast, the 2 Tesla, 100 cm bore (whole body) superconducting magnet has a useful imaging volume of approximately a 25 cm sphere, whereas a 7 Tesla, 9 cm bore magnet (such as that used for NMR spectroscopy) has a useful imaging volume of approximately a 2 cm sphere. In the Herchel Smith Laboratory for Medicinal Chemistry, such

[18] P. G. Morris, "Nuclear Magnetic Resonance Imaging In Medicine and Biology." Oxford Science Publications, Oxford, 1986.

[19] J. J. Attard and L. D. Hall, *in* "Encyclopedia of Applied Physics." (G. L. Trigg, ed.), Vol. 9, p. 47. Wiley-VCH, Berlin, 1994.

[20] S. C. Bushong, "Magnetic Resonance Imaging: Physical and Biological Principles." Mosby, London, 1996.

FIG. 8. (*Top*) End view of the 31 cm bore magnet; the white cylinder inside the magnet bore houses a 10 cm diameter set of gradient coils. (*Bottom*) Photograph of a 31 cm bore magnet (Oxford Instruments, Oxford, U.K.) inside a copper mesh Faraday cage, and connected to an MRI console (Bruker, Karlsruhe, Germany) outside the cage.

magnets are each connected to Bruker BMT (Bruker Medzintechnik Biospec II, Karlsruhe, Germany) imaging consoles (Fig. 8, bottom). The data set of a 2D image is on the order of tens of kilobytes and that of a 3D image is in the order of tens of megabytes. Since a standard PC computer has a storage capacity of at least 4.3 gigabytes, those can be used to store and process MRI data. Usually, the data are compressed and archived on a CD with storage capacity of 650 megabytes.

Gradient Set and Spatial Resolution

A number of factors should be taken into account when choosing a gradient set; those include the size of the sample, ease of access to it, the required spatial resolution, and for shorter echo time (TE) imaging or complex pulse sequences, fast gradient switching.

Since gradient power supplies are complex and expensive, smaller gradient sets are generally used to produce the large gradient strengths required for higher spatial resolution. Since that limits the size of the object that can be accommodated within the gradient set, high resolution is generally achievable for smaller samples. In addition, smaller gradient sets tend to be more suited for the fast gradient switching required for short echo time imaging (for short T_2 samples; see below). The manufacture of the gradient sets needs to be robust since the pulses of electric current through the wires generate large forces that exert large pressure on the formers that hold the coils.

The spatial resolution achievable for modest scan times (minutes) on the 9, 31, and 100 cm bore magnets is submillimeter, ca. 0.02, 0.1, and 0.5 mm, respectively. It is known from previous work that the depth of hydrated *Citrobacter* biofilm (grown on glass slides) ranges between 20 and 150 μm, measured using a confocal laser microscope.[10] Hence resolution of 20 μm could be achievable for a biofilm using a 9 cm bore magnet, provided that a small sample was used with a tightly fitting RF coil. However, for MRI of larger volumetric elements (e.g., bulk flow volumes) a 31 cm bore magnet is required (e.g., Figs. 1a, 2a, and 2b). For instance, Lewandowski *et al.*[3] used a 1.89 Tesla, 31 cm bore magnet to determine the flow velocity distributions in a reactor with and without biofilm, but imaging of the flow patterns around and within the biofilm itself would require a 9 cm bore magnet.

Practicalities of Image Acquisition

Spatial vs Temporal Resolution and Signal-to-Noise

In general, the time required to scan an object is given by Eq. (3):

$$\text{Acquisition time} = \text{TR } N_y N_A N_z \qquad (3)$$

where TR is the repetition time; N_y is the number of pixels in the second dimension; N_A is the number of averages; and N_z is the number of pixels in the third dimension (the number of pixels in the first dimension, usually 128 or 256, has no bearing on the scan time). Therefore, a two-dimensional image must have N_y "line scans," each separated by the repetition time, TR; a three-dimensional image must have $N_y \times N_z$ "line scans," each separated by TR.

Typically a two-dimensional, 256×128 pixel, gradient echo image can be acquired for a defined slice thickness of 2–5 mm, in 26 sec (TR = 100 msec, $N_y = 128$, $N_A = 2$, $N_z = 1$). A three-dimensional $256 \times 128 \times 128$ pixel matrix can be acquired over the sample of interest in 13 min and 39 sec (TR = 50 msec, $N_y = 128$, $N_A = 1$, $N_z = 128$). The experimental times can be reduced further at the expense of accuracy (lowering TR and N_A) and/or spatial resolution (lowering N_y and N_z).

The signal-to-noise ratio (SNR) of a spin echo image can be related to the MRI sampling parameters by the following expression:

$$\text{SNR} \propto \frac{F_x F_y F_z}{N_x N_y N_z} \sqrt{N_y} \sqrt{N_z} \sqrt{N_A} \, [1 - \exp(-\text{TR}/\text{T}_1)] \tag{4}$$

where F_x, F_y, and F_z are the field of view in the x, y, and z spatial coordinates, respectively; N_x, N_y, and N_z are the number of pixels in the x, y, and z spatial coordinates, respectively; and T_1 is the longitudinal relaxation time. Thus, the signal-to-noise of an image increases on decreasing the spatial resolution (N_y and/or N_z), increasing the repetition time, and increasing the number of averages. Reducing TR and N_A increases temporal resolution but reduces signal-to-noise. Reducing spatial resolution also increases temporal resolution, but increases signal-to-noise, since this increases the voxel (volumetric element) size and consequently the signal within.

Image Contrast

MRI contrast is based on the spin density M_0 (governed by the water content) and the relaxation times, T_1 and T_2 (an indicator of molecular mobility), as well as the experimental time parameters, TE and TR (echo time and repetition time, respectively). The magnetization available for successive line scans is governed by TR (the time left for the longitudinal magnetization to recover), and this is an important contributor to the total time required for acquisition of an image (see above). To obtain maximum signal detection it is necessary to use a short TE to capture the transverse signal before it decays, and a long TR (5T_1) for full longitudinal recovery. Under ideal conditions, this gives a proton density (or water content) image in which the signal contrast is based effectively on the number of the water nuclei in each pixel (voxel).

Since the water in the biofilm has a shorter T_2 relaxation time than pure water,[4] increasing the echo time (TE) will result in T_2 contrast since the signal will differentiate pixels that are at a different timepoint in the transverse relaxation decay (Fig. 9). The differential increases with the increases in TE; however, at the same time, the signal-to-noise ratio of the images decreases. In addition, the water in the biofilm has a shorter T_1 relaxation time than pure water[4]; hence, decreasing TR (from 5T_1) will result in T_1 contrast, as the scan will differentiate pixels that are at a different timepoints in their T_1 saturation recovery. Thus, the differential increases with decreasing TR; however, the signal-to-noise ratio will also decrease as the magnetization will not be fully recovered.

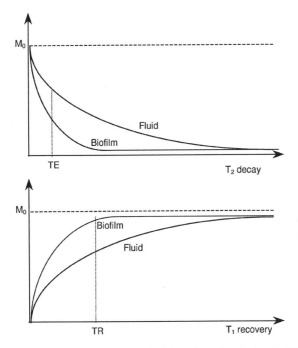

FIG. 9. Image contrast. The signal from each pixel depends on the spin density (M_0), longitudinal magnetization recovery (T_1), and transverse magnetization decay (T_2) of the nuclear species, as well as the repetition time (TR) and echo time (TE) of the scan protocols. A proton density image is one where the difference in the numbers of protons per unit volume is the main determining factor in forming image contrast. In order to achieve proton density weighting, the effects of T_1 and T_2 contrast must be diminished. A long TR allows both water within biofilm and perfusing fluid to fully recover their longitudinal magnetization, and therefore diminishes T_1 weighting. A short TE does not give the water within the biofilm and perfusing fluid time to decay and therefore diminishes T_2 weighting.

Of the two basic MRI sequences, spin echo and gradient echo,[16] most protocols used for 3D visualization are based on the latter, which allow the acquisition of a 3D image in minutes, rather than the hours required for conventional spin echo protocols. This is because the longitudinal magnetization is not fully inverted by the RF observation pulse and therefore takes a shorter time for full recovery. In addition, the transverse magnetization in a gradient echo sequence is also dephased by magnetic field inhomogeneities; therefore, the decay constant is $T_2{}^*$. These field inhomogeneities can be caused by variations in the magnetic susceptibility between different materials; a common example encountered in this work is the interface between air and water in air bubbles.

Bioreactor and Biofilm Visualization

The 3D image shown in Fig. 1a was acquired using a 31 cm bore magnet, a gradient echo sequence[16] with $TR/TE/\theta = 200$ msec/7.5 msec/70°, and spatial resolution 312.5 μm × 312.5 μm × 312.5 μm; the total acquisition time was 54 min and 37 sec. The 3D images shown in Figs. 2a and 2b were acquired with $TR/TE/\theta = 50$ msec/7.5 msec/90°; the total acquisition time was 13 min and 39 sec.

Figure 1a shows areas of high signal intensity from water. There is little or no signal from the porous support because of its lower water content and, more importantly, the high magnetic susceptibility differential between water and the porous support, leading to such a short T_2^* that the water magnetization decays before it can be detected.

A 2D slice from the 3D image data set of the packed-bed reactor containing biofilm immobilized onto six cubes of polyurethane foam is shown in Fig. 10a; this image was acquired using the same conditions as for Figs. 2a and 2b. The pore size of the foam is ca. 1.3 mm and the width of the matrix of the foam structure is ca. 150 μm; values were obtained from the scanning electron micrograph shown in Fig. 10b. MRI is able to visualize the polyurethane foam because it gives zero signal compared to the surrounding water. However, as expected under the imaging conditions used in this study with a spatial resolution of 312.5 μm, which is larger than the width of the foam matrix (ca. 150 μm), the 3D MR image of the matrix itself is not well defined (image not shown). To obtain image data of the matrix and biofilm more closely comparable to those in Fig. 10b would require the use of a higher field magnet and far longer scan times, to build up the required signal-to-noise and resolution.

Examination of the *Citrobacter* biofilm (dehydrated during sample preparation), using scanning electron microscopy, shows clusters of biomass of uneven shapes attached to the foam structure (Figs. 10b and 10c). It was not expected, using the conditions and magnet described in this example, to be able to visualize details of the structure of the biofilm (see Gradient Set and Spatial Resolution section). However, as shown in Fig. 10a the foam is more distinct when it is coated with biofilm, since the signal from the water within the biofilm also gives a low signal. Therefore Fig. 10a and an image equivalent to Fig. 10b could not both be achieved using any single magnet bore system. A high-resolution image (2D slice taken from the 3D MRI data set) of biofilm immobilized onto a single foam cube, obtained using the 9 cm bore magnet, is shown in Fig. 10d. This image was acquired using a gradient echo sequence with $TR/TE/\theta = 1000$ msec/7.5 msec/45° and $N_A = 2$; a longer TR was used because the surrounding fluid (isotonic saline solution, 8.5 g sodium chloride/liter) has a long T_1, and consequently the total acquisition time was 9 hr and 6 min. This compares well with the photographic image of reticulated foam (not itself visible by MRI) and following biofilm development (Figs. 2c and 10e, respectively). The foam and biofilm are clearly visible

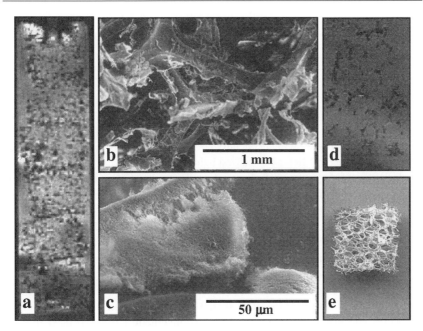

FIG. 10. (a) A 2D slice taken from the 3D MRI data set of a packed-bed reactor, containing biofilm immobilized onto cubes of polyurethane foam and challenged with a solution sodium glycerol 2-phosphate (5 mM) in sodium citrate buffer (2 mM), pH 6.0. The image was obtained using a 31 cm bore magnet. (b) Scanning electron micrograph of biofilm immobilized onto the polyurethane foam (50× magnification) and (c) the same as for (b), but 1000× magnification. (d) A 2D slice taken from the 3D image data set of biofilm immobilized onto a single foam cube, imaged using the 9 cm bore magnet. Note that this permits high resolution imaging of the biofilm gross structure. (e) Photograph of biofilm growth on polyurethane foam cube. Note that although the biofilm appears patchy by scanning electron microscopy because of dehydration during sample preparation (b), the fully hydrated biofilm (e) is visible in the native state by MRI (d).

by MRI at a resolution of 156 μm (Fig. 10d). The image contrast is better despite only halving the resolution because of the higher signal-to-noise, associated with a higher field strength.

The removal of metals, e.g., UO_2^{2+}, La^{3+}, Cd^{2+}, Pb^{2+}, and Sr^{2+}, from aqueous solutions by immobilized *Citrobacter* cells in packed-bed reactors has been well documented.[5,9] The processes of metal uptake by *Citrobacter* biofilm can be summarized as follows: (1) nucleation of metals by phosphate groups present within the extracellular polymeric matrix, and (2) production of enzymatically liberated phosphate, which consolidates the growth of the metal phosphate crystal.[21,22] For

[21] K. M. Bonthrone, J. Quarmby, C. J. Hewitt, V. J. M. Allan, M. Paterson-Beedle, J. F. Kennedy, and L. E. Macaskie, *Environmental Technol.* **21,** 123 (1999).
[22] L. E. Macaskie, K. M. Bonthrone, P. Yong, and D. T. Goddard, *Microbiol.* **14,** 635 (2000).

simplicity, the metal uptake shown in Fig. 2b is confined to the former process. The addition of a metal solution changes the MR properties of both the biofilm and solution causing a different, in this case higher, contrast.

Many atoms exhibit magnetic properties, which are not due to the properties of the nucleus, but to the structure of their circulating clouds of electrons. The prediction of the magnetic behavior of these orbiting electrons is complicated because they also possess spin, which creates a magnetic field. In practice, many atoms are paramagnetic to some degree, with the transition elements in the middle of the periodic table showing especially strong paramagnetism; these can be used as contrast agents in MRI, as discussed below.

Paramagnetic compounds enhance the proton relaxation of the water in which they are dissolved; as a result, T_1 and T_2 are both decreased, which affects the signal intensity. The measured MRI signal intensity is a complex function of the contrast agent used, the materials under study, and the scan protocol and its parameters. In general practice, any decrease in these two relaxation times has the opposite effect on the signal intensity; T_1 decreases tend to increase the observed signal intensity, whereas T_2 decreases will decrease the signal intensity.[23]

Manganese and gadolinium complexes have been successfully used as contrast agents in clinical MRI, because of their strong paramagnetic properties. Other transition and rare earth metals that are paramagnetic have also been used as contrast agents, e.g., ferric and ferrous salts, their complexes, and those of copper, chromium, europium, and dysprosium.[24]

The lanthanum present within the biofilm (Fig. 2b) increases the perceived thickness of the foam matrix when compared to the same biofilm without metal (Fig. 10a). The uptake of metal between the foam cubes is noticeable as black areas (Fig. 2b). Also, the presence of these salts in the aqueous solutions alters the intensity of the interstitial aqueous fluid relative to the biofilm and support matrix.

Small air bubbles are problematic in MRI, since they significantly disturb the magnetic field homogeneity; fortunately, they are easily identified because they appear in MRI scans as a cross-shaped void with white dots around the edges (see top of columns, Figs. 2b and 10a). Since air bubbles interfere with studies of bioreactor flow and its modeling, it is important that MRI is able to visualize them; they are not always evident from visual observation and it is impossible to visualize them directly within a packing matrix.

The 3D gradient echo image may appear to give inherently better contrast because of the higher resolution in the third dimension; a 2D gradient echo

[23] M. F. Tweedle, H. G. Brittain, W. C. Eckelman, G. T. Gaughan, J. J. Hagan, P. W. Wedeking, and V. M. Runge, *in* "Magnetic Resonance Imaging" (C. L. Partain, R. R. Price, J. A. Patton, M. V. Kulkarni, and A. E. James, Jr., ed.), 2nd ed., Vol. 1, p. 793. W. B. Saunders Company, Philadelphia, 1988.

[24] L. D. Hall and P. G. Hogan, *in* "Functional Studies Using NMR" (V. R. McCready, M. Leach, and P. J. Ell, ed.), p. 107. Springer-Verlag, London, 1987.

image is "volume averaged" across a multi-millimeter slice, whereas via 3D that thickness is submillimeter. This can be particularly important in distinguishing the polyurethane foam support and biofilm from the bulk water. Alternatively, 2D quantitation of the T_1 and T_2 relaxation times may be used to distinguish the different constituents.[4] Thus, acquisition of a set of T_1- and T_2-weighted images can produce high signal-to-noise T_1 and T_2 maps, which provide distinct values for the individual constituents. Such quantitation experiments are very time consuming, and the increased time required for similar 3D quantitation may not be practicable (it would take hours to acquire images with the signal-to-noise necessary for accurate quantitation).

A major benefit of the MRI technique is that in addition to visualization of the packing (support plus immobilized biofilm) and fluid distributions within bioreactors, it can be used to monitor systems over a period of time. This technique is sensitive to motion and temperature (among other variables), and hence it can be used on-line to measure quantitatively dynamic changes, such as flow, diffusion, and mass and heat transfer. Consequently MRI can be used to obtain the experimental data required for validation of the modeling of complex systems (such as the metal accumulation process described in this work) and, importantly, for studying the effects of perturbation on the processes of recovery back to the steady-state baseline. In the case of industrial waste treatment, for example, transient toxic shock overload, or changes in the flow composition or pH, may occur. That particular system is characterized by temporal and spatial variations of physical and chemical properties that occur within the reactor: for example, metal adsorption and precipitation onto the biofilm, channelling, variation of flow rates and pH, metal overload, and eventually blockage of the reactor.

Applications of the MRI Technique

An increasing amount of research has used the MRI technique to study physical and chemical properties of products and chemical engineering processes. Here we outline some of the applications of the MRI technique, which are relevant to the study of bioreactors.

MRI has been used to determine the spatial distribution of fluid saturation and flow within porous media[25,26] and columnar reactors.[27,28] Baldwin and Gladden[29]

[25] M. A. Horsfield, E. J. Fordham, C. Hall, and L. D. Hall, *J. Magn. Reson.* **81**, 593 (1989).

[26] P. A. Osment, K. J. Packer, M. J. Taylor, J. J. Attard, T. A. Carpenter, L. D. Hall, N. J. Herrod, and S. J. Doran, *Phil. Trans. R. Soc. Lond. A* **333**, 441 (1990).

[27] L. D. Hall and V. Rajanayagam, *J. Chem. Soc., Chem. Commun.* **8**, 499 (1985).

[28] K. Potter, "Magnetic Resonance Imaging of Columnar Reactors." Ph.D. thesis, University of Cambridge, U.K. (1993).

[29] C. A. Baldwin and L. F. Gladden, *AIChE J.* **42**, 1341 (1996).

demonstrated that MRI could be used to quantify the volume of nonaqueous phase liquid present in an otherwise water-saturated porous medium. Baldwin et al.[30] used MRI to determine and characterize the structure of the pore space of a packed bed of uniform spheres (6 mm diameter).

The MRI technique is capable of measuring fluid velocities in three spatial dimensions (velocimetry)[31]; consequently, it is uniquely suitable for the analysis of opaque samples, including complex dispersions. Amin et al.[32] used MRI to obtain displacement and velocity spectra of saturated steady-state flow through a column packed with glass beads (0.3–0.36 mm average particle diameter). Rigby and Gladden[33] have used MRI to study the modeling of transport in porous media. Sederman et al.[34] used MRI volume- and velocity-measurement techniques to probe structure–flow correlation within the interparticle space of a packed bed of ballotini (5 mm diameter). Manz et al.[35] used a modified version of the echo-planar imaging technique to study the drying process in an initially water-saturated model porous medium.

MRI is also being widely used to probe diffusion. In addition to direct studies of water diffusion, techniques have been developed to visualize and quantify diffusion of metal ions through gels and cartilage.[36–38]

MRI is also known to be sensitive to pH[39]; however, a strong acid is needed for direct observation of a MR effect.[40] "Molecular amplifiers" can be used to increase the sensitivity of a measurable MR effect. Generally, paramagnetic metal complexes alter the magnetism of the water protons because of their exchange with the hydration shell surrounding paramagnetic ions; thus, the MR effect is sensitive to changes in the concentration or nature of these ions. For instance, the degree with which ethylenediaminetetraacetic acid (EDTA) binds Cu^{2+} ions depends on the protonation of the EDTA complexing agent; therefore, the proportion of free and bound Cu^{2+} ions, and in turn the MR relaxation properties, are pH dependent.[41]

[30] C. A. Baldwin, A. J. Sederman, M. D. Mantle, P. Alexander, and L. F. Gladden, J. Colloid Interf. Sci. 181, 79 (1996).

[31] B. Newling, S. J. Gibbs, J. A. Derbyshire, D. Xing, L. D. Hall, D. E. Haycock, W. J. Frith, and S. Ablett, J. Fluids Eng. 119, 103 (1997).

[32] M. H. G. Amin, S. J. Gibbs, R. J. Chorley, K. S. Richards, T. A. Carpenter, and L. D. Hall, Proc. R. Soc. London 453, 489 (1997).

[33] S. P. Rigby and L. F. Gladden, Chem. Eng. Sci. 51, 2263 (1996).

[34] A. J. Sederman, M. L. Johns, A. S. Bramley, P. Alexander, and L. F. Gladden, Chem. Eng. Sci. 52, 2239 (1997).

[35] B. Manz, P. S. Chow, and L. F. Gladden, J. Magn. Reson. 136, 226 (1999).

[36] A. E. Fischer, B. J. Balcom, E. J. Fordhami, T. A. Carpenter, and L. D. Hall, J. Phys. D: Appl. Phys. 28, 384 (1995).

[37] A. E. Fischer, T. A. Carpenter, J. A. Tyler, and L. D. Hall, Magn. Reson. Imag. 13, 819 (1995).

[38] A. E. Fischer and L. D. Hall, Magn. Reson. Imag. 14, 779 (1996).

[39] L. D. Hall and S. L. Talagala, J. Magn. Reson. 65, 501 (1985).

[40] A. E. Fischer and L. D. Hall, MAGMA 2, 203 (1994).

[41] B. J. Balcolm, T. A. Carpenter, and L. D. Hall, Can. J. Chem. 70, 2693 (1992).

In a similar fashion the Fe(II)↔Fe(III) equilibrium can be used to monitor reduction–oxidation reactions since there is a 50-fold difference in the MR signal intensity between these two ions.[42]

3D MRI has been used to study a filtration process. A paramagnetic material used as contaminant, together with a gradient-echo imaging sequence, enabled the location of areas on a filter surface where particles are deposited, and also showed the earliest stages of blocking.[43] Biofiltration systems are a logical extension to these studies.

Conclusions

The three-dimensional MRI technique discussed in this article has great potential for studies not only of biofilms immobilized onto supports, but also of the overall functioning of bioreactors. Study of the structure of a biofilm at a resolution level of 20 μm will always require long scan times and, possibly, use of higher field strength on smaller samples. For larger volumetric elements, such as the packed-bed reactors tested in this chapter, a 31 cm bore magnet (2.35 Tesla) can easily produce 300 μm resolution, and far higher resolution (ca. 70 μm) can be readily achieved. Since the uptake of metal salts by biofilm enhances the image contrast, the biofilm becomes more distinct.

Perhaps most importantly, MRI can also be used to monitor systems over a period of time. This technique is sensitive to motion and to temperature and it can be used to measure quantitatively in three dimensions dynamic changes, such as flow, diffusion, and mass- and heat-transfer. Beside the obvious importance of such experimental data in their own right, such data have unique potential for validation of the computer software used to model bioreactor processes.

Acknowledgments

This work was supported by a BBSRC (Grant No. EO 9214) and by the Herchel Smith Endowment. The authors thank Norton Chemical Process Products Corporation (USA) and Recticel (Belgium) for the supports for growth of biofilm.

[42] B. J. Balcolm, T. A. Carpenter, and L. D. Hall, *J. Chem. Soc., Chem. Commun.* **4,** 312 (1992).

[43] C. J. Dirckx, S. A. Clark, L. D. Hall, B. Antalek, J. Tooma, and J. M. Hewitt, *AIChE J.* **46,** 6 (2000).

[21] Detachment, Surface Migration, and Other Dynamic Behavior in Bacterial Biofilms Revealed by Digital Time-Lapse Imaging

By PAUL STOODLEY, LUANNE HALL-STOODLEY, and
HILARY M. LAPPIN-SCOTT

Introduction

Over the past decade much biofilm research has focused on the ultrastructural complexity of biofilms and the implications that the observed organization may have on biofilm processes.this interest was largely sparked by the development of confocal microscopy (CM), which allowed biofilms to be observed at high resolution in 3D with no disruption to structure.[1] These studies revealed that biofilms, both in the laboratory and in nature, are often heterogeneous and consist of microcolonies or cell clusters (aggregates of microbial cells in an extracellular polysaccharide slime matrix) separated by interstitial voids and channels.[2] Further, with the use of tracer particles[3] and microelectrodes,[4] it was found that the channels allowed water to flow around the biofilm structures. These channels, depending on the hydrodynamic conditions, could significantly increase the supply of nutrients to bacterial cells in the biofilm. Detailed descriptions of the principles and applications of CM to the imaging of biofilms can be found in Caldwell *et al.*[1] and Lawrence and Neu[5]. Evidence that biofilm structure may influence growth and activity, as well as the possibility that the structure may represent an optimal organizational arrangement for growth in a particular environment, has prompted increased interest in this area of biofilm research. Other microscopic techniques such as deconvolution microscopy[6] and differential interference contrast microscopy,[7] have also been used to image biofilm structure at high resolution.

Although much work has been done, and is currently being undertaken, to determine the spatial complexities of biofilm systems, the dynamic behavior in biofilms has been less well studied. Time lapse studies are generally limited to the tracking of individual cells attaching, moving over, and detaching from surfaces in the initial stages of biofilm formation.these studies have allowed the

[1] D. E. Caldwell, D. R. Korber, and J. Lawrence, *in* "Advances in Microbial Ecology," Vol. 12 (K. C. Marshall, ed.), p. 1. Plenum Press, New York, 1992.

[2] P. Stoodley, D. deBeer, J. D. Boyle, and H. M. Lappin-Scott, *Biofouling* **14,** 75 (1999).

[3] P. Stoodley, D. deBeer, and Z. Lewandowski, *Appl. Environ. Microbiol.* **60,** 2711 (1994).

[4] D. deBeer and P. Stoodley, *Water Sci. Tech.* **32,** 11 (1995).

[5] J. R. Lawrence and T. R. Neu, *Methods Enzymol.* **310,** 131 (1999).

[6] D. Phipps, G. Rodriguez, and H. Ridgway, *Methods Enzymol.* **310,** 178 (1999).

[7] J. Rogers and C. W. Keevil, *Appl. Environ. Microbiol.* **58,** 2326 (1992).

direct measurement of the accumulation rate of cells on a surface[8] and given insight into the dynamic behavior of attached cells. Dalton *et al.*[9] and O'Toole and Kolter[10] have demonstrated that biofilm microcolonies can repeatedly form and disperse from the coordinated movement of single attached cells to specific loci on a surface. Without time lapse imaging it might be assumed that in these cases the microcolonies were generated from growth, not migration. However, it is much more difficult to keep track of single cells within mature, thick biofilms, since they are easily lost against a background of cells of similar appearance. This task is made more difficult if the biofilm cells are motile or "twitching." To overcome these problems we have used time lapse imaging at lower power magnification to reveal dynamic behavior in biofilms by tracking entire biofilm microcolonies rather than individual bacterial cells. These methods and some results that demonstrate the utility of these methods are presented in this paper.

Biofilm Reactor Systems, Flow Cells, and Digital Imaging

Two components are required to image the development and behavior of living biofilms in real time. First, a reactor system is required that incorporates flow cells (sometimes called perfusion chambers) that have a transparent wall or window for microscopic observation. Second, a microscope and imaging system are required to collect and manipulate digital images.

Reactor Systems

There are many types of flow cells, which are often designed and constructed "in house." Some of these have been described in detail elsewhere.[11-13] Some flow cells are also commercially available (BioSurface Technologies Corporation, Bozeman, MT, http://www.imt.net/~mitbst/flowcell.html). The flow cells are incorporated into a flow system that is generally either "once through" or "recirculating." In "once through" systems, which are the most commonly used in biofilm studies, sterile nutrients are pumped through the flow cell into a waste container at the effluent end.[12] In these setups the flow rate cannot be adjusted independently of

[8] A. Escher and W. G. Characklis, *in* "Biofilms" (W. G. Characklis and K. C. Marshall, eds.), p. 445. Wiley, New York, 1990.

[9] H. M. Dalton, A. E. Goodman, and K. C. Marshall, *J. Ind. Microbiol.* **17,** 228 (1996).

[10] G. A. O'Toole and R. Kolter, *Mol. Microbiol.* **30,** 295 (1998).

[11] L. Hall-Stoodley, J. C. Rayner, P. Stoodley, and H. M. Lappin-Scott, *in* "Methods in Biotechnology," Vol. 12: "Environmental Monitoring of Bacteria" (C. Edwards, ed.), p. 307. Humana Press, Totowa, NJ, 1999.

[12] R. J. Palmer, *Methods Enzymol.* **310,** 160 (1999).

[13] M. S. Zinn, R. D. Kirkegaard, R. J. Palmer, and D. C. White, *Methods Enzymol.* **310,** 224 (1999).

the residence time and the cost and time required to prepare sterile media usually limits experiments to low, laminar flows. In "recirculating" systems the flow cells are incorporated into a recirculating loop attached to a mixing chamber (Fig. 1). Sterile nutrients are added to the mixing chamber and the residence time (θ) can be controlled by the nutrient flow rate (Q_n) according to $\theta = V/Q_n$, where V is the volume of the mixing chamber *plus* the recirculation loop. These systems have the advantage that the flow rate in the flow cells can be adjusted independently of the nutrient flow rate so that much higher flow rates can be achieved in the flow cells without using impractical volumes of media. Also, the Q_n can be adjusted so that the dilution rate ($D = Q_n/V$) is greater than the specific growth rate of the organism being investigated so planktonic cells are continually "washed out" and only the attached biofilm population is retained.[14] Therefore, it can be assumed that any cells or microcolonies in the bulk liquid must have resulted from detachment and not planktonic growth. However, these systems have several disadvantages. First, they generally have large surface areas and there can be significant biofilm accumulation in areas other than those being observed in the flow cells (mixing chamber, connective tubing). Second, the biofilms in the flow cells are continually exposed to detached cells and waste products that build up to steady-state concentrations in the reactor system. A comprehensive mass balance analysis of various biofilm reactor systems may be found in Characklis.[14]

Image Capture, Enhancement, and Analysis

Digital microscopic image analysis requires four key components: a microscope, a digital camera, a computer with an installed framestore board, and software for capture, image enhancement, and analysis. Often the same software can be used for image capture and analysis. There are many different types of systems available with a variety of features and a range of prices. In this paper we describe the system found to be economical and relatively easy to operate.

Biofilm Reactor System for Growing Biofilms in Laminar or Turbulent Flow

For some experiments we wished to determine the influence of flow velocity on biofilm structure and behavior and required a flow system that could operate over a wide range of laminar and turbulent flows and in which the hydrodynamics were well characterized. To achieve this we designed a recirculating flow system based on a system previously described by Bryers and Characklis.[15] The reactor

---

[14] W. G. Characklis, *in* "Biofilms" (W. G. Characklis and K. C. Marshall, eds.), p. 17. Wiley, New York, 1990.

[15] J. D. Bryers and W. G. Characklis, *Water Res.* **15**, 483 (1981).

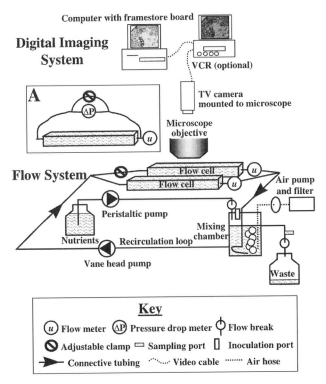

FIG. 1. Biofilm reactor system designed to image biofilms grown under laminar and turbulent flows. The schematic shows two parallel flow cells, although four have also been used. In our experiments we usually maintain laminar flow in one flow cell and turbulent flow in the other. The flow rate in the turbulent flow cell was controlled by the pump RPM and an adjustable clamp was used to reduce flow in the laminar flow cell. The inset "A" shows how the pressure drop meter was connected to the flow cell. A bypass loop facilitates flushing of air bubbles from the lines but must be clamped (as shown) during measurements.

system consists of either two or four parallel flow cells that are incorporated into a recycle loop with a mixing chamber for aeration and nutrient addition (Fig. 1). For flow cells we use 20 cm sections of square glass tubing (S-103 Camlab Ltd., Cambridge, U.K.) that are 3 mm wide (W) and 3 mm deep (D). The sections are cut from longer stock tubing with a diamond knife and the ends flame polished. Nutrients are delivered by peristaltic pump (Masterflex, Cole Parmer, Niles, IL) and the recycle flow rate is controlled with a vane head pump (Masterflex, Cole Parmer). Peristaltic pumps should be avoided because they provide a pulsatile flow in which there can be a wide range of shears. The volume (V) of the mixing chamber and recycle loop, including two flow cells, is approximately 175 ml. The nutrient influent flow rate (Q_n) is 4.3 ml/min, for a resulting residence time ($\theta = V/Q_n$) of

40 min. The flow rate through each of the flow cells (Q_f) is measured using flow meters (McMillan Flo-sensor model 101T # 3724 and 3835, supplied by Cole-Parmer). The flow meters are calibrated by timed volumetric displacement. To minimize disruption to the flow by sudden expansions and contractions in the recirculating loop, we use connective tubing (silicone or Phar-Med size 16, Cole-Parmer) and connectors with similar hydraulic diameters to the flow cells (3 mm). The tubing is connected directly to the flow cells and crimped using cable ties. To characterize the flow in the flow cells we measure the pressure drop (ΔP) across each flow cell (using differential pressure transducers RS Components, Corby, Northants, U.K., model 286-686) at different average flow velocities ($u_{(ave)} = Q_f/WD$). The pressure transducers are calibrated using a manometer. The flow velocity Q_f, through each flow cell can be controlled independently by tightening or loosening clamps on the inlet tubing. The flow cells are positioned on a polycarbonate holder that is mounted on the stage of an Olympus BH2 upright microscope with epifluorescence capabilities. The thickness of the glass wall of the flow cells limits observations to low numerical aperture long and ultralong working distance (LWD) objectives. However, 100 and 50× LWD objectives can resolve single cells on the top surface of the flow cells. To observe the biofilm growing on the bottom surface, objectives of 10× and lower are required to focus through the thickness of the entire flow cell. Under operating conditions the water temperature in the reactor system is 28° at a room temperature of 22°.

Hydrodynamic Characteristics of Flow Cells

We determined the hydrodynamic characteristics of the flow cells using the relationship between the Fanning friction factor (f) and Reynolds number (Re). The f and Re are found from ΔP, $u_{(ave)}$, and flow cell geometry. The Reynolds number (Re) is calculated from Eq. (1):

$$\text{Re} = \frac{u_{(ave)}D_h}{\nu} \tag{1}$$

where D_h is the hydraulic diameter and ν is the kinematic viscosity. The D_h is (3 mm) from $D_h = 4CSA/WP$ where CSA is the cross-sectional area (CSA $= WD$) and WP the wetted perimeter ($WP = 2W + 2D$) of the square flow cells. The f is found from[16] Eq. (2):

$$f = \frac{\Delta P \times D_h}{2l_f \, \rho_w u_{(ave)}^2} \tag{2}$$

where ρ_w is the density of water and l_f is the distance between pressure ports. For ρ_w and ν we use values for water, since there is no significant difference between the

---

[16] W. G. Characklis, M. H. Turakhia, and N. Zelver, *in* "Biofilms" (W. G. Characklis and K. C. Marshall, eds.), p. 265. Wiley, New York, 1990.

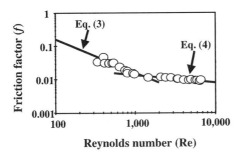

FIG. 2. Friction factor chart showing the measured ($\bigcirc$) and predicted (—) values of f for a clean flow cell. The curve described by Eq. (3) shows the laminar region and Eq. (4) shows the turbulent region. The transition point occurred at approximately Re 1200.

viscosity (using a falling ball viscometer) of the sterile minimal salts medium[4,17] and distilled water ($p = 0.1$, $n = 6$), and the viscosity of the spent medium is only 1.2% greater than that of distilled water. Results show that the transition between laminar and turbulent flow occurs at Re 1200 ($u = 0.35$ m /sec) (Fig. 2). The measured f is in close agreement with that predicted by the Hagen–Poiseuille equation[18] in the laminar region:

$$f = \frac{16}{\text{Re}} \tag{3}$$

and the Blasius formula[18] in the turbulent region:

$$f = 0.0791/\text{Re}^{0.25} \tag{4}$$

for flow through smooth pipes. The transition point between flows occurs at a Re of approximately 1200. This has been confirmed by dye tracer studies.

In many experiments it is useful to report attachment or detachment of cells or microcolonies as a function of the wall shear stress caused by the moving fluid. In turbulent flow the shear stress at the wall (τ_w) can be estimated from[16,18] Eq. (5).

$$\tau_w = \frac{f \rho_w u_{(ave)}^2}{2} \tag{5}$$

In laminar flow:

$$\tau = \frac{4 \eta u_{(max)}}{D_h} \tag{6}$$

where η is the absolute viscosity and $u_{(max)}$ is the maximum velocity. For a circular pipe $u_{(max)} = 2u_{(ave)}$ and for a flat plate reactor (when $W \approx 100D$) $u_{(max)}$ can be

[17] P. Stoodley, I. Dodds, J. D. Boyle, and H. M. Lappin-Scott, *J. Appl. Microbiol.* **85**, 19S (1999).

[18] W. L. McCabe and J. C. Smith, "Unit Operations of Chemical Engineering," 3rd ed., Chemical Engineering Series. McGraw-Hill, New York, 1976.

approximated to $\frac{3}{2}u_{(ave)}$. However, for square flow cells the edge effects of the corners will cause τ_w to vary along the width of the wall, and Eqs. (5) and (6) will only give approximate estimates of τ_w. When correlating observed biofilm structure or dynamic phenomena with τ_w, images should be taken in the center area of the channel and the edges avoided. The shear distribution for laminar flow in square rectangular pipes can be calculated using analytical solutions described elsewhere.[19] However, these calculations are not trivial and there are no analytical solutions for turbulent flow. An alternative approach is to measure the velocity profile using particle image velocimetry[3] and calculate τ_w directly from Eq. (7):

$$\tau_w = \frac{\eta\,du}{dz} \tag{7}$$

where z is the distance from the wall; du/dz at the wall (i.e., when $z = 0$) can be determined graphically.

Reactor Sterilization

The reactor system and nutrients are sterilized by autoclaving at 121° for 15 min. The flow meters are sterilized using a method adapted from Fisher and Petrini[20] in which they are exposed to 70% ethanol for 15 min, 40% NaOCl solution (approximately 12% available chlorine when undiluted) for 15 min, and 70% ethanol for 15 min. The flow meters are then drained and positioned in the recycle loop aseptically by flame sterilized glass tubing connectors. When the pressure drop meters are used they are also sterilized by exposure to bleach and ethanol solutions. To check for sterility we operate the flow system with sterile media for 1 to 2 days prior to inoculation before plating three 0.1 ml effluent samples onto nutrient agar plates.

Image Capture and Analysis

A COHU 4612-5000 CCD camera (Cohu, Inc., San Diego, CA) is used to collect gray-scale images and a Scion VG-5 PCI framestore board (Scion Inc., Frederick, MD) is used for image capture. The framestore board is controlled with either NIH-Image (developed at the National Institutes of Health and available for free download at http://rsb.info.nih.gov/nih-image/) for use with a Macintosh or Scion Image (available at http://www.scioncorp.com/) for use with a PC. These software packages are also used for image enhancement and analysis. The VG-5 framestore board allows video rate imaging of 30 frames per second (fps). However, at this rate the computer RAM is quickly used up and we recommend at least 128 MB. It is also important to purchase a computer with adequate hard disk space for storage as well as having access to a CD writer or some other large-capacity

[19] F. M. White, "Viscous Fluid Flow," 2nd ed. McGraw-Hill, New York, 1991.
[20] P. J. Fisher and O. Petrini, *New Phytol.* **120,** 1370 (1992).

portable storage device such as Zip disks (Iomega, http://www.iomega.com/). Time lapse sequences are made by capturing images at programmed time intervals. Distance and area measurements are calibrated from pixels to microns using a 1 mm graticule with 10 μm divisions (Ref. # CS990, Graticules Ltd., Tonbridge, Kent, U.K.).

Application of Time Lapse Digital Imaging to Analyze Dynamic Behavior in Biofilms

Downstream Migration of Biofilm Ripples

A four species biofilm consisting of *Pseudomonas aeruginosa, P. fluorescens, Klebsiella pneumoniae,* and *Stenatrophomonas maltophilia* is grown in the previously described recirculating flow reactor[21] (Fig. 1) under turbulent flow of 1 m/sec. The biofilm is initially grown on a minimal salts medium (MSM) with 40 ppm glucose as the sole carbon source. After approximately 10 days the biofilm develops ripple-like structures on the walls of the flow cells. Low power (4 to 20× objectives) time-lapse images taken at 0.5 hr intervals over a 20 hr period have revealed that the biofilm appears to flow along the flow cells in a fluid-like manner as the ripples and other biofilm microcolonies migrate downstream along the walls of the flow cell (Figs. 3 and 4). The migration velocity is determined by measuring the location of individual ripples at different times (Fig. 4). We found the migration velocity is a function of the bulk liquid velocity and that the ripples have a maximum velocity of approximately 1 mm/hr when the bulk liquid velocity is approximately 0.5 m/sec. The time lapse images also revealed that, in some areas, ripples were continually detaching into the bulk liquid from the front (downstream edge) of the ripple bed, although it was not clear what initiated this detachment.[21]

Detachment of Biofilm Microcolonies

In a replicate set of experiments the four species biofilm was grown under turbulent flow for 21 days on MSM + 40 ppm glucose. The biofilm developed ripple structures similar to those observed in replicate experiments.[17] After 21 d the glucose concentration was increased to 400 ppm. Within 24 h the biofilm had increased in thickness from approximately 30 μm to 130 μm and the biofilm structure had markedly changed. The ripple structures had disappeared and the biofilm consisted of large oval shaped microcolonies separated by interstitial channels.[17] The structure was monitored for a further 2 days with no observed changes, and so it was assumed that the new structure was stable after this period. In this case low power time-lapse images did not show significant downstream motion but revealed

[21] P. Stoodley, Z. Lewandowski, J. D. Boyle, and H. M. Lappin-Scott, *Environ. Microbiol.* **1,** 447 (1999).

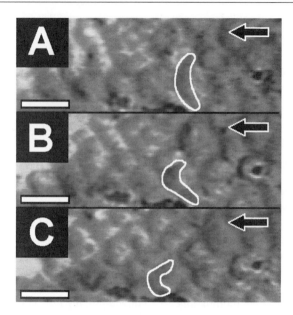

FIG. 3. Magnified region in a biofilm ripple bed showing an individual ripple (outlined in white) migrating downstream. The time interval between panels was 30 min. Scale bar = 80 μm. Note how the ripple changes shape as it moves along the surface.

that biofilm microcolonies were continually growing and detaching over a 20 hr monitoring period (Figs. 5 and 6). When each image was subtracted from the image taken previously using the "Image Math" routine in Scion Image (for settings we selected "real result" and set "x" at "1" and "+" at "0"), clusters that had detached appeared white. The image was then inverted and an appropriate threshold manually applied so that the clusters appeared black against a white background.

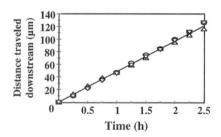

FIG. 4. The distance traveled downstream of three individual ripples in the ripple bed shown in Fig. 3 as a function of time. Over a 2.5 hr period the ripple velocity was a constant 49 μm/hr ($r^2 = 0.997$, $n = 33$) and there was very little variability between the velocity of the individual ripples.

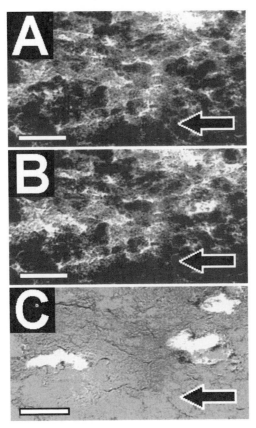

FIG. 5. Selected area from a larger field of view (area = 1.02 mm^2) showing the detachment of microcolonies from a 24 day old four species biofilm growing on MSM + 400 ppm glucose, Image "B" was subtracted from the image taken an hour earlier (A). Three large microcolonies that had detached during the elapsed time interval appeared as white areas on the resultant image "C." The direction of the bulk fluid flow is indicated by arrow. Scale bar = 200 μm.

The "Analyze Particles" routine was used to measure the number and area of clusters detaching between each time interval. These data can then be used to determine the detachment rate (Fig. 6) and the size distribution of detaching microcolonies.

Surface Area Coverage and Attachment of Microcolonies to a Surface

Mycobacterium fortuitum biofilms are grown on silastic or HDPE surfaces in a "once through" reactor system.[22] Since these experiments require surfaces

---

[22] L. Hall-Stoodley, C. W. Keevil, and H. M. Lappin-Scott, *J. Appl. Microbiol.* **85,** S60 (1999).

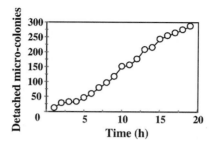

FIG. 6. The detachment rate of microcolonies from the area containing Fig. 5 was 14 mm^{-2} hr^{-1} ($r^2 = 0.96$, $n = 19$) calculated using the "Analyze Particles" routine on each subtracted image. The detachment rate was relatively constant over the 20 hr monitoring period. Because of limited resolution from the low power microscopy used (objective = 10×) only microcolonies with a projected area >140 μm (approx. diameter 14 μm) were included in the data set.

other than glass, we use a flat plate flow cell (BioSurface Technologies Corporation, http://www.imt.net/~mitbst/flowcell.html) with a recessed well in which coupons of different material can be fitted. The flow cell incorporates an overlying glass coverslip, for observing the developing biofilm. These flow cells allow the study of biofilm formation on a wide range of metallic and nonmetallic materials. However, they have the disadvantage that the channel depth is limited to 100 or 200 μm by the working distance of most high-resolution objectives. This in turn limits the flow regime to low, laminar flows. Also, because these types of flow cells generally have sudden expansions and contractions into and out of the coupon area, the hydrodynamics are often less well characterized. If the coupon is translucent, bright-field microscopy may be used for imaging, but reflected interference contrast or epifluorescence microscopy generally yield better images. Using this system with the same image analysis system as previously described, we have monitored the development and dynamic behavior of *M. fortuitum* biofilms. The progression of biofilm accumulation is monitored by measuring the percent surface area covered as a function of time. Images of the biofilm are captured using a low-power (20×) objective to increase the field area and give a better overall picture of the biofilm morphology. A threshold is then manually applied using the "Map" tool so the biofilm microcolonies appear black and the surrounding channels white. The total black area is recorded to a text file and the percent coverage calculated from the known field area. Single cells on the surface between the cell clusters cannot be observed at the low magnification and are therefore not included in the surface coverage measurement.

In addition to the routine monitoring of biofilm growth and accumulation, digital time lapse imaging is used to look for dynamic behavior in the biofilms. Although extensive ripple beds were not observed in these biofilms, some

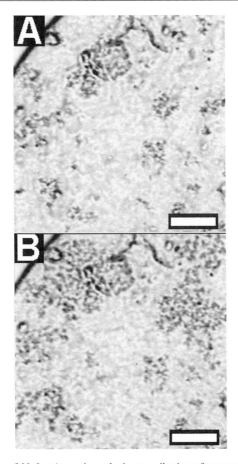

FIG. 7. Attachment of *M. fortuitum* microcolonies to a silastic surface over a 1 hr interval. Image B was subtracted from the previous image (A). The individual cocci in the three outlined microcolonies that had attached during the 1-hr interval appeared black in the resultant image (C). Scale bar = 10 μm.

ripple-shaped and round microcolonies were seen moving over the surface in the downstream direction (unpublished results). Furthermore, image subtraction using "Image Math" showed that microcolonies up to 20 μm in length could reattach to the surfaces from the bulk liquid (Fig. 7). We estimated from these images that such microcolonies could each harbor on the order of several hundred cells. It is not yet known if such detached microcolonies have the antibiotic resistance often found in attached biofilms.

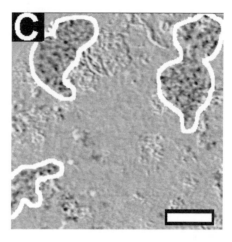

FIG. 7. (*continued*)

Conclusions

Digital time-lapse microscopic imaging by ourselves and others has revealed that biofilms not only are spatially complex but can also be temporally complex. Although this is more apparent during the initial stages of surface colonization, it can be more difficult to see in mature biofilms. It is often not evident that thicker, older biofilms are changing significantly, and an image taken one day may look very similar to an image taken on a different day. *In situ* microscopic measurements such as surface coverage and thickness, as well as scraping and conventional enumeration techniques used for monitoring biofilms, would indicate that the biofilm was in "steady state" with respect to biomass. However, with the use of time lapse imaging such dynamic process of growth, attachment, detachment, and motion of single cells and microcolonies can be observed and monitored. This work demonstrates that transient behavior in biofilms can cover a wide range of time scales, ranging from the high-frequency oscillations of biofilm filaments[23] to the slow surface migration of ripples,[21] which can be on the order of micrometers per hour. The notion of biofilms as dynamic, moving structural entities rather than static "coats of slime" has significant ramifications not only for how we view and model biofilms, but also as to how we may successfully control them. Furthermore, the detachment, reattachment, and movement of biofilm microcolonies over surfaces may give new insight into mechanisms by which infection or contamination may be disseminated in medical and industrial settings. As time lapse imaging continues to be used in the study of biofilms it is certain that other fascinating dynamic behavior will be revealed over a spectrum of time scales.

[23] P. Stoodley, Z. Lewandowski, J. Boyle, and H. M. Lappin-Scott, *Biotech. Bioeng.* **57**, 536 (1998).

Acknowledgments

Work in the laboratories of P.S. and H.M.L.-S. was supported by grants from the National Institutes of Health (1 RO1 GM60052-01), the cooperative agreement EEC-8907039 between the National Science Foundation and Montana State University—Bozeman, and by the Wellcome Trust Ref. 050950/MJM/APH.

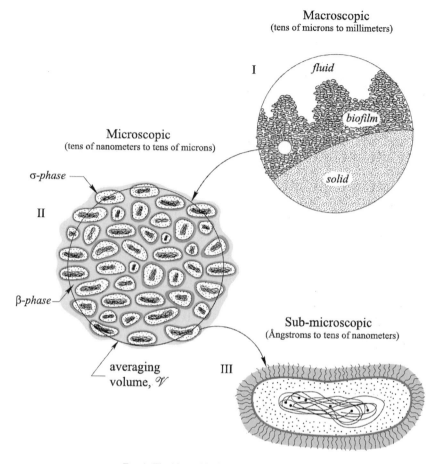

FIG. 1. The hierarchical structure of a biofilm.

This article begins with a description of the transport and reaction processes in a biofilm at the micro scale indicated by level II in Fig. 1. Next, the macroscopic equations describing solute transport in biofilms is presented and we briefly describe how these equations can be obtained through volume averaging. Finally, some analytical solutions for predicting the effective diffusivity are presented, and some special cases that are routinely used in practice are considered.

The Microscopic Description

Solute transport in biofilms is a multiphase process, and in order for this diffusion and reaction process to be fully understood, biofilms need to be analyzed beginning at the micro scale.[2] In level II of Fig. 1, we have illustrated the simplest type of biofilm containing only two phases: an extracellular phase and an intercellular phase. The extracellular phase (denoted as the β-phase in Fig. 1) is composed of highly hydrated polymeric substances whose diffusivity is usually close to that of water. The intercellular phase (denoted as the σ-phase in Fig. 1) represents chemical species bound by the cell membrane, and this region contains enzymes and other macromolecules (RNA, DNA, etc.) involved in cell metabolism.

Transport across the cell membrane is of primary importance to the understanding of diffusion in biofilms and cellular systems. There has been confusion in the literature about the process of species transport across the cell membrane. It is often suggested that chemical species *diffuse* across the cell membrane, but this process is generally only important for small nonionic molecules such as water, oxygen, and carbon dioxide.[3] The transmembrane transport of ions and larger molecules (such as sugars and amino acids) is usually facilitated by specific proteins.[4] Examples of such proteins include carriers that transport molecules across cell membranes with an associated conformational change (e.g., glucose)[5] or porins that form protein channels across the cell membrane. With the input of energy, some of these specific proteins can maintain concentration ratios between the two sides of the membrane of up to 10^5. In any event, the transport of molecules across the cell membrane is not generally well represented as a passive diffusion process, and the presence of a specific transport mechanism can make a significant difference in the effective diffusivity of the biofilm. We have used two different models to describe solute transport across the cell membrane. For small nonionic molecules we have adopted a simple permeability model.[3] For all other solutes we have used a simple carrier model.[6] The simple carrier model is illustrated in Fig. 2, and this model is known to represent a wide spectrum of protein-mediated transport processes.

For the two-phase model of biofilms that we have adopted, the description of the micro-scale transport must include (1) diffusion of chemical species in the extracellular phase, (2) transport of chemical species across the cell membrane, (3) diffusion of chemical species in the intercellular phase, and (4) reactions among species in the intercellular phase. For a single reactive species with transport mediated by a specific protein, the micro-scale description of transport can be written as[7]

[2] B. D. Wood and S. Whitaker, *Chem. Eng. Sci.* **55**, 3397 (2000).

[3] S. J. Marrink and H. J. C. Berendsen, *J. Phys. Chem.* **100**, 16729 (1996).

[4] L. Stryer, *Biochemistry*, 2nd ed. Freeman and Company, New York, 1981.

[5] A. Carruthers, *Prog. Biophys. Mol. Biol.* **43**, 33 (1984).

[6] P. Läuger, *Science (N.Y.)* **178** (1972).

[7] B. D. Wood and S. Whitaker, *Chem. Eng. Sci.* **53**, 397 (1998).

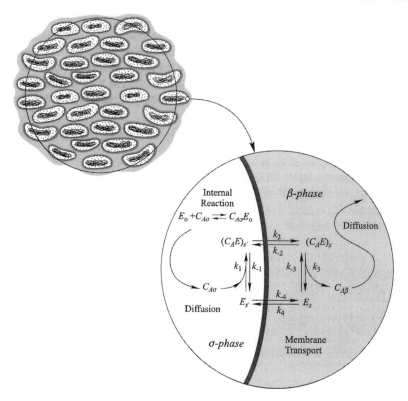

FIG. 2. The simple carrier model for protein-mediated membrane transport.

Diffusion in the extracellular phase:

$$\underbrace{\frac{\partial c_{A\beta}}{\partial t}}_{\text{accumulation}} = \underbrace{\nabla \cdot (\mathscr{D}_{A\beta} \cdot \nabla c_{A\beta})}_{\text{diffusion}} \text{ in the } \beta\text{-phase} \tag{1}$$

Continuity of flux at the boundary:

B.C.1 $$\underbrace{\mathscr{D}_{A\beta} \cdot \nabla c_{A\beta} \cdot \mathbf{n}_{\beta\sigma}}_{\substack{\text{interfacial flux,}\\\text{extracellular phase}}} = \underbrace{\mathscr{D}_{A\sigma} \cdot \nabla c_{A\sigma} \cdot \mathbf{n}_{\beta\sigma}}_{\substack{\text{interfacial flux,}\\\text{intercellular phase}}} \text{ at } \mathscr{A}_{\beta\sigma} \tag{2}$$

Transport across membrane (specific protein mediated):

B.C.2 $$\underbrace{\mathscr{D}_{A\beta} \cdot \nabla c_{A\beta} \cdot \mathbf{n}_{\beta\sigma}}_{\substack{\text{interfacial flux,}\\\text{extracellular phase}}} = \underbrace{\frac{E_0(c_{A\beta} - \alpha_1 c_{A\sigma})}{\alpha_2 + \alpha_3 c_{A\beta} + \alpha_4 c_{A\sigma} + \alpha_5 c_{A\beta} c_{A\sigma}}}_{\substack{\text{rate of transport across}\\\text{cell membrane}}} \text{ at } \mathscr{A}_{\beta\sigma} \tag{3}$$

Diffusion and reaction in the intercellular phase

$$\underbrace{\frac{\partial c_{A\sigma}}{\partial t}}_{\text{accumulation}} = \underbrace{\nabla \cdot (\mathscr{D}_{A\sigma} \cdot \nabla c_{A\sigma})}_{\text{diffusion}} - \underbrace{\frac{c_{A\sigma}}{c_{A\sigma} + K_{A\sigma}}}_{\substack{\text{intercellular reaction} \\ \text{rate}}} \quad \text{in the } \sigma\text{-phase} \quad (4)$$

The parameters α_1 through α_5 can be related to the intrinsic kinetic rate parameters illustrated in Fig. 2, and the details of these relationships can be found elsewhere.[7,8]

For cases in which the diffusion of small nonionic species is of interest, the boundary condition given by Eq. (3) should be replaced by a permeability model of the form

Transport across membrane (diffusion of small nonionic solutes):

B.C.2 $$\underbrace{\mathscr{D}_{A\beta} \cdot \nabla c_{A\beta} \cdot \mathbf{n}_{\beta\sigma}}_{\substack{\text{interfacial flux,} \\ \text{extracellular phase}}} = \underbrace{k(c_{A\beta} - K_{eq}c_{A\sigma})}_{\substack{\text{rate of transport across} \\ \text{cell membrane}}} \quad \text{at } \mathscr{A}_{\beta\sigma} \quad (5)$$

A complete description of the transport of small, nonionic solutes can be found elsewhere.[2] In the developments that follow we will assume that either Eq. (3) or Eq. (5) will be used as the boundary condition, as dictated by the membrane transport mechanism for the solute of interest.

Equations (1) through (5) provide a description of transport and reaction in biofilms at the micro scale. In principal, these equations could be solved directly for the biofilm illustrated in Fig. 1. Such a solution, however, has several drawbacks. First, a detailed structure of biofilms is rarely available, and such characterization itself represents an enormous challenge. Second (and more importantly), one is generally not interested in the detailed solutions of solute transport in biofilms at the micro scale. Rather, it is usually more useful to have a macroscopic description of biofilms in which the complex structure of the two-phase system is accounted for in terms of effective transport and reaction rate parameters.

Volume averaging[9] is a method that allows one to formally connect the microscopic and macroscopic representations of the biofilm. In this method, one averages the micro-scale transport processes given by Eqs. (1)–(5) over a region that is large enough so that it can be considered *representative* in a statistical sense. The result is a set of averaged equations that apply at the macro scale. The details of the volume averaging of Eqs. (1)–(5) are presented elsewhere[2,7] and will not be reproduced here. However, some basic information about the process of volume averaging is presented in the next section.

---

[8] N. Lakshminarayanaiah, "Equations of Membrane Biophysics." Academic Press, New York, 1984.
[9] S. Whitaker, "The Method of Volume Averaging." Kluwer, Dordrecht, 1999.

The Macroscopic Description

Our goal is to relate the *microscopic* process of solute diffusion in biofilms to an effective *macroscopic* representation, and this is accomplished by averaging Eqs. (1)–(5) over the representative volume indicated by $\mathscr{V}$ in Fig. 1. One begins by defining the *intrinsic volume averages* associated with the β- and σ-phases by

$$\langle c_{A\beta} \rangle^{\beta} = \frac{1}{V_{\beta}} \int_{V_{\beta}(t)} c_{A\beta} \, dV \tag{6}$$

$$\langle c_{A\sigma} \rangle^{\sigma} = \frac{1}{V_{\sigma}} \int_{V_{\sigma}(t)} c_{A\sigma} \, dV \tag{7}$$

Here $c_{A\beta}$ represents the molar concentration of species A in the β-phase, and V_{β} is the volume of the β-phase contained in the averaging volume, $\mathscr{V}$. Using these definitions, the intrinsic average concentration given in Eq. (6) is interpreted as the moles of species A in the β-phase divided by the volume of the β-phase; an analogous interpretation holds for the σ-phase. The volume fractions of the β- and σ-phases are defined by

$$\varepsilon_{\beta} = V_{\beta}/\mathscr{V} \tag{8}$$

$$\varepsilon_{\sigma} = V_{\sigma}/\mathscr{V} \tag{9}$$

Using the definitions indicated by Eqs. (6) and (7), one can form the average of Eqs. (1)–(5) to develop volume averaged transport equations for both the β- and σ-phases. This procedure provides a macroscopic *two-equation model* for diffusion and reaction in biofilms, with separate transport equations governing the distribution of solute in each phase. Although two-equation models have been applied to problems in heat and mass transfer,[10,11] biofilms have typically been described by *one equation models*.[12] It has been shown that there are in fact three regimes that can be used to classify mass transport in biofilms at the macro-scale.[7] These regimes are described by (1) a one-equation, local mass equilibrium model, (2) a two-equation model, and (3) a pseudo-one equation model. More detail about each of these macroscopic models is presented below.

Regime 1: Local Mass Equilibrium

When the principal of *local mass equilibrium* is valid, the volume averaged concentrations can be related by

$$\langle c_{A\beta} \rangle^{\beta} = \alpha_1 \langle c_{A\sigma} \rangle^{\sigma} \quad \text{at local mass equilibrium} \tag{10}$$

[10] M. Quintard and S. Whitaker, *Adv. Heat Transfer* **23**, 369 (1993).

[11] M. Quintard and S. Whitaker, *Int. J. Heat Mass Trans.* **38**, 2779 (1995).

[12] W. Gujer and O. Wanner, *in* "Biofilms" (W. G. Characklis and K. C. Marshall, eds.) Wiley, New York, 1990.

Strictly speaking, this condition is achieved only at thermodynamic equilibrium; however, there are conditions for which $\langle c_{A\beta}\rangle^\beta$ and $\alpha_1 \langle c_{A\sigma}\rangle^\sigma$ are *close enough* so that Eq. (10) is an acceptable approximation. These conditions are referred to as *local mass equilibrium*. In a previous analysis, it has been shown that local mass equilibrium is valid when the following constraints are satisfied[7]:

Capacity constraint:

$$\varepsilon_\beta \varepsilon_\sigma \left(1 - \alpha_1^{-1}\right) \frac{\ell_{\beta\sigma}^2}{\mathsf{D}_{\text{eff}}' t^*} \ll 1 \tag{11}$$

Diffusion constraint:

$$\varepsilon_\beta \varepsilon_\sigma \frac{\left(\mathscr{D}_{A\beta} - \alpha_1^{-1} \mathscr{D}_{A\sigma}\right)}{\mathsf{D}_{\text{eff}}'} \left(\frac{\ell_{\beta\sigma}}{L}\right)^2 \ll 1 \tag{12}$$

Reaction constraint:

$$\varepsilon_\beta \varepsilon_\sigma \frac{\alpha_1 K_A}{(\{c_A\} + \alpha_1 K_A)} \frac{\ell_{\beta\sigma}^2 \mu_A}{(\{c_A\} + \alpha_1 K_A)\mathsf{D}_{\text{eff}}'} \ll 1 \tag{13}$$

Here D_{eff}' represents an estimate of the effective diffusivity that is defined in the next section, t^* is a characteristic process time associated with changes in the average concentrations, and L is a characteristic length scale associated with spatial variations of the average concentration. The small length scale $\ell_{\beta\sigma}$ is a derived quantity defined by

$$\ell_{\beta\sigma}^2 = \frac{1}{a_\nu} \frac{\varepsilon_\beta \varepsilon_\sigma \mathscr{D}_{\beta\sigma}}{(\varepsilon_\sigma + \alpha_1 \varepsilon_\beta) E_0} \left(\alpha_2 + \alpha_3\{c_A\} + \alpha_4 \alpha_1^{-1}\{c_A\} + \alpha_5 \alpha_1^{-1}\{c_A\}^2\right) \tag{14}$$

where $a_\nu = A_{\beta\sigma}/\mathscr{V}$. Note that similar constraints can be developed for the case of small nonionic solutes for conditions when the boundary condition given by Eq. (5) is used, and these results have been presented elsewhere.[2]

When the constraints given by Eqs. (11)–(13) are satisfied, the equilibrium condition given by Eq. (10) is a valid approximation. This condition allows one to *add* the volume-averaged transport equations for the β- and σ-phases to obtain a one-equation model that takes the form[7]

One equation, equilibrium model:

$$\underbrace{\frac{\partial}{\partial t}\left[(\varepsilon_\beta + \alpha_1^{-1}\varepsilon_\sigma)\{c_A\}\right]}_{\text{accumulation}} = \underbrace{\nabla \cdot \left(\mathbf{D}_{\text{eff}}' \cdot \nabla\{c_A\}\right)}_{\text{diffusion}} - \underbrace{\mu_{\text{eff}}'\langle \rho_\sigma\rangle \frac{\{c_A\}}{\{c_A\} + \alpha_1 K_A}}_{\text{reaction}} \tag{15}$$

Because the β- and σ-phase concentrations are related by Eq. (10), one can define the *equilibrium-weighted* concentration, $\{c_A\}$, by

$$\{c_A\} = \varepsilon_\beta \langle c_{A\beta}\rangle^\beta + \varepsilon_\sigma \alpha_1 \langle c_{A\sigma}\rangle^\sigma \tag{16}$$

The coefficients that appear in Eq. (15) are effective parameters whose values can be related to the microscopic parameters and geometry. The coefficient μ'_{eff} is defined by

$$\mu'_{eff} = \frac{\mu_A}{\langle \rho_\sigma \rangle^\sigma} \tag{17}$$

where $\langle \rho_\alpha \rangle^\sigma$ is the (intrinsic) density of the microbial phase. The quantity indicated by $\langle \rho_\sigma \rangle$ is the superficial volume averaged microbial density, and it is defined as the mass of microbes in the averaging volume divided by the volume of the averaging volume.[2,7] The superficial density is related to the intrinsic density by $\langle \rho_\sigma \rangle = \varepsilon_\sigma \langle \rho_\sigma \rangle^\sigma$. The parameter $\mathbf{D}'_{eff}$ is an effective diffusivity tensor that depends on the diffusion coefficients of the extracellular and intercellular phases, the membrane transport parameters, and the microscopic geometry. The definition of the effective diffusivity will be discussed in more detail in the following sections.

When variations in $\left(\varepsilon_\beta + \alpha_1^{-1} \varepsilon_\sigma \right)$ can be neglected everywhere in the biofilm, Eq. (15) can be expressed in the form

$$\underbrace{\frac{\partial \{c_A\}}{\partial t}}_{\text{accumulation}} = \underbrace{\nabla \cdot (\mathbf{D}_{eff} \cdot \nabla \{c_A\})}_{\text{diffusion}} - \underbrace{\mu_{eff} \langle \rho_\sigma \rangle \frac{\{c_A\}}{\{c_A\} + \alpha_1 K_A}}_{\text{reaction}} \tag{18}$$

where the relations between the effective parameters in Eqs. (15) and (18) are given by

$$\mathbf{D}_{eff} = \frac{\mathbf{D}'_{eff}}{\left(\varepsilon_\beta + \alpha_1^{-1} \varepsilon_\sigma \right)}, \quad \mu_{eff} = \frac{\mu'_{eff}}{\left(\varepsilon_\beta + \alpha_1^{-1} \varepsilon_\sigma \right)} \tag{19}$$

In the literature, $\mathbf{D}'_{eff}$ is usually referred to as the *effective diffusivity* and $\mathbf{D}_{eff}$ is often called the *effective diffusive permeability*. The fact that there are two distinct definitions for the effective diffusivity has created some confusion in the literature. Often, Eq. (18) is presented as the species conservation equation with the effective diffusivity, $\mathbf{D}'_{eff}$, in place of the effective diffusive permeability, $\mathbf{D}_{eff}$. This representation is technically valid only for the limiting condition $\alpha_1 = 1$ (or $K_{eq} = 1$). Such an assumption should be applied with caution, since it precludes the possibility that the equilibrium concentrations in the σ- and β-phases are different. Although this assumption may be reasonable for cell constituents whose concentrations are nearly identical on both sides of the membrane (e.g., water), other solutes are concentrated in the cell against a concentration gradient by the expenditure of energy (e.g., potassium). It is important to note there are many chemical species for which the assumption $\alpha_1 = 1$ (or $K_{eq} = 1$) is not valid.

In certain circumstances it may be more convenient to express the transport equation in terms of the β-phase (extracellular) concentration rather than the equilibrium-weighted concentration. When local mass equilibrium is valid, Eq. (15) can be rewritten in the form

One equation, equilibrium model (alternate expression):

$$\frac{\partial}{\partial t}\left[\left(\varepsilon_\beta + \alpha_1^{-1}\varepsilon_\sigma\right)\langle c_{A\beta}\rangle^\beta\right] = \underbrace{\nabla \cdot \left(\mathbf{D}'_{\text{eff}} \cdot \nabla\langle c_{A\beta}\rangle^\beta\right)}_{\text{diffusion}} - \underbrace{\mu'_{\text{eff}}\langle \rho_\sigma\rangle \frac{\langle c_{A\beta}\rangle^\beta}{\langle c_{A\beta}\rangle^\beta + \alpha_1 K_A}}_{\text{reaction}}$$

$$\underbrace{\phantom{\frac{\partial}{\partial t}\left[\left(\varepsilon_\beta + \alpha_1^{-1}\varepsilon_\sigma\right)\langle c_{A\beta}\rangle^\beta\right]}}_{\text{accumulation}}$$

(20)

Again, note that when variations in $(\varepsilon_\beta + \alpha_1^{-1}\varepsilon_\sigma)$ can be neglected, Eq. (20) can be put in a form analogous to Eq. (18). Under these conditions, Eqs. (15), (18), and (20) are all equivalent. It is important to note that available analytical models usually predict the effective diffusivity, $\mathbf{D}'_{\text{eff}}$, rather than the effective diffusive permeability, $\mathbf{D}_{\text{eff}}$.

Regime 2: Two-Equation Model

If the constraints associated with local mass equilibrium are not valid, it may be necessary to use a *two-equation model* for describing mass transport in biofilms. Such a model may be required, for example, when the rate of mass transport across the cell membrane is much slower than the rate of diffusion in the extracellular and intercellular phases. Under these conditions, the concentrations in the β- and σ-phases cannot be assumed to be near equilibrium, and separate transport equations must be used to describe the mass transport in each phase. Although there may be instances where two-equation models are required for describing mass transport in biofilms, the theory has not yet been worked out in detail. However, much can be inferred about the structure of such a representation from previous studies of heat and mass transfer in other two-region systems.[10,11]

Regime 3: Pseudo One-Equation Model

A third situation arises when the intercellular reaction rate is sufficiently rapid that the concentration in the intercellular phase is negligible relative to the concentration in the extracellular phase. This condition is met when the following constraint is valid[7]:

Pseudo one-equation model constraint:

$$\frac{\alpha_2\sqrt{\mathscr{D}_{A\sigma}\mu_A/(c_{A\sigma} + K_A)}}{\alpha_1 E_0} \gg 1 \tag{21}$$

Under these conditions, a *pseudo one-equation model* analogous to Eq. (15) applies. Although the two models have a similar mathematical form, the definition of the effective parameters and the dependency on the biomass concentration are somewhat different. The pseudo one-equation model takes the form

Pseudo one equation model

$$\frac{\partial}{\partial t}\left(\varepsilon_\beta \langle c_{A\beta}\rangle^\beta\right) = \nabla \cdot \left(\mathbf{D}'_{\text{eff}} \cdot \nabla \langle c_{A\beta}\rangle^\beta\right) - \mu_{\text{eff,pseudo}}\, a_v \frac{\langle c_{A\beta}\rangle^\beta}{\langle c_{A\beta}\rangle^\beta + \alpha_2/\alpha_3} \quad (22)$$

$$\langle c_{A\sigma}\rangle^\sigma = 0 \quad (23)$$

where

$$\mu'_{\text{eff,pseudo}} = \frac{E_0}{\alpha_3} \quad (24)$$

When variations in ε_β can be neglected, Eq. (22) can be put in a form that is analogous to Eq. (18).

There are important differences in the reaction terms of the two one-equation models. For the one-equation, equilibrium model [Eq. (20)] the reaction rate is proportional to the *mass* of microorganisms per unit volume, $\langle \rho_\sigma \rangle$, whereas for the pseudo one-equation model [Eq. (22)], the reaction rate is proportional to the *area* of microorganisms per unit volume, a_v. Although Eqs. (15) and (22) have similar forms, their interpretation is very different.

Calculation of the Effective Diffusivity

The primary results from the process of volume averaging are (1) the derivation of the mathematical form of the macroscopic transport equations [Eqs. (15) and (22)], and (2) the definitions of the effective diffusion and reaction rate parameters in terms of microscopic properties of the biofilm. For the case of the effective reaction rate parameters, these results take the simple algebraic forms given by Eqs. (17) and (24). In contrast, the effective diffusion coefficient depends on both the microscopic transport parameters and the geometry of the biofilm. To obtain predictive expressions for this quantity, one must develop and solve a *closure problem* that relates the effective diffusion coefficient to the geometric structure of the biofilm as well as to the microscopic transport parameters. Such a development is beyond the scope of the work reported here, and the relevant details have been provided elsewhere.[2,7] For completeness, a brief account of the closure problem is given below in Appendix A.

The closure problem relates the microscopic properties of the biofilm to the macroscopic effective diffusion coefficient, and this scheme is most useful for predicting the effective diffusivity in a representative (statistically homogeneous) volume of biofilm. Intuitively, this means that if the size of the averaging volume is increased or its location is changed, the calculated value of the effective diffusion coefficient is not altered.

There are two methods that have been used to solve the closure problem required to predict the effective diffusivity. The first involves finding an analytical solution for a particularly simple geometry. Although this may represent a significant abstraction from the actual biofilm, for some problems simple geometries can capture much of the important structure that affects the solution to the closure problem. When such solutions exist, they can provide closed-form expressions for predicting the effective parameters that are easy to use in practice. For prediction of the effective diffusion coefficient, Chang's unit cell[13] provides a simple geometry that is tractable by analytical methods. This method leads to results that are identical to some of the classical solutions of Maxwell.[14]

The second approach that can be used to predict the effective diffusion coefficient is to solve the closure problem numerically in a complex unit cell. The idea in this approach is to fully resolve all of the important structural details in a representative volume of biofilm. Then, when the closure problem is solved over this representative volume, the predicted effective diffusion coefficient includes all of the effects of the complex structure. Although this method may lead to more realistic values for the effective diffusion coefficient, it has the drawback that it is a numerical scheme requiring substantial computational resources. We will discuss the prediction of the effective diffusion coefficient using both approaches.

Chang's Unit Cell

Experimental investigations have shown that in systems with an impermeable σ-phase, it is primarily the volume fraction ε_β that determines the effective diffusion coefficient.[9] For biofilms, it is reasonable to assume that a simple geometry might be used to predict the effective diffusion coefficient provided that it reproduced the key elements of the biofilm structure. In Fig. 3a we have illustrated the simple spherical geometry associated with Chang's unit cell, and Fig. 3b illustrates the structure of a complex unit cell that is patterned after a real biofilm. For the use of Chang's unit cell, one must accept the idea that the cell geometry can be represented by concentric spheres. The volume fraction and the surface area per unit volume of the cell are chosen to match the experimental system. Given these conditions, one can develop an analytical solution for the closure problem in order to predict the effective diffusion coefficient under the conditions of local mass equilibrium or for conditions for which the pseudo one-equation model is valid. The details are presented elsewhere,[2,15] and the closure problem for this simple geometry is listed in Appendix A. Here we simply list the solution as

Chang's unit cell solution:

$$\frac{D'_{eff}}{\mathscr{D}_{A\beta}} = \frac{3\kappa - 2\varepsilon_\beta(\kappa - 1) + 2\varepsilon_\beta(3\varepsilon_\sigma)^{-1}a_v\kappa\psi}{3 + \varepsilon_\beta(\kappa - 1) + (3 - \varepsilon_\beta)(3\varepsilon_\sigma)^{-1}a_v\kappa\psi} \tag{25}$$

[13] H.-C. Chang, *Chem. Eng. Commun.* **15**, 13 (1982).
[14] J. C. Maxwell, "Treatise on Electricity and Magnetism," 3rd ed., Vol. I, reproduction of the Claridon Press imprint of 1891, p. 506. Dover Publications, New York, 1954.
[15] J. A. Ochoa, S. Whitaker, and P. Stroeve, *Chem. Eng. Sci.* **41**, 2999 (1986).

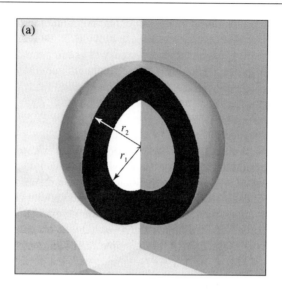

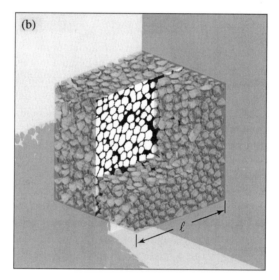

FIG. 3. (a) Simple spherical (Chang's) unit cell. (b) Complex unit cell.

where the notation is

$$\kappa = \frac{\mathscr{D}_{A\sigma}}{\alpha_1 \mathscr{D}_{A\beta}}, \quad \psi = \frac{\Gamma \mathscr{D}_{A\beta}}{E_o} \tag{26}$$

For the case of small nonionic solutes [and the associated boundary condition given by Eq. (5)] these definitions must be expressed in the form

$$\kappa = \frac{\mathscr{D}_{A\sigma}}{K_{eq}\,\mathscr{D}_{A\beta}}, \quad \psi = \frac{\mathscr{D}_{A\beta}}{k} \tag{27}$$

Note that these expressions contain (1) an effect due to mass transfer within the two phases, and (2) an effect due to mass transfer across the cell membrane. In particular, the parameter κ gives an indication of the relative rate of solute transport in the σ- and β-phases, whereas the parameter ψ provides an indication of the relative rates of diffusive mass transfer in the β-phase to the rate of mass transport across the membrane.

In many conventional representations, the influence of the membrane on the effective diffusion coefficient is neglected entirely. Strictly speaking, this approximation is valid only when the rate of mass transfer across the membrane is very rapid relative to the diffusive redistribution of mass. For this case, one can set ψ equal to zero, and this leads to the classical solution of Maxwell[14] given by

Maxwell's solution:

$$\frac{\mathrm{D}'_{eff}}{\mathscr{D}_{A\beta}} = \frac{3\kappa - 2\varepsilon_\beta(\kappa - 1)}{3 + \varepsilon_\beta(\kappa - 1)} \tag{28}$$

Equation (28) represents a substantial simplification of Eq. (25) and is only valid when the mass transfer resistance of the membrane is negligible. In applications to biofilms, it is often further assumed that the equilibrium coefficient α_1 or K_{eq} is unity. As discussed above, this assumption is only reasonable for constituents that are not specifically concentrated or excluded from cells. Because many chemical species of interest (e.g., potassium) may be concentrated in cells by specific proteins, this assumption must be applied with caution. When it can be assumed that the equilibrium coefficient α_1 or K_{eq} is identically unity, Eq. (28) can be expressed in the commonly used form

Maxwell's solution (permeable spheres, $\alpha_1 = 1$):

$$\frac{\mathrm{D}'_{eff}}{\mathscr{D}_{A\beta}} = \frac{2\,\mathscr{D}_{A\beta} + \mathscr{D}_{A\sigma} - 2\varepsilon_\sigma(\mathscr{D}_{A\beta} - \mathscr{D}_{A\sigma})}{2\,\mathscr{D}_{A\beta} + \mathscr{D}_{A\sigma} + \varepsilon_\sigma(\mathscr{D}_{A\beta} - \mathscr{D}_{A\sigma})} \tag{29}$$

Note that for this case, the effective diffusivity and the effective diffusive permeability are identical, as shown by Eq. (19).

In some cases, the diffusivity of solutes that are not transported across the cell wall is of interest. This situation occurs when there is no mechanism for the solute to be transported across the cell wall, or when the enzymes in the cell have been denatured so that enzyme-mediated processes are not active. If transport across the cell membrane is prevented by the destruction of specific proteins by heat or chemicals, then the effective diffusivity will not be representative of the value that applies when these proteins are active. This suggests that a certain amount of caution is required in the preparation of cells for measurement of the effective

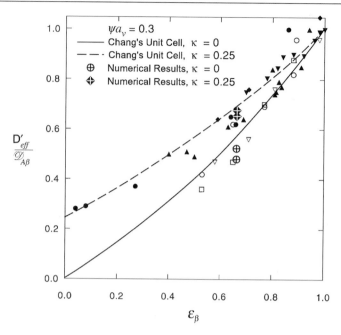

FIG. 4. Comparison between experimental data, analytical results, and numerical results for the effective diffusivity. Open symbols indicate data collected from systems where solute was excluded from cells; filled symbols indicate data collected from systems in which solutes were transported across the cell membrane.

diffusion coefficient. For conditions where no transport across the cell membrane occurs, Eq. (29) simplifies to the form

Maxwell's solution (impermeable spheres, $\kappa = 0$):

$$\frac{D'_{eff}}{\mathscr{D}_{A\beta}} = \frac{2\varepsilon_\beta}{3 - \varepsilon_\beta} = \frac{2(1 - \varepsilon_\sigma)}{2 + \varepsilon_\sigma} \tag{30}$$

This result also represents the correct expression for the effective diffusivity to be used with the pseudo one-equation model given by Eq. (22).

In Fig. 4 a plot comparing the analytical solution to Chang's unit cell with measured values of the effective diffusion coefficient is shown. These data are from measurements reported previously in the literature[16–19] and are tabulated in a previous paper.[2] The data presented in Fig. 4 are separated into two groups: (1) chemical species that *were not* transported across the cell membrane (open symbols), and (2) chemical species that *were* transported across the cell membrane by either diffusion or by a specific protein system (filled symbols). Solutions to Chang's unit cell are presented for the same two cases. The solid curve appearing in Fig. 4 is the effective diffusivity predicted by Eq. (30) for the case in which solute is excluded from the σ-phase, and the predictions are in reasonable agreement with the experimental data. Predictions for this case are relatively straightforward because knowledge of only the volume fraction ε_β is required.

For a case in which solutes are transported into the σ-phase, the prediction of the effective diffusivity is complicated by the fact that estimates of two additional parameter groups (κ and ψa_v) are required. A review of the literature suggests that reasonable values for κ are between 0 and 1; however, for solutes that are highly concentrated inside cells (relative to the extracellular concentration) this parameter could be much larger. The value of $\mathscr{D}_{A\sigma}/\mathscr{D}_{A\beta}$ can be estimated to be between 0.2 and 0.3,[16] and this is helpful for estimating κ. We have adopted a moderate value, $\kappa = 0.25$, for the curve plotted in Fig. 4 (dashed line). Appropriate values for ψa_v are much harder to estimate, but the literature suggests that reasonable values are between 0 to 30. For the plot shown in Fig. 4 (dashed line), a value of $\psi a_v = 0.3$ is used, and this provides reasonable agreement with experimental data from a large variety of sources.

Numerical Solutions of Complex Unit Cells

Chang's unit cell provides a convenient and simple method for predicting the effective diffusion coefficient in a biofilm on the basis of a few parameters. However, for some biofilms this approach may not be suitable. We have completed a study of diffusion in biofilms with complex structure,[20] such as that illustrated in Fig. 3b. For these systems, no analytical solution is available and one must use numerical techniques to solve the closure problem. The numerical solution to the closure problem is described elsewhere[20] and will not be discussed in detail here.

We have found that the numerical solution to the closure problem is in reasonable agreement with the solution provided by Chang's unit cell. In Fig. 4 plotted values for the effective diffusivity calculated numerically for the structure illustrated in Fig. 3b are shown. It was assumed that κ took the values of 0 and 0.25 and that $\psi a_v = 0.3$ so that these calculations correspond to the predictions of the effective diffusivity obtained using Chang's unit cell. Because only one three-dimensional biofilm structure was used, numerically calculated effective diffusivities are available only for a single β-phase volume fraction, $\varepsilon_\beta = 0.67$. The structure illustrated in Fig. 3b is not spatially isotropic, so there are some small differences in the diffusivity in each of the three principal directions of the

[16] C. S. Ho and L.-K. Ju, *Biotechnol. Bioeng.* **32**, 313 (1988).

[17] T. J. Chresand, B. E. Dale, S. L. Hanson, and R. J. Gillies, *Biotechnol. Bioeng.* **32**, 1029 (1988).

[18] S. B. Libicki, P. M. Salmon, and C. R. Robertson, *Chem. Eng. Sci.* **32**, 68 (1988).

[19] A. Axelsson and B. Persson, *Appl. Biochem. Biotechnol.* **18**, 231 (1988).

[20] B. Wood, M. Quintard, and S. Whitaker, *submitted to Biotechnol. Bioeng.* (2001).

unit cell. For each of the numerical simulations considered, three points are plotted in Fig. 4 representing the separate values of the effective diffusivity obtained for the x-, y-, and z-directions. Because the structure in the x- and y-directions is very similar, the values calculated for the diffusivity in these directions are nearly indistinguishable. All of the values calculated for the effective diffusivity are in reasonable agreement with the corresponding values predicted by Chang's unit cell (dashed line). Although in many instances it may be more accurate to calculate the effective diffusivity by a numerical solution to the closure problem, in practice this is generally not feasible. Because of the good correspondence between the two approaches, the analytical solution provided by Chang's unit cell [Eqs. (25), (28), (29), or (30)] will provide reasonable estimates for many applications.

Comments

A closed-form solution that can be used to predict the effective diffusion coefficient for a wide variety of systems has been presented. At the core of this solution is the assumption that a complex biofilm structure can be approximated by a spherical unit cell with the same surface area per unit volume and volume fractions as the complex biofilm. For the complex biofilm shown in Fig. 3b, the simple geometry of Chang's unit cell provides an estimate for the effective diffusivity that is in reasonable agreement with the estimate from the complex unit cell. This suggests that the effective diffusivity can, in some instances, be predicted using the solution for Chang's unit cell, as presented in Eq. (25). However, even for the simplified analytical solutions provided here, the prediction of effective diffusivities requires careful analysis. In particular, we note the following:

1. The effective diffusivity predicted by Eq. (25) requires that the condition of *local mass equilibrium* be valid. This leads to a one-equation model for solute transport of the form of Eq. (15) or (20). For local mass equilibrium to be valid, the constraints given by Eqs. (11)–(13) must be met.

2. The effective diffusivity can also be predicted for the case where the *pseudo one-equation model* applies. In this case, Eq. (30) can be used to predict the effective diffusivity. For the pseudo one-equation model to be valid, the intercellular reaction rate must be large enough so that the constraint given by Eq. (21) is satisfied.

3. The effective diffusivity depends on the *chemical species* that are diffusing. For many solutes of interest, transport across the membrane may be mediated by specific proteins, and these transporter proteins can have solute-specific properties. Therefore, the effective diffusivity may depend strongly on the particular solute of interest. One must be extremely careful when attempting to extrapolate the effective diffusivity of one solute on the basis of the values for a different solute. Although sufficient evidence may exist under many circumstances to suggest that

the transport properties of two solutes are similar, this issue must be considered carefully.

4. The effective diffusivity of a biofilm or cellular system may depend on the *metabolic state* of the cells. Again, if a specific protein is responsible for transmembrane transport of a particular solute, then the transport properties of that solute will change if the specific protein transport system is deactivated. This has potentially serious ramifications for measurements of the effective diffusivity in systems that have been enzymatically deactivated by heat or chemicals. The effective diffusivity measured in such systems may not be representative of the conditions that are of interest for applications.

5. Numerical simulations suggest that the solution to Chang's unit cell provides a reasonable estimate of the effective diffusivity for some cases. In particular, the cases that have been tested so far suggest that the correspondence is best for values of κ less then 1; when ψa_v is either very large ($\psi a_v \gg 1$) or zero, good correspondence is also observed for a wide range of values of κ. The numerical simulations have tested only systems in which the cells are nearly spherical (such as those illustrated in Fig. 3b); the effective diffusivity predicted by Eq. (25) will probably not give good results for cell geometries that are not well represented by spheres. An analytical solution to an elliptical version of Chang's unit cell has been developed,[21] and this solution may provide estimates of the effective dispersivity in cases where the structure is anisotropic. For more general cases where the biofilm structure is not well represented by spheres or ellipsoids, a numerical solution of the closure problem remains the only tractable means of determining the effective diffusivity.

Appendix A: The Closure Problem

Wood and Whitaker[7] have shown that the effective diffusion tensor can be defined by

$$\mathbf{D}'_{\text{eff}} = \left(\varepsilon_\beta \, \mathscr{D}_{A\beta} + \alpha_1^{-1}\varepsilon_\sigma \, \mathscr{D}_{A\sigma}\right)\mathbf{I} + \frac{\mathscr{D}_{A\beta}}{\mathscr{V}} \int_{A_{\beta\sigma}} \mathbf{n}_{\beta\sigma}\mathbf{b}_{A\beta} \, dA$$

$$+ \frac{\alpha_1^{-1} \, \mathscr{D}_{A\sigma}}{\mathscr{V}} \int_{A_{\sigma\beta}} \mathbf{n}_{\sigma\beta}\mathbf{b}_{A\sigma} \, dA \qquad (\text{A.1})$$

where the quantities $\mathbf{b}_{A\beta}$ and $\mathbf{b}_{A\sigma}$ are defined by

$$\tilde{c}_{A\beta} = \mathbf{b}_{A\beta} \cdot \nabla \langle c_{A\beta}\rangle^\beta \qquad (\text{A.2})$$

$$\alpha_1 \tilde{c}_{A\sigma} = \mathbf{b}_{A\sigma} \cdot \nabla \langle c_{A\sigma}\rangle^\sigma \qquad (\text{A.3})$$

[21] J. A. Ochoa-Tapia, *Chem. Eng. Sci.* **49**, 709 (1994).

In Eqs. (A.2) and (A.3), the quantities $\tilde{c}_{A\beta}$ and $\tilde{c}_{A\sigma}$ are *deviation concentrations* defined by

$$\tilde{c}_{A\beta} = c_{A\beta} - \langle c_{A\beta} \rangle^{\beta} \tag{A.4}$$

$$\tilde{c}_{A\sigma} = c_{A\sigma} - \langle c_{A\sigma} \rangle^{\sigma} \tag{A.5}$$

Equations (A.4) and (A.5) suggest that transport equations for predicting $\tilde{c}_{A\beta}$ and $\tilde{c}_{A\sigma}$ can be found by subtracting the volume averaged transport equations for the β- and σ-phases, respectively, from the point transport equations given by Eqs. (1) and (4). Substituting the representation of the deviation concentrations given by Eqs. (A.2) and (A.3) into the closure problem yields the following differential equation that must be solved to predict the effective diffusion tensor:

$$\nabla^2 \mathbf{b}_{A\beta} = \varepsilon_{\beta}^{-1} \mathbf{c} \quad \text{in the } \beta\text{-phase} \tag{A.6}$$

B.C.1 $\quad \mathbf{n}_{\beta\sigma} \cdot \nabla \mathbf{b}_{A\beta} = \kappa\, \mathbf{n}_{\beta\sigma} \cdot \nabla \mathbf{b}_{A\sigma} + \underbrace{\mathbf{n}_{\beta\sigma}(\kappa - 1)}_{\text{source}} \quad \text{at } A_{\beta\sigma} \tag{A.7}$

B.C.2 $\quad \mathbf{b}_{A\beta} - \mathbf{b}_{A\sigma} = -\underbrace{\kappa\psi\, \mathbf{n}_{\beta\sigma}}_{\text{source}} - \kappa\psi\, \mathbf{n}_{\beta\sigma} \cdot \nabla \mathbf{b}_{A\sigma} \quad \text{at } A_{\beta\sigma} \tag{A.8}$

$$\nabla^2 \mathbf{b}_{A\sigma} = -\frac{\varepsilon_{\sigma}^{-1}}{\kappa} \mathbf{c} \quad \text{in the } \sigma\text{-phase} \tag{A.9}$$

Periodicity: $\quad \mathbf{b}_{A\beta}(\mathbf{r}) = \mathbf{b}_{A\beta}(\mathbf{r} + \ell_i), \quad i = 1, 2, 3 \tag{A.10}$

Periodicity: $\quad \mathbf{b}_{A\sigma}(\mathbf{r}) = \mathbf{b}_{A\sigma}(\mathbf{r} + \ell_i), \quad i = 1, 2, 3 \tag{A.11}$

where

$$\mathbf{c} = \frac{1}{\mathcal{V}} \int_{A_{\beta\sigma}(t)} \mathbf{n}_{\beta\sigma} \cdot \nabla \mathbf{b}_{A\beta}\, dA = -\kappa \frac{1}{\mathcal{V}} \int_{A_{\sigma\beta}(t)} \mathbf{n}_{\sigma\beta} \cdot \nabla \mathbf{b}_{A\sigma}\, dA \tag{A.12}$$

$$\Gamma = \alpha_2 + \alpha_3 \langle c_{A\beta} \rangle^{\beta} + \alpha_4 \langle c_{A\sigma} \rangle^{\sigma} + \alpha_5 \langle c_{A\beta} \rangle^{\beta} \langle c_{A\sigma} \rangle^{\sigma} \tag{A.13}$$

For the simplified geometry specified for Chang's unit cell, the closure problem reduces to

$$\nabla^2 \mathbf{b}_{A\sigma} = 0 \quad \text{in the } \sigma\text{-phase} \tag{A.14}$$

B.C.1 $\quad -\mathbf{n}_{\beta\sigma} \cdot \nabla \mathbf{b}_{A\beta} = -\mathbf{n}_{\beta\sigma} \cdot \kappa \nabla \mathbf{b}_{A\sigma} + \mathbf{n}_{\beta\sigma}(1 - \kappa) \quad \text{at } r = r_1 \tag{A.15}$

B.C.2 $\quad \kappa\mathbf{n}_{\beta\sigma} + \kappa\mathbf{n}_{\beta\sigma} \cdot \nabla \mathbf{b}_{A\sigma} = \frac{1}{\psi}(\mathbf{b}_{A\sigma} - \mathbf{b}_{A\beta}) \quad \text{at } r = r_1 \tag{A.16}$

$$\nabla^2 \mathbf{b}_{A\beta} = 0 \quad \text{in the } \beta\text{-phase} \tag{A.17}$$

B.C.3 $\quad \mathbf{b}_{A\beta} = constant \quad \text{at } r = r_2 \tag{A.18}$

Similar results hold when the boundary condition given by Eq. (5) is used. Additional details can be found in previously published work.[2,7]

Appendix B: Nomenclature

Roman Letters

a_v	$A_{\beta\sigma}/\mathcal{V}$, area per unit volume, m^{-1}
$\mathcal{A}_{\beta\sigma}$	interfacial area contained within the biofilm, m^2
$A_{\beta\sigma}$	interfacial area contained within a representative volume of the biofilm, m^2
$\mathbf{b}_{A\beta}$	vector field that maps $\nabla\{c_A\}$ onto $\tilde{c}_{A\beta}$, m
$\mathbf{b}_{A\sigma}$	vector field that maps $\nabla\{c_A\}$ onto $\alpha_1\tilde{c}_{A\sigma}$, m
$c_{A\beta}$	concentration of species A in the β-phase, mol m^{-3}
$c_{A\sigma}$	concentration of species A in the σ-phase, mol m^{-3}
$\langle c_{A\beta}\rangle^\beta$	intrinsic average concentration of species A in the β-phase, mol m^{-3}
$\langle c_{A\beta}\rangle$	superficial average concentration of species A in the β-phase $(=\varepsilon_\beta\langle c_{A\beta}\rangle^\beta)$, mol m^{-3}
$\langle c_{A\sigma}\rangle^\sigma$	intrinsic average concentration for the σ-phase, mol m^{-3}
$\langle c_{A\sigma}\rangle$	superficial average concentration of species A in the σ-phase $(=\varepsilon_\sigma\langle c_{A\sigma}\rangle^\sigma)$, mol m^{-3}
$\langle c_A\rangle$	spatial average concentration of species A $(=\varepsilon_\beta\langle c_{A\beta}\rangle^\beta + \varepsilon_\sigma\langle c_{A\sigma}\rangle^\sigma)$, mol m^{-3}
$\{c_A\}$	equilibrium average weighted average concentration of species A $(=\varepsilon_\beta\langle c_{A\beta}\rangle^\beta + \varepsilon_\sigma\alpha_1\langle c_{A\sigma}\rangle^\sigma)$, defined by Eq. (16), mol m^{-3}
$\tilde{c}_{A\beta}$	spatial deviation concentration for species A in the β-phase $(=\langle c_{A\beta}\rangle^\beta - c_{A\beta})$, mol m^{-3}
$\tilde{c}_{A\sigma}$	spatial deviation concentration for species A in the σ-phase $(=\langle c_{A\sigma}\rangle^\sigma - c_{A\sigma})$, mol m^{-3}
d_p	effective particle diameter, m
$\mathcal{D}_{A\beta}$	mixture diffusivity for species A in the β-phase, $m^2\ s^{-1}$
$\mathcal{D}_{A\sigma}$	mixture diffusivity for species A in the σ-phase, $m^2\ s^{-1}$
$\mathbf{D}'_{\mathrm{eff}}$	effective diffusivity for the one-equation, equilibrium model or for the pseudo one-equation model, $m^2\ s^{-1}$
$\mathbf{D}_{\mathrm{eff}}$	alternate effective diffusivity for the one-equation, equilibrium model $(\mathbf{D}'_{A,\mathrm{eff}}/[\varepsilon_\beta + \alpha_1^{-1}\varepsilon_\sigma])$, $m^2\ s^{-1}$
E_o	surface concentration of transporter proteins, mol m^{-2}
$\mathbf{I}$	unit tensor
k	membrane permeability for a small nonionic species, m s^{-1}
K_{eq}	equilibrium coefficient for a small nonionic species, m s^{-1}
K_A	half saturation constant, mol m^{-3}
$\ell_{\beta\sigma}$	small-scale characteristic length defined by Eq. (14), m
ℓ	lattice vector used in defining the periodic boundary condition in the closure problem, m

L characteristic length defined by $L^2 = L_c L_{c1}$, m

$\mathbf{n}_{\beta\sigma}$ $-\mathbf{n}_{\sigma\beta}$, unit normal vector directed from the β-phase toward the σ-phase

$\mathbf{O}$ Order-of-magnitude symbol

t time, s

t^* characteristic process time, s

$\mathscr{V}$ averaging volume, m^3

V_β volume of the β-phase contained in the averaging volume, m^3

V_σ volume of the σ-phase contained in the averaging volume, m^3

Greek Letters

α_1 equilibrium coefficient

α_2 reaction rate parameter, moles-s m^{-3}

α_3 reaction rate parameter, s

α_4 reaction rate parameter, s

α_5 reaction rate parameter, m^{3}s moles^{-1}

ε_β volume fraction of the β-phase ($= V_\beta/\mathscr{V}$)

ε_σ volume fraction of the σ-phase ($= 1 - \varepsilon_\beta$)

ψ dimensionless parameter defined by Eq. (26) or Eq. (27)

κ dimensionless parameter defined by Eq. (26) or Eq. (27)

ρ_σ density of the σ-phase, kg m^{-3}

$\langle \rho_\sigma \rangle^\sigma$ intrinsic average σ-phase density (mass of cells per unit volume of cells), kg m^{-3}

$\langle \rho_\sigma \rangle$ superficial average σ-phase density ($= \varepsilon_\sigma \langle \rho_\sigma \rangle^\sigma$), kg m^{-3}

μ_A maximum specific substrate utilization rate parameter, moles m^{-3} s^{-1}

μ'_{eff} effective maximum specific substrate utilization rate parameter, pseudo one-equation model [defined by Eq. (17)], mol kg^{-1} s^{-1}

μ_{eff} effective maximum specific substrate utilization rate parameter, one-equation equilibrium model [defined by Eq. (19)], mol kg^{-1} s^{-1}

$\mu'_{\text{eff,pseudo}}$ effective maximum specific substrate utilization rate parameter, pseudo one-equation model [defined by Eq. (24)], mol m^{-2} s^{-1}

Γ membrane kinetic rate term ($= \alpha_2 + \alpha_3 \langle c_{A\beta} \rangle^\beta + \alpha_4 \langle c_{A\sigma} \rangle^\sigma + \alpha_5 \langle c_{A\beta} \rangle^\beta \langle c_{A\sigma} \rangle^\sigma$), mol-s m^{-3}.

Acknowledgments

The first author received support from the Department of Health and Environmental Research (EMSL Intermediate Flow Cells project), U.S. Department of Energy, and from the ECRI LDRD program at Pacific Northwest National Laboratory. Pacific Northwest Laboratory is operated for the U.S. Department of Energy by Battelle Memorial Institute under contract DE-AC06-76RLO 1830.

[23] Limiting-Current-Type Microelectrodes for Quantifying Mass Transport Dynamics in Biofilms

By Zbigniew Lewandowski and Haluk Beyenal

Overview

Quantifying factors affecting mass transport in biofilms help in understanding the mechanics of biofilm processes. Nutrients are delivered to the microorganisms embedded in extracellular polymers by mass transport and waste products are removed the same way. Since biofilm thickness is conveniently expressed in microns, the tools to study the intrabiofilm mass transport dynamics have to provide comparable resolution. We use microelectrodes, often in conjunction with confocal scanning laser microscopy. Although the use of microelectrodes for measuring concentration profiles of various substances in biofilms is well known, quantifying factors affecting mass transport in biofilms define a set of new challenges for making these measurements. These challenges are determined by the fact that the measured parameters are not simple physical quantities (e.g., oxygen concentration) but complex ones (e.g., mass transport coefficient or effective diffusivity) that appear as proportionality constants in respective equations describing mass transport dynamics. To evaluate these parameters the entire measurement system has to be designed in such way that the microelectrode response reflects the magnitude of these complex parameters. This chapter demonstrates the use of microelectrodes for evaluating the mass transport coefficient, effective diffusivity, and flow velocity in space occupied by biofilms.

Many of the concepts and all of the experimental results presented here were generated by the Biofilm Structure and Function Research Group at the Center for Biofilm Engineering. There are many excellent measurements of mass transport in biofilms reported by other research groups; however, limited space only allows referring to those that can be directly compared with our techniques. At several places in the text we indicate the types of instruments we use. This is not meant as an endorsement or recommendation of any specific piece of hardware, because many other instruments and tools are available that perform the same functions. Specifying what we use should serve as guidance, a point of departure for those intending to build such measurement systems.

Microelectrodes for Quantifying Factors Affecting Mass Transport in Biofilms

Microelectrodes are often used to measure local nutrient concentration profiles across biofilms.[1-3] The most commonly evaluated parameters of such profiles are

METHODS IN ENZYMOLOGY, VOL. 337

the fluxes of nutrients consumed by the biofilm, e.g., oxygen, which are calculated by multiplying the slope of the concentration profile at the biofilm surface by the molecular diffusivity of the respective nutrient. However, difficulties arise in interpreting the results of such measurements. For instance, the shape of the nutrient concentration profile is affected by many factors, the most influential being those that affect mass transport dynamics (mass transport coefficient, effective diffusivity, hydrodynamics) and biofilm activity (substrate concentration, presence of inhibitory substances).[4] If the biofilm respiration rate or the flow velocity increase, the overall effect is the same: mass transport rate to the biofilm increases. At the macro scale of observation, it is not possible to distinguish the effects of mass transport dynamics from the effects of microbial activity on the overall nutrient consumption rate of the biofilm. Because biofilm reaction rates are generally limited by mass transport rates, an increase in flow velocity and an increase in microbial respiration rate both produce steeper nutrient concentration gradients.

To gain an insight into the dynamics of biofilm processes, microelectrode measurements of nutrient concentration profiles are often supplemented by the measurements of factors affecting mass transport dynamics. Combining results of these two groups of measurements allows us, in simple experimental systems, to distinguish the effects of mass transport from the effects of microbial activity on the overall biofilm activity. Although other procedures to evaluate the parameters characterizing mass transport in biofilms are available—microinjections combined with confocal scanning laser microscopy[5,6] and fluorescent recovery after photobleaching[7] (FRAP)—using microelectrodes to quantify mass transport dynamics is often the superior method because it can be integrated with the microelectrode measurements of nutrient concentration profiles. The purpose of this chapter is to demonstrate the utility of the limiting-current-type microelectrodes for quantifying factors affecting mass transport in biofilms.

Quantifying the factors affecting mass transport in biofilms is complicated by the fact that biofilms are structurally heterogeneous, i.e., microorganisms in biofilms are clustered in microcolonies separated by voids.[8–12] As a consequence,

[1] N. P. Revsbech, *J. Microbiol. Meth.* **9**, 11 (1989).

[2] Z. Lewandowski, G. Walser, and W. G. Characklis, *Biotechnol. Bioeng.* **38**, 877 (1991).

[3] Y. C. Fu, T. C. Zhang, and P. L. Bishop, *Wat. Res.* **29**, 455 (1994).

[4] K. Rasmussen and Z. Lewandowski, *Biotechnol. Bioeng.* **59**, 302 (1998).

[5] D. de Beer, P. Stoodley, and Z. Lewandowski, *Biotechnol. Bioeng.* **53**, 151 (1997).

[6] J. R. Lawrence, G. M. Wolfaardt, and D. R. Korber, *Appl. Environ. Microbiol.* **60**, 1166 (1994).

[7] J. D. Bryers and F. Drummond, *Biotechnol. Bioeng.* **60**, 562 (1998).

[8] D. de Beer, P. Stoodley, and Z. Lewandowski, *Biotechnol. Bioeng.* **44**, 636 (1994).

[9] D. de Beer, P. Stoodley, F. Roe, and Z. Lewandowski, *Biotechnol. Bioeng.* **43**, 1131 (1994).

[10] D. R. Noguera, S. Okabe, and C. Picioreanu, *Wat. Sci. Tech.* **39**, 273 (1999).

[11] P. L. Bishop and B. E. Rittmann, *Wat. Sci. Tech.* **32**, 263 (1995).

[12] J. R. Lawrence, D. R. Korber, B. D. Hoyle, J. W. Costerton, and D. E. Caldwell, *J. Bacteriol.* **173**, 6558 (1991).

chemistries and mass transport rates vary among points in the biofilm,[13–15] and measuring a single concentration profile at an arbitrarily selected location cannot describe the complex, three-dimensional effects of mass transport dynamics on biofilm activity. It is apparent that more complex protocols must be applied for such measurements.

In our laboratory, we have developed tools and experimental protocols to measure parameters that quantify factors affecting mass transport dynamics in three dimensions, as well as to generate maps of the spatial distribution of the measured parameters in biofilms. Although using these protocols made studying the mass transport in heterogeneous biofilms in three dimensions possible, it also makes the interpretation of the results complex because they are in three dimensions and because there are no simple mathematical models available to facilitate this process.

Factors affecting mass transport in biofilms can be monitored at two levels of observation: macro and micro scale. It is difficult to prescribe a specific dimension separating these scales of observation because any magnitude selected to separate them would have to be chosen arbitrarily. Therefore, it is better to base such distinction on what is expected from each scale of observation and what tools are routinely applied at the macro and micro scale. Macro-scale observations refer to the average overall properties of the entire system measured by chemical analysis of the bulk solution with the aid of mass balances. Micro-scale observations refer to the spatial distribution of factors affecting the local microbial activity and the local mass transport dynamics, e.g., local biofilm structure, local hydrodynamics, and local mass transport rates. Tools used for micro-scale observations must principally provide resolution high enough to distinguish the distribution of the measured quantities at a scale comparable with the smallest dimension of the biofilm—biofilm thickness. A variety of microscopy and microsensor tools can provide the micron-sized resolution necessary for these measurements.

Results generated at the micro scale are difficult to compare with results generated at the macro scale. Clearly, the effects measured at the macro scale have to be related to the underlying biofilm processes taking place at the micro scale. It is a major challenge in biofilm engineering to integrate the results of the micro-scale observations for predicting the results measured at the macro scale. Thus far these efforts of "scaling up" the micro-scale observations have not been very successful, demonstrating that our knowledge of biofilm processes is still inadequate.

For system analysis, it is convenient to subdivide the space influenced by biofilm activity into two zones: external (the bulk solution) and internal (the biofilm). Micro-scale measurements of mass transport dynamics in both external and internal zones of bacterial biofilms are the subject of this chapter. To simplify

[13] B. R. Rittmann, M. Pettis, W. H. Reeves, and D. Stahl, *Wat. Sci. Tech.* **39,** 99 (1999).
[14] C. Picioreanu, M. C. M. van Loosdrecht, and J. J. Heijnen, *Biotechnol. Bioeng.* **57,** 718 (1998).
[15] C. Picioreanu, M. C. M. van Loosdrecht, and J. J. Heijnen, *Biotechnol. Bioeng.* **58,** 101 (1998).

the considerations, we assume that (1) the mass transport dynamics in the external zone is entirely defined by the bulk liquid hydrodynamics, and also that there is no substrate consumption in that zone, and (2) the mass transport dynamics in the internal zone is entirely defined by the intrabiofilm hydrodynamics, by biofilm structure, and by biofilm activity.

Limiting Current and Limiting-Current-Type Sensors

The limiting-current-type sensors belong to a large group of amperometric sensors whose principle use is measuring current generated by reducing (or oxidizing) electroactive materials at surfaces of electrically polarized electrodes.[16] The current measured by these devices is equivalent to the rate of the electrode reaction, which is determined by the applied potential and the rate at which the reactant arrives at the electrode (mass transport rate). In this reaction, increasing the potential increases the current up to a point termed "limiting current" beyond which any further increase of the potential does not produce more current. This condition exists because the reaction rate is now determined by the rate at which the electroactive material arrives at the electrode, i.e., the mass transport rate. Factors determining the rate are diffusivity of the material surrounding the electrode, bulk liquid reactant concentration, and hydrodynamics.

The amperometric microelectrodes used for measuring concentrations of selected nutrients in biofilms (e.g., oxygen) are covered with membranes that determine the mass transport resistance of the measured substance to the electrode. In principle, such microelectrodes are not sensitive to the conditions determining the mass transport rate of the measured substance in the bulk solution (e.g., flow rate), because the majority of mass transport resistance to the microelectrode is confined within the membrane. In other words, the membrane shields the device from the variations in external mass transport rates. To better quantify factors affecting mass transport dynamics in biofilms, we use amperometric sensors without membranes, which makes them sensitive to the mass transport rates in the solution. As long as the bulk-liquid reactant concentration remains constant, the limiting current measured by such devices depends strictly on the mass transport rate of the electroactive material to the microelectrode. Devices that build on the limiting current principle are well known for determining mass transport rates and studying the effects of hydrodynamics on mass transport rates in various systems.[17–21] Two differences distinguish our sensors from those described in the literature: our sensors have tip diameters less than 10 μm, and they are mobile.

[16] T. Mizushina, *Adv. Heat Transfer* **7**, 87 (1971).

[17] S. Yapici, M. A. Patrick, and A. A. Wragg, *J. Appl. Electrochem.* **24**, 685 (1994).

[18] N. M. Juhasz and W. M. Deen, *AIChE. J.* **39**, 1708 (1993).

[19] S. J. Konopka and B. McDuffie, *Anal. Chem.* **42**, 1741 (1970).

[20] P. H. Vogtlander and C. A. P. Bakker, *Chem. Eng. Sci.* **18**, 583 (1963).

[21] W. E. Ranz, *AIChE. J.* **4**, 338 (1958).

To apply the limiting-current-type microelectrodes in biofilm systems, two conditions need to be satisfied: (1) the only sink for the electroactive substance must be the microelectrode tip and (2) there can be only one electroactive substance available for the electrode to process at a time. After a suitable electroactive substance has been selected and introduced to a biofilm, a microelectrode, polarized to an extent that satisfies the limiting current conditions, reduces (or oxidizes, depending on the procedure) this substance. The electroactive substance is then locally removed (converted to another substance) at the tip of the microelectrode, thereby locally decreasing the concentration of this substance in the vicinity of the microelectrode tip. After a few seconds, an equilibrium is established between the rate at which the electroactive substance is locally removed by the electrode reaction and the rate at which it is supplied by the mass transport. The limiting current reflects the position of this equilibrium. If, for any reason, the rate of mass transport of the electroactive substance to the tip of the microelectrode changes, the limiting current changes. For our experiments we use mobile microelectrodes and expose their tips at various locations in biofilms. As the electroactive reactant we use a 25 mM solution of potassium ferricyanide, $K_3Fe(CN)_6$, in 0.2 M KCl. With proper experimental control, the ferricyanide is reduced to ferrocyanide, $Fe(CN)_6^{4-}$, at the surface of cathodically polarized microelectrodes:

$$Fe(CN)_6^{3-} + e^- \rightarrow Fe(CN)_6^{4-} \tag{1}$$

Increasing the polarization potential applied to the microelectrode increases the rate of the electrode reaction (that is, the rate of reduction of ferricyanide to ferrocyanide) until the rate reaches its limit for the existing set of conditions. At limiting current, the concentration of ferricyanide at the electrode surface is zero, and the concentration gradient cannot increase any further. Figure 1 shows a microelectrode immersed in a solution of ferricyanide, polarized with respect to a reference electrode, and measuring limited current. For such conditions, the concentration of ferricyanide at the electrode surface is zero ($C_s = 0$) while in the bulk solution it remains C_0. The concentration gradient is said to be confined entirely within the mass transfer boundary layer, δ. The flux of the ferricyanide, J, to the exposed surface of the microelectrode with the sensing area, A, is related to the limiting current, I, by Eq. (2):

$$J = \frac{I}{nAF} \tag{2}$$

Here F is Faraday's constant, and n is the number of electrons transferred in the balanced reaction. For practical reasons, the results of current measurement are reported as limiting current density, (I/A), which describes the ratio of the limiting current to the reactive surface area of the microelectrode tip. The flux of ferricyanide to the electrode can also be expressed as Fick's first law:

$$J = \frac{D(C_0 - C_s)}{\delta} = k(C_0 - C_s) \tag{3}$$

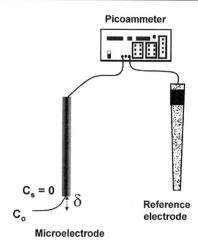

FIG. 1. At limiting current, ferricyanide concentration at the electrode surface is zero ($C_s = 0$) while the concentration of ferricyanide in the bulk solution remains C_0. Concentration gradient is confined within the boundary layer of δ thickness.

Here, D is the diffusivity of ferricyanide; concentration of the ferricyanide in the bulk solution is C_0, and at the surface of the microelectrode it is C_s; k is the mass transport coefficient; and δ designates the thickness of the diffusion boundary layer, controlled by the local shear stress and flow velocity.[16,22–25] In stagnant water, δ is controlled only by the diffusivity of ferricyanide.[26–28] A properly arranged experimental system using a mobile microelectrode allows for measuring rates of transport of ferricyanide to the microelectrode tip, and this transport rate is equivalent of the nutrient transport rate in biofilms at that location.

We use several modes of measurements to quantify factors influencing mass transport dynamics in biofilms. The simplest to measure is the local mass transfer coefficient.[29] More complicated are local effective diffusivity[30,31] and local flow

[22] T. J. Hanratty and J. A. Campbell, in "Fluid Mechanics Measurements" (R. J. Goldstein, ed.), p. 616. Taylor and Francis, Washington, D.C., 1996.

[23] J. R. Selman and C. W. Tobias, in "Advances in Chemical Engineering" (B. Drew, G. R. Cokelet, J. W. Hoopes, and T. Vermeulen, Jr., eds.), p. 211. Academic Press, New York, 1978.

[24] H. G. Dimopoulos and T. J. Hanratty, Fluid, Mech. 33, 303 (1968).

[25] J. A. Campbell and T. J. Hanratty, AIChE. J. 28, 988 (1982).

[26] X. Gao, J. Lee, and H. S. White, Anal. Chem. 67, 1541 (1995).

[27] C. Amatore and B. Fosset, Anal. Chem. 65, 2311 (1993).

[28] Z. Galus and J. Osteryoung, J. Phys. Chem. 92, 1103 (1988).

[29] S. Yang and Z. Lewandowski, Biotechnol. Bioeng. 48, 737 (1995).

[30] H. Beyenal, A. Tanyolaç, and Z. Lewandowski, Wat. Sci. Tech. 38, 171 (1998).

[31] H. Beyenal and Z. Lewandowski, Wat. Res. 34, 528 (2000).

velocity.[32] The greatest benefits of these measurements can be expected when the results are compared with the local nutrient consumption rates measured by other microelectrodes. Such a combination gives us an image of the relations between local mass transport dynamics and local microbial activity. Since application of limiting-current-type sensors requires introducing ferricyanide, which inactivates metabolic reactions in biofilms, measurements of mass transport dynamics have to be conducted separately for the measurement of the primary substrate utilization rate by the biofilm. Local substrate utilization rates are measured first, then the nutrient solution is replaced by the electrolyte and the factors affecting mass transport dynamics are quantified. It has been demonstrated that replacing the nutrient solution with the electrolyte does not change the biofilm structure.[29] To combine the results of local microbial respiration rate and local mass transport dynamics, the microelectrode measurements have to be conducted at exactly the same location, which is a challenging task. To find the location using different microelectrodes, we grow biofilms on microscope coverslips that are marked with a grid (Fisher Scientific). The slides are positioned at the bottom of the flat plate reactor, and the exact location for microelectrode measurements is then found using the inverted microscope positioned below the reactor (Fig. 2).

Technically, all microelectrode measurements are similar: a cathodically polarized microelectrode measures the limiting current at different locations in a biofilm. The differences in experimental conditions and the differences in microelectrode calibration procedures define what is actually measured. For example: to measure a local mass transfer coefficient, the measurements are conducted in flowing electrolyte because the convective mass transport component needs to be included in the final result. To measure local effective diffusivity we use the same system, but the flow is stopped and the measured limiting current reflects only the diffusional component of mass transport. In the former case the electrode is not calibrated; however, in the latter case the microelectrode has to be calibrated using gels of known densities and effective diffusivities.

The mass transfer coefficient can be found[29] by solving Eqs. (2) and (3) for $C_s = 0$.

$$k = \frac{I}{nAFC_0} \qquad (4)$$

If the current is measured directly and all other factors are known, the mass transport coefficient is calculated from Eq. (4). The surface area of the microelectrode needs to be known *a priori*, but the microelectrodes that measure the mass transport coefficient do not need to be calibrated.

In contrast, microelectrodes used to measure effective diffusivity need to be calibrated. The reason that the local effective diffusivity cannot be directly

[32] F. Xia, H. Beyenal, and Z. Lewandowski, *Wat. Res.* **32**, 3637 (1998).

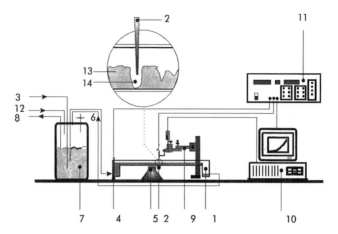

FIG. 2. Experimental setup used to grow biofilms and perform microelectrode measurements: 1, open channel, flat plate reactor; 2, microelectrode; 3, air conduit; 4, reference electrode; 5, inverted microscope/confocal scanning laser microscope; 6, air filter; 7, mixing chamber; 8, effluent; 9, micromanipulator; 10, data acquisition system; 11, picoammeter/DC voltage source; 12, fresh feed conduit; 13, biofilm cluster; 14, biofilm void.

calculated from Eq. (3) is that the limiting current depends on the thickness of the diffusion boundary layer, δ, around the microelectrode tip. However, this value is unknown and depends on the diffusivity, D, in a stagnant solution. For that reason microelectrodes used to measure effective diffusivity have to be calibrated in gels with characteristics similar to those of the extracellular biopolymers in biofilms and of known effective diffusivities.

Microelectrodes used to measure local flow velocity have to be calibrated as well. The major factor defining mass transport rates in the flowing electrolyte is convection, which makes it possible to quantitatively relate the limiting current measured by microelectrodes in ferricyanide solution to local flow velocities in the vicinity of the microelectrode tip. We have tested that the procedure works at low and very low velocities not exceeding a few centimeters per second. To calibrate such microelectrodes, the local flow velocities in the vicinity of the microelectrode tip are measured using velocimetry combined with confocal scanning laser microscopy as described later in this chapter.

Procedures

Growing Biofilms

An open channel biofilm reactor (Fig. 2) made of transparent polycarbonate, 2 cm deep, 4 cm wide, and 75 cm long, is used.[29] The nutrient solution consists of

KH_2PO_4 (2.2 mM), K_2HPO_4 (4.8 mM), $(NH_4)SO_4$ (0.25 mM), $MgSO_4$(0.04 mM), and yeast extract (0.01 g/liter), and it is recycled through the reactor at a rate of 10 ml/s by a magnet driven vane pump (Cole-Parmer, Chicago, IL). Oxygen is supplied by preaerating the nutrient solution. The operating volume of the recycle loop, including the reactor, is 450 ml. The inoculum consists of *Pseudomonas aeruginosa* (ATCC 700829), *Pseudomonas fluorescens* (ATCC 700830), and *Klebsiella pneumoniae* (ATCC 700831). After inoculation, the reactor is operated as a batch for 12 hr, and then in a continuous flow mode. Nutrient solution is added by a peristaltic pump (Masterflex; Cole-Parmer, Niles, IL) at a rate of 15 ml/min to achieve a hydraulic residence time of 30 min. A short retention time prevents buildup of suspended growth. The biofilm is allowed to grow for 4 to 7 days before the measurements are taken.

Replacing Nutrients with Electrolyte

Just before the measurements are taken, the nutrient delivery is interrupted and the nutrient solution is slowly drained from the reactor and replaced with the electrolyte solution properly formulated for the measurements: 25 mM $K_3Fe(CN)_6$ in 0.2 M KCl. When measurements of mass transport parameters are conducted in flowing electrolyte, the solution is recycled through the biofilm reactor using a gravitational flow system to ensure that the flow rate is constant and without pulsation. Flow rates are adjusted by a control valve and continuously monitored by a flowmeter (Model GF-1500, Gilmont Instruments, Barrington, IL). The recirculating electrolyte is filtered through a quartz wool plug to remove large aggregates of cells. As a final step prior to the measurements, the electrolyte solution is allowed to flow through the biofilm reactor for at least 30 min to equilibrate with the biomass.

Constructing and Testing Microelectrodes

To make the microelectrodes, we use a platinum wire (California Wire Company, Grover Beach, CA), 100 μm in diameter (pure TC grade). The tip of the wire is etched electrochemically (7 volts AC against graphite counterelectrode) to a diameter between 2 and 6 μm in a saturated KCN solution. The wire is rinsed with distilled water and carefully inserted into a capillary made of soda-lime glass. The capillary is positioned in a microelectrode puller (Stoelting Co., Wood Dale, IL) in such way that the tip of the platinum wire is about 1.5 cm above the heating coil. The heat is slowly increased until the glass around the platinum wire melts and adheres to the wire, while the entire capillary drops down, separates, and cools in the air. The tip of the wire is then exposed by grinding it on a rotating diamond wheel (Model EG-4, Narishige Co., Tokyo) and then cleaned in a sonication bath, first in distilled water, then in acetone. The diameter of the microelectrode tip is measured under a microscope, and its surface area is then calculated. To prevent

(A)

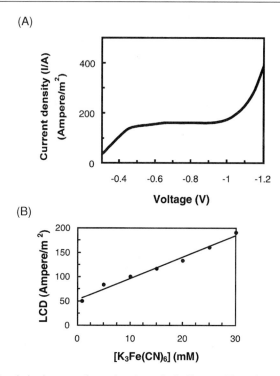

(B)

FIG. 3. (A) A polarization curve for a microelectrode tip diameter 10 μm in an unstirred solution of 25 mM K₃Fe(CN)₆ in 0.2 M KCl. (B) The limiting current is linearly correlated with the bulk water ferricyanide concentration, demonstrating that the reaction at the tip of the microelectrode is truly diffusion controlled.

damaging the biofilm structure, we use electrodes with an outside tip diameter less than 10 μm. The electrical connection with the external circuit is made by inserting a small piece of solder wire and a bare copper wire into the capillary and melting the solder in a stream of hot air to form a metallic droplet connecting the platinum and the copper wires. A voltammetry test is performed on each microelectrode to check its performance. The polarization potential is scanned from 0.0 to −1.2 V in a solution of 0.2 M KCl and 25 mM potassium ferricyanide, $K_3Fe(CN)_6$. Potassium chloride, KCl, is used as supporting electrolyte to suppress the contribution of electromigration to the microelectrode.[23,29] The microelectrodes having stable limiting current in the voltage range between −0.6 and −0.9 V are considered usable (see Fig. 3A). The actual concentrations of $Fe(CN)_6^{3-}$ are determined by iodimetric titration before each measurement.[33] Occasionally, the dependence of

[33] A. Vogel, "A Textbook of Quantitative Inorganic Analysis Including Elementary Instrumental Analysis." Longman, London, 1978.

the limiting current on the ferricyanide concentration is tested; if it is linear, it confirms that the reaction at the tip of the microelectrode is truly diffusion controlled[34] (see Fig. 3B).

As reference/counterelectrodes we use commercial calomel electrodes (Model 13-620-51, Fisher Scientific, Pittsburgh, PA). A Hewlett Packard 4140B multimeter is used as a voltage source and picoammeter. During the biofilm measurements the microelectrodes are polarized cathodically to -0.8 V against the reference electrode, which is placed next to it within the reactor in the downstream direction.

Figure 3A shows typical results of microelectrode testing. When the applied potential increases from 0 to -0.5 V, current increases as well. For potentials between approximately -0.6 and -1.0 V, the current remains constant, which reflects the limiting current conditions. As a rule, if the current is stable between -0.6 and -0.9 V, the tested microelectrode is accepted. Increasing the potential beyond -1.0 V causes chemical reduction of water, an additional reaction that increases the current. Based on the results similar to those in Fig. 3A, we decided to polarize the microelectrodes at -0.8 V during the measurements in biofilm rectors. Since at -0.8 V oxygen can be reduced, we have considered the need for removing oxygen from the system prior to taking the measurements. However, tests demonstrated that the limiting current generated by reducing ferricyanide to ferrocyanide was not affected by the presence of oxygen. Two reasons may, hypothetically, explain this result: (1) the concentration of ferricyanide in the system, 25 mM, exceeds the solubility of oxygen in water, 0.5 mM, by a factor of 50, and (2) the reduction of ferricyanide is kinetically preferred.

An important condition that needs to be satisfied by measuring systems using limiting-current-type microelectrodes is that the electrode reaction must be the only sink (or source) of electrons for the selected electroactive substance. This condition prevents using some convenient cathodic reactants, such as oxygen, as electroactive materials for limiting current measurements in biofilms. The reason is that oxygen would be reduced by the microorganisms, and the electrode reaction would not be the only source of electrons to reduce oxygen. Effectively, in such a system the microelectrode would compete for oxygen with the microorganisms.

The condition that the electrode reaction is the only sink (or source) of electrons needs to be verified for each selected reactant. For example, in our measurements the biomass can serve as a source of electrons to reduce ferricyanide, therefore competing with the electrode reaction. Our tests show that it takes at least 30 min for the ferricyanide to equilibrate with the biomass. To test if the system is equilibrated, we measure profiles of ferricyanide across the biofilm using ferricyanide microelectrodes. Ferricyanide microelectrodes are constructed in the same way as

[34] A. D. Dawson and O. Trass, *J. Heat Mass Transfer* **15**, 1317 (1972).

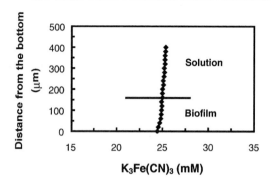

FIG. 4. Profiles of ferricyanide concentration measured to evaluate whether the system was already equilibrated. The ferricyanide in the bulk solution was 25 mM $K_3Fe(CN)_6$ in 0.2 M KCl. The slight decrease of the ferricyanide concentration near the bottom shows that the system may need more time to equilibrate.

the microelectrodes used to measure the mass transport dynamics, with one exception: we cover their tips with a 5% cellulose acetate membrane. An appropriate amount of cellulose acetate to make 5% solution is dissolved in acetone and the tip of the microelectrode dipped into the solution and dried overnight. The presence of the membrane makes the electrodes useless for measuring the mass transport coefficients, but makes them convenient tools to measure the concentration of ferricyanide within the biofilm.

Figure 4 shows results of such measurements in a biofilm that was equilibrated with the ferricyanide for 30 min. If the system is properly equilibrated, the concentration of ferricyanide across the biofilm does not change. The results in Fig. 4 show a slight bend on the profile just above the bottom, indicating that the system could benefit from having more time to equilibrate. For some measurements (e.g., in thick and dense biofilms) we equilibrate the system for 2 hr.

Calibrating Effective Diffusivity Microelectrodes

To calibrate the effective diffusivity microelectrodes, we use agar gels of known effective diffusivities. To prepare such gels, one needs to determine the effective diffusivity of ferricyanide in agar gels of different densities. For that purpose we use a diffusion cell consisting of two tanks, each 11 cm deep, 10 cm wide, and 10 cm long, separated by a dialysis membrane (Spectrum, 132709, La Cadena, CA); a fresh membrane is used for each experiment. A suspension of agar (BBL 11853; Becton Dickinson Microbiology Systems, Cockeysville, MD) in 0.2 M KCl is heated to 100° (in boiling water), then cooled to 40–60° and carefully poured onto the dialysis membrane. The lower tank (1.040 liter working volume) is filled with 25 mM K$_3$Fe(CN)$_6$ in 0.2 M KCl and stirred by a magnetic stirrer to avoid mass transport limitations to the membrane, while the upper tank (0.560 liter working volume) is filled with 0.2 M KCl. The ferricyanide diffuses from the lower tank through the membrane and the agar layer into the upper tank. The entire diffusion cell operates in a temperature controlled (25 ± °) water bath. The concentration of ferricyanide in the upper tank is measured spectroscopically at a wavelength of 430 nm. The effective diffusivity is measured both through the membrane alone and through the membrane and agar together, and the difference is used to calculate the effective diffusivity in agar. The procedure of calculating the effective diffusivity from such results is described in detail in Beyenal and Lewandowski.[31]

To prepare agar layers of known density and effective diffusivity, the known amount of agar is dissolved in a solution of K$_3$Fe(CN)$_6$ and 0.2 M KCl. The solution is slowly heated until the agar dissolves, then it is slowly cooled to 40–60° and poured as a layer 250–1000 μm thick on the bottom of the flat plate reactor (without biofilm). After the agar layer solidifies, the reactor is carefully filled with K$_3$Fe(CN)$_6$ in 0.2 M KCl electrolyte solution. The solution is recirculated for 2 hr to equilibrate the ferricyanide with the agar. Before the measurements are taken, the recycling process is stopped, and the solution remains stagnant for the duration of the measurements. The microelectrode is mounted on a micromanipulator (Model M3301L, World Precision Instruments, New Haven, CT), which is equipped with a stepper motor (Model 18503, Oriel, Stratford, CT), controlled by the Oriel Model 20010 interface, and introduced perpendicularly to the agar layer from the top of the reactor. The microelectrode is moved at a 20 μm step length across the agar layer, and the limiting current density, (I/A), is measured at different positions within the agar. The distance from the tip of the microelectrode to the bottom of the reactor is determined at each step. Figure 5A shows typical results of limiting current densities measured in agar gels of different densities and effective diffusivities.

Figure 5 shows that, in agar gels of constant density and constant effective diffusivity, the limiting current density is constant at a distance from the bottom of the reactor but decreases suddenly just above the bottom of the reactor. This variance is most likely because of the mass transfer limitations near the wall. This effect, also noticed by Beyenal et al.[30] and Yang and Lewandowski,[29] limits the application of microelectrodes at distances less than 60 μm from the bottom. Consequently, the results of the limiting current measurements closer than 60 μm from the bottom have to be ignored. The relation between the limiting current density measured by the microelectrodes and the effective diffusivity in the agar gel is linear within the range of tested agar densities (Fig. 5B). Using such calibration curves, the results of limiting current measurements can be expressed in terms of local effective diffusivities in the vicinity of the microelectrode tip at a selected location in the biofilm.

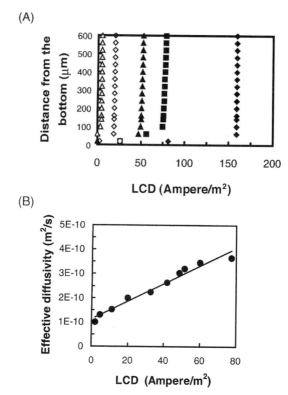

FIG. 5. (A) Limiting current density (LCD) in agar layers. ◆, Electrolyte solution. D_{ag} (m²/s): ■, 3.66×10^{-10}; ▲, 3.21×10^{-10}; ◇, 1.98×10^{-10}; △, 1.29×10^{-10}. (B) The relation between the LCD and the effective diffusivity of ferricyanide within the range of tested agar densities and effective diffusivities. [Reproduced from H. Beyenal and Z. Lewandowski, *Wat. Res.* **34,** 528 (2000).]

Calibrating Local Flow Velocity Microelectrodes

The limiting current microelectrodes respond to the rate of mass transport to the electrode tip. For flowing liquids the major factor that determines the mass transport rate is the thickness of the mass transfer boundary layer, which, in its turn, is determined by the local flow velocity (diffusivity of ferricyanide is constant). Since the response of the microelectrode depends on local flow velocity, we relate the limiting current measured by an ammeter to the local flow velocity. The difficult part here is determining the local flow velocity in a flow velocity field. We use the Bio-Rad model MRC600 confocal scanning laser microscope (CSLM) integrated with the Olympus model BH2 light microscope for particle tracking. Detailed procedures of the measurement of local flow velocities in biofilms are given by Stoodley et al.[35] and De Beer et al.[8] Neutral-density fluorescent latex spheres [density at 20°, 1055 kg/m^3; excitation wavelength (Ex), 580 nm; emission wavelength (Em), 605 nm; diameter, 0.216 μm; Molecular Probes, Eugene, OR] are added to the reactor. Local velocity profiles in the reactor are obtained by raising or lowering the motorized stage and capturing images at different focal depths. The CSLM is used to capture images by using the Bio-Rad COMOS operating software. Particles traveling across the field of view appear as dashed lines. The velocity of a particle is calculated from the distance between the dashes and the time taken to scan the number of frame lines crossed by the particle. We then insert the microelectrode at the exact point where the CSLM images were taken and determine the limiting current. Principles of the calibration procedure are exemplified in Fig. 6.

Modes of Measurements and Interpretation of Results

Appropriate experimental setups are arranged in such way that the results of microelectrode measurements are collected as (1) profiles of local mass transport coefficients, effective diffusivities, or flow velocities across the biofilm, or (2) maps of spatial distribution of those parameters at different levels in the biofilm.

Profiles of Measured Parameters

The microelectrode is attached to a linear, computer-controlled micropositioner (Model CTC-322-20, Micro Kinetics) positioned above a selected location in the biofilm reactor. The micropositioner moves the electrode across the biofilm in predetermined intervals, e.g., 10 μm, taking measurement of the limiting current at each position. Figure 7 shows a profile of local mass transfer coefficients calculated from such results.[4]

Concentration profiles measured across the biofilm are site specific, meaning they are different at different locations (see Yang and Lewandowski[29]). Mathematical modeling of biofilms predicts biofilm activity over a certain surface. To verify biofilm models, the measurements need to evaluate an average biofilm activity over the same surface for which the activity was predicted using the model. Concentration profiles measured at different locations can show surprising variability, and it is not quite clear how to average these profiles. To quantify the extent of this variability, and to average the concentrations measured at different locations, it is often desirable to construct maps of concentration distribution at different levels in biofilms, instead of reconstructing concentration profiles across these biofilms.

[35] P. Stoodley, D. De Beer, and Z. Lewandowski, Appl. Environ. Microbiol. **60**, 2711 (1994).

(A)

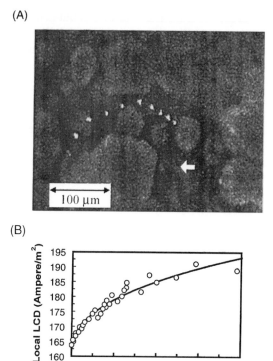

(B)

FIG. 6. Calibrating local flow velocity microelectrodes. (A) CSLM of fluorescent beads is used to measure local flow velocities. Reproduced from P. Stoodley, D. De Beer, and Z. Lewandowski, *Appl. Environ. Microbiol.* **60**, 2711 (1994). (B) Local flow velocities correlated with the local limiting current density measurements indicate that for flow velocities not exceeding 2 cm/s, which are the flow velocities in the voids of many biofilms, the limiting current density correlates well with the local flow velocity. [Reproduced from F. Xia, H. Beyenal, and Z. Lewandowski, *Wat. Res.* **32**, 3637 (1998).]

Maps of Spatial Distribution of Measured Parameters

To calculate the average mass transport coefficient and average effective diffusivity for the entire biofilm, we map spatial distribution of these parameters in the biofilm. Generating such maps is more complicated than reconstructing profiles of these parameters. To describe the spatial distribution of the measured parameters, and to find meaningful average mass transport coefficients and effective diffusivities, we use a different protocol from that used to measure profiles of these parameters. First, we design a grid above a selected part of the biofilm surface and make the measurements at the grid points. Second, instead of measuring the selected parameter across the biofilm in predesigned intervals (as is done when

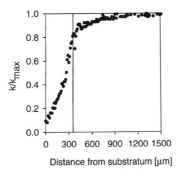

FIG. 7. Profile of the local mass transport coefficient across a cell cluster in a biofilm. The vertical line designates the approximate position of the biofilm surface. The results were normalized with respect to the maximum mass transport coefficient measured in the bulk solution. [Reproduced from K. Rasmussen and Z. Lewandowski, *Biotechnol. Bioeng.* **59**, 302 (1998).]

measuring profiles), the linear micropositioner lowers the microelectrode to a pre-selected distance from the bottom of the reactor to make a single measurement. Then, the microelectrode is withdrawn from the biofilm, moved to the next grid point, and lowered to exactly the same distance from the bottom as for the previous measurement; another measurement is taken; and so on. Once these measurements are completed, we repeat the cycle of measurements at another distance from the bottom, using the same grid points. When all measurements are completed we have several sets of results of the measured parameters collected at selected distances from the bottom, and these values are then converted to maps of spatial distribution of the measured parameter at different levels in the biofilm. An advantage of this protocol is that such maps of spatial distributions of the measured parameters—local mass transfer coefficient, local effective diffusivity, and local flow velocity—at selected distances from the bottom are directly comparable with the CSLM images collected at the same distances from the bottom, making it possible to compare biofilm structure with the distribution of factors affecting mass transport dynamics.

For practical purposes, we use a combination of linear and X–Y micropositioners. The flat plate biofilm reactor is fixed to the X–Y micropositioner stage (Model CTC-462-2S, Micro Kinetics, Laguna Hills, CA), while the microelectrode is attached to the linear micropositioner (Model CTC-322-20, Micro Kinetics) positioned above the reactor. Both micropositioner are manipulated by a computerized controller (CTC-283-3, Micro Kinetics) with a positioning precision of 0.1 μm. Custom software is used to simultaneously control the X–Y stage movement, the microelectrode movement, and the data acquisition (Model CIO-DAS08PGL, Computer Boards, Inc., Mansfield, MA). For typical measurements a 10×10 grid matrix with a step size of 100 μm in both X and Y directions is used.

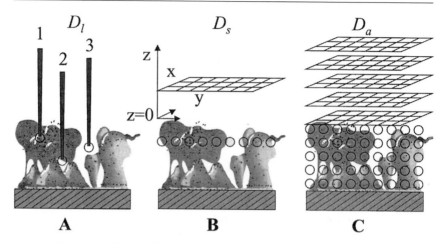

FIG. 8. A system of effective diffusivity measurements. (A) Local relative effective diffusivity (D_l) measured by microelectrodes at arbitrarily selected locations, and at different distances from the bottom. (B) The D_l values are measured at grid points equally distant from the bottom. The measured D_l values are then averaged, yielding the surface averaged relative effective diffusivity, $D_s = \sum_{n=1}^{k} D_{ln}/k$. (C) The average relative effective diffusivity, $D_a = \sum_{n=1}^{p} D_{sn}/p$, is the average of all local measurements for the entire biofilm. [Reproduced from H. Beyenal and Z. Lewandowski, *Wat. Res.* **34**, 528 (2000).]

The microelectrode is manually positioned above the first grid point and the linear micropositioner moves the microelectrode into the biofilm to a predetermined distance from the bottom. After the limiting current has been measured and the data have been accepted, the linear micropositioner moves the microelectrode out of the biofilm and into the bulk liquid. The motorized X–Y stage moves the reactor to the next grid point, and then the linear micropositioner lowers the microelectrode to the same distance from the bottom as for the previous measurement. Maps of limiting current densities measured at the grid points equidistant from the bottom of the reactor are displayed on the computer's monitor in real time during the measurements. Figure 8 shows the experimental arrangement. Individual profiles of *local relative effective diffusivity, D_l*, vary from location to location and do not reveal the structured pattern of relative effective diffusivities found in biofilms (Fig. 8A). Local measurements are then arranged in a way that demonstrates that organization: we measure D_l at grid locations equidistant from the bottom and take the average of the results (Fig. 8B). The average of these D_l measurements is termed *surface averaged relative effective diffusivity, D_s*, and is different at different distances from the bottom. The set of these D_s averages contains as many individual results as the number of levels that were selected for effective diffusivity measurements. The average of all the D_l measurements taken at different distances from the bottom is termed *average relative effective diffusivity, D_a*, and represents the average effective diffusivity for the entire biofilm (Fig. 8C).

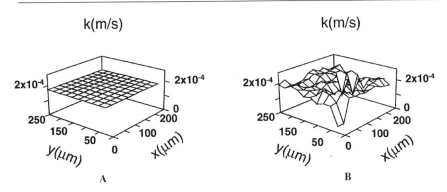

FIG. 9. Horizontal distributions of local mass transfer coefficient. Bulk flow velocity was 1.58 cm/s. Distances from the bottom: (A) $z = 1000\,\mu\text{m}$ (outside the biofilm), (B) $z = 200\,\mu\text{m}$ (inside the biofilm). [Reproduced from S. Yang and Z. Lewandowski, *Biotechnol. Bioeng.* **48**, 737 (1995).]

When results are arranged as maps of spatial distribution, one can estimate average local mass transport coefficient and average local effective diffusivities at different levels within the biofilm over the surface area covered by the grid. Figure 9 shows typical results of such measurements: a map of mass transport coefficient distribution at selected levels above and within a biofilm.

Using this system of measurements one can, for example, quantify the effects of biofilm growth conditions on the average relative effective diffusivities in heterogeneous biofilms. For that purpose, we grow biofilms at different conditions, evaluate their average relative effective diffusivities, D_a, and present the results in a coordinate system that visualizes the effects of growth conditions on the average relative effective diffusivity. As an example, the effect of flow velocity and glucose concentration on average relative effective diffusivity of biofilms is visualized in Fig. 10.[31] The results show that concentration of glucose had a stronger effect on the average effective diffusivity than did flow velocity.

Local Biofilm Density

Biofilm density cannot be directly measured by microelectrodes. However, Fan *et al.*[36] suggested, based on extensive literature studies of the effective diffusivity measured in biofilms, activated sludge flocs, and mycelial pellets, the following empirical relation between average biofilm density and average relative effective diffusivity:

$$D_a = 1 - \frac{0.43 X_f^{0.92}}{11.19 + 0.27 X_f^{0.99}} \tag{5}$$

[36] L. S. Fan, R. L. Ramos, K. D. Wisecarver, and B. J. Zehner, *Biotechnol. Bioeng.* **35**, 279 (1990).

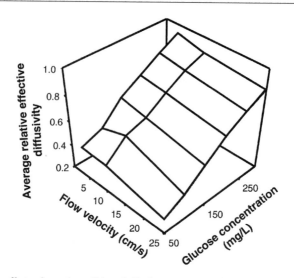

FIG. 10. The effects of growth conditions (bulk glucose concentration and average flow velocity) on the average relative effective diffusivity in biofilms. [Reproduced from H. Beyenal and Z. Lewandowski, *Wat. Res.* **34**, 528 (2000).]

Here D_a is average relative effective diffusivity of ferricyanide in the electrolyte or average relative effective diffusivity of growth limiting substrate in the medium, and X_f is average biofilm density (g/liter). This equation can be solved for the biofilm density as a function of the effective diffusivities measured in biofilm. Figure 11 shows that the average density of the biofilm near the bottom was on the order of 100 g/liter, which is consistent with observations reported by Zhang et al.[37] and Zhang and Bishop.[38]

Concluding Remarks

Limiting-current-type microelectrodes are remarkably versatile when applied to the measurement of mass transport rates in biofilms. Local mass transport co-efficient, local effective diffusivity, and local flow velocity can all be measured. When the results of these measurements are superimposed on the results of nutrient uptake dynamics, the intricate interplay between mass transport dynamics and microbial activity in biofilms becomes visible, at least in simple experimental systems.

The most controversial part of the limiting current techniques, when applied to biological systems, is the choice of cathodic reactants. Ideally, such a reactant

[37] T. C. Zhang, Y. C. Fu, and P. L. Bishop, *Wat. Environ. Res.* **67**, 992 (1995).
[38] T. C. Zhang and P. L. Bishop, *Wat. Res.* **28**, 2279 (1994).

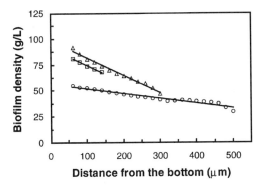

FIG. 11. Profiles of biofilm density in mixed population and pure culture *Pseudomonas aeruginosa* biofilms. $\bigcirc$, Mixed culture ($v = 1.6$ cm/s); $\triangle$, mixed culture ($v = 3.2$ cm/s); $\square$, *Pseudomonas aeruginosa* ($v = 3.2$ cm/s). [Reproduced from H. Beyenal, A. Tanyolaç, and Z. Lewandowski, *Wat. Sci. Tech.* **38**, 171 (1998).]

would not be toxic to the microorganisms, would not serve as a substrate in metabolic reactions, and would not react chemically with the biomass. At the same time it should exhibit acceptable electrochemical properties, reversible kinetics, and low overpotential on the selected electrode. It is difficult to find such a reagent! Potassium ferricyanide, the cathodic reactant that we use, is not an ideal reagent mainly because it reacts with the biomass. We bypass this problem by waiting long enough for the biomass to equilibrate with the ferricyanide, which essentially means until the reaction between the ferricyanide and biomass is completed. As a result, we need to assume that this reaction does not change the measured properties of the biofilms significantly.

Acknowledgments

The research was supported by the cooperative agreement EED-8907039 between the National Science Foundation and Montana State University.

Section V

Susceptibility Testing of Biofilm Microbiota

[24] Development of a Standardized Antibiofilm Test

By NICK ZELVER, MARTIN HAMILTON, DARLA GOERES, and
JOANNA HEERSINK

Introduction

Existing standard methods for evaluating biocides, disinfectants, or antibiotics are inadequate for analyzing the performance of antibiofilm products. Methods designed for testing antibiofilm treatments are needed because biofilm bacteria respond differently to antimicrobials compared to suspended bacteria.[1-5]
These differences include the following:

1. Biofilm bacteria may express different genes than do suspended bacteria. Some of these genes may express antimicrobial resistance.[6,7]
2. Antimicrobials may not fully penetrate a biofilm.[8,2]
3. Biofilms exhibit heterogeneity such as gradients in physiology where slower growing organisms may be more resistant to antimicrobials.[8-11]
4. Biofilms in the field may accumulate contaminants from the surrounding medium. These contaminants can present a demand on the antimicrobial.[10]

For the preceding reasons, current methods (based on suspended cultures) for evaluating antimicrobials may exaggerate performance of the treatment. In addition to determining the ability to kill bacteria, which is typically the end point of standard antimicrobial tests, biofilm tests may have other end points that are unique to biofilm, such as the following:

1. The application of an antibiofilm product may be designed to remove the biofilm rather than kill the bacteria.

[1] J. W. Costerton, K. J. Cheng, G. G. Geesey, T. I. Ladd, J. C. Nickel, M. Dasgupta, and T. J. Marrie, *Annu. Rev. Microbiol.* **41,** 435 (1987).
[2] D. R. de Beer, R. Srinivasan, and P. S. Stewart, *Appl. Environ. Microbiol.* **60,** 4339 (1994).
[3] D. J. Hassett, J. G. Elkins, J.-F. Ma, and T. R. McDermott, *Methods Enzymol.* **310,** 599.
[4] C. T. Huang, F. P. Yu, G. A. McFeters, and P. S. Stewart, *Appl. Environ. Microbiol.* **61,** 2252 (1995).
[5] P. S. Stewart and J. B. Raquepas, *Chem. Eng. Sci.* **50,** 3099 (1995).
[6] J. G. Elkins, D. J. Hassett, P. S. Stewart, H. P. Schweizer, and T. R. McDermott, *Appl. Environ. Microbiol.* **65,** 4594 (1999).
[7] D. J. Hassett, J. G. Elkins, J.-F. Ma, and T. R. McDermott, *Methods Enzymol.* **310,** 599.
[8] J. W. Costerton, A. K. Camper, P. S. Stewart, N. Zelver, and M. E. Dirckx, *Analyst* **6,** 18 (1999).
[9] C.-T. Huang, P. S. Stewart, and G. A. McFeters, *in* "Digital Image Analysis of Microbes" (M. H. F. Wilkinson and F. Schut, eds.), p. 411. Wiley, New York, 1998.
[10] R. Srinivasan, P. S. Stewart, T. Griebe, and C.-I. Chen, *Biotechnol. Bioeng.* **46,** 553 (1995).
[11] P. S. Stewart, L. Grab, and J. A. Diemer, *J. Appl. Microbiol.* **85,** 495 (1998).

2. The antibiofilm product may be designed to prevent regrowth or resuscitation of the bacteria. The heterogeneity of a natural biofilm provides a rich and diverse ecological community of bacteria. This heterogeneity allows microniches where bacteria can reside and resist the antimicrobial application. Following application of the antimicrobial, the resistant biofilm bacteria may exhibit rapid regrowth.[10]
3. An antibiofilm product may be designed to prevent the biofilm from initially colonizing a surface rather than killing an established biofilm.

In summarizing the above issues, standard antibiofilm methods must use relevant biofilm as the test sample. Depending on the manufacturers' claims for antibiofilm products, methods are also needed to evaluate (1) kill of the biofilm bacteria, (2) removal of biofilm slime layer, and (3) rate of biofilm regrowth following antibiofilm application.

In this paper, we use the rotating disk reactor protocol with sodium hypochlorite treatment as a model for developing a standard *tier-one* antibiofilm test. A tier-one test is a quick, inexpensive method that is relevant to generic biofilm conditions. Since the tier-one test does not represent a specific field application, it may be most applicable to general screening of antibiofilm products. The tier-one test may also be used as the starting point in developing a *tier-two* test. The tier-two test is a more stringent test designed to represent a particular field system, such as a cooling tower, a swimming pool, or a medical implant device. For tier-two tests, defining parameters of the field system such as water chemistry, water temperature, water filtration, and types of inoculum must be represented in the test apparatus.

The antibiofilm test demonstration uses the rotating disk reactor (RDR) system described previously by the authors.[12] The RDR, shown in Fig. 1, consists of a 1 liter glass beaker fitted with a drain spout. The bottom of the reactor contains a magnetically driven rotor with six 1.27 cm^2 biofilm test-surface coupons. The rotor is constructed from a star-head magnetic stir bar to which a Teflon and Viton rubber disk is attached that holds the coupons. The coupons rotate continuously to provide fluid shear and mixing. Biofilm growth nutrients are continuously pumped into the reactor.

Continuous Stirred Tank Reactor Chemostat Approach to Growing Biofilm

The RDR is operated as a continuous stirred tank reactor (CSTR) chemostat. As a chemostat, the RDR is a perfectly mixed environment with continuous feed

[12] N. Zelver, M. Hamilton, B. Pitts, D. Goeres, D. Walker, P. Sturman, and J. Heersink, *Meth. Enzymol.* **310**, 608.

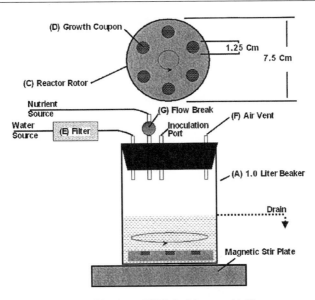

FIG. 1. Rotating disk reactor (RDR) for laboratory biofilm assays.

rate matching the effluent rate. The rate at which the nutrients are added to a chemostat is referred to as the dilution rate.

Controlling the dilution rate controls the rate of bacterial growth. Above a critical dilution rate, when nutrients are flowing through the reactor faster than the bacteria can divide, suspended bacteria are flushed out of the reactor. Operating a chemostat beyond the critical dilution rate selects for biofilm growth since suspended organisms are not available to compete for the nutrients.[13] The RDR is a biofilm growth reactor because it is operated at a high dilution rate. The RDR is operated until a steady-state biofilm bacterial cell density is reached (biofilm growth equals biofilm sloughing). The steady-state assumption provides a repeatable model biofilm.

Impact of Inoculation of the RDR on Repeatability

In the protocol presented previously by the authors, the RDR was inoculated at the beginning of an experiment by injecting a thawed culture of test organisms into the reactor and then operating the RDR in batch mode for 24 hr to establish a biofilm. Subsequently, we found this inoculation procedure to be a source of variability and we established a more repeatable inoculation protocol where the

---

[13] W. G. Characklis, *in* "Biofilms" (W. G. Characklis and K. C. Marshall, eds.), p. 37. John Wiley and Sons, New York, 1990.

thawed organisms are cultured to a known bacterial cell density prior to inoculation of the RDR. In this new protocol, a colony of *Pseudomonas aeruginosa (ERC-1)* is inoculated into a flask containing 100 ml of sterile tryptic soy broth (TSB) at 300 mg/liter. The flask is incubated in an orbital mixer at 35° for a maximum of 24 hr to attain a viable bacterial cell density of $\sim 10^8$ CFU/ml. The inoculation culture is serially diluted and plated to ensure that the cell bacterial cell density is correct for each RDR test. We believe the new inoculation procedure is a microbiological improvement because of the following reasons:

1. The initial method of injecting frozen inoculum was associated with additional nutrients being added from nutrients in the inoculum.
2. The new inoculation method allows the organisms to be acclimated to the type of environment found in the RDR so that the organisms are not shocked during inoculation.
3. The new inoculation method allows the technician to check for a pure culture when inoculating.

Biofilm Sampling and Enumeration in the RDR

Biofilm samples from the RDR are obtained by aseptically removing the test coupon from the rotor and removing the biofilm with a sterile wooden applicator stick. The stick is stirred vigorously into a test tube containing 9 ml of sterile buffered water. The entire coupon surface is scraped approximately three times for 1–2 min. The coupon is rinsed with 1 ml of sterile buffered water into the original 9 ml, bringing the final volume of the tube to 10 ml. Prior to enumeration, the cells are disaggregated by homogenization at a speed of 20,500 rpm for 30 sec to eliminate biofilm clumps.

Alternatively, a Teflon scraper can be used to remove the biofilm. Figure 2 shows comparisons of RDR biofilm coupons following the 24 hr CSTR growth phase. Figure 2a shows a coupon prior to sampling, Fig. 2b shows a coupon following sampling using a wooden scraper, and Fig. 2c shows a coupon following sampling using a Teflon scraper. With the wooden scraper, we measured biofilm (untreated) on 13 coupons sampled from the RDR and obtained a standard deviation of 0.47. With the Teflon scraper, we measured biofilm (untreated) on seven coupons sampled from the RDR and obtained a standard deviation of 0.57.

Enumeration of the biofilm bacteria is achieved using standard microbiological practices. Samples are diluted and plated using a modification of the spread plate method called the drop plate method.[12,14] If a mixed culture was used, viable cell enumeration would occur via the spread plate method.[15]

[14] A. A. Miles and S. S. Misra, *J. Hygiene* **38**, 732 (1938).
[15] A. L. Koch, *in* "Methods for General and Molecular Bacteriology" (P. Gerhardt, ed.), p. 255. ASM Press, Washington, D.C., 1994.

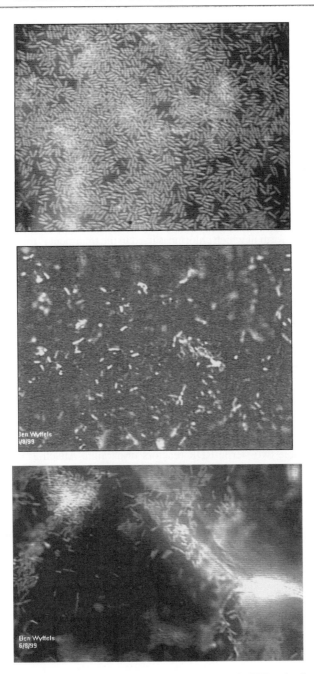

FIG. 2. Digital epifluorescent images of biofilm samples grown in the RDR and stained with DAPI (1000× magnification).

Treatment Protocol using Sodium Hypochlorite as a Model Antibiofilm Treatment

To demonstrate the ability of the RDR to test the efficacy of disinfectants on biofilm, we conducted a series of tests to determine the log reduction in biofilm viable bacterial cell density resulting from a 10 min treatment of 5 mg/liter total chlorine added as NaOCl (bleach). Chlorine was chosen as the model antibiofilm treatment since the efficacy of the biocide has been previously studied on biofilm, chlorine water chemistry is understood, and a standard method exists for measuring bulk fluid concentrations.[2,8,11,16–18] Sodium thiosulfate will neutralize chlorine, and neutralization is required so that the biocide reaction does not continue following the applied dosing time.

Chlorine is a nondiscriminating oxidizing agent; therefore, the protocol includes a rinse step to help reduce the chlorine demand in the reactor prior to the treatment. The form in which chlorine is used in the treatment is important because various types of chlorines affect the water chemistry differently.

In this protocol, the chlorine is added as NaOCl, which will increase the alkalinity of the bulk water to the extent that the OCl^- reacts with the H_2O.[16] A change in the alkalinity is important because at pH < 7.5 HOCl dominates and at pH > 7.5 OCl^- dominates. HOCl is known to have the greater efficacy. In the protocol, the bulk water is buffered with K_2HPO_4 and KH_2PO_4 to a pH of 7.2.

The following treatment protocol was used to evaluate the performance of chlorine in killing biofilm organisms using the RDR system as described previously by the authors.[12] For the treatment protocol, two parallel RDRs are set up such that one is used as a control with no treatment and the other receives the chlorine treatment.

Setup of RDR Reactor Vessels

1. Assemble two sterile RDR reactors in parallel. Attach a "Y" joint to the water filter on the effluent end so both reactors can be run off of the same filter and pump. Adjust the flow rate to 16 ml/min.
2. Set up a single carboy for nutrient supply to both RDRs.
3. Set up separately pumped nutrient lines to each RDR.
4. Set flow rate for nutrients at 1 ml/min.

Batch Nutrients Preparation

The 24-hr batch phase is used to establish the biofilm.

1. Mix 0.15 g TSB, 3.5 g K_2HPO_4, and 1.0 g KH_2PO_4 in 500 ml reverse osmosis (RO) water for each reactor. K_2HPO_4 and KH_2PO_4 added to buffer the

[16] V. L. Snoeyink and D. Jenkins, "Water Chemistry." Wiley, New York, 1980.
[17] A. Camper, M. Burr, B. Ellis, P. Butterfield, and C. Abernathy, *J. Appl. Microbiol. Symposium Suppl.* **85**, 1S (1999).

solution are at the following concentrations: 7 g/liter K_2HPO_4 and 2 g/liter KH_2PO_4.

2. Check pH and adjust to 7.2.
3. Autoclave on liquid setting for 25 minutes.

CSTR Nutrients

These nutrients are used during the 24-hr CSTR growth period following batch operation. Buffer and nutrient concentrations are determined using Eq. (1):

$$C_1 \times Q_1 = C_2 \times Q_2 \tag{1}$$

(Q, flow rate; C, concentration; 1, influent; 2, effluent)
Using a 4 liter carboy, the following calculations are made:

$$C_1 \times (1 \text{ ml/min}) = (30 \text{ mg/liter TSB}) \times (17.0 \text{ ml/min})$$
$$= 510 \text{ mg/liter TSB}$$
$$= (2040 \text{ mg/liter TSB}) \times (1 \text{ g/1000 mg})$$
$$= 2 \text{ g TSB added to 4 liter RO water}$$

$$K_2HPO_4 \text{ buffer}: C_1 \times (1 \text{ ml/min}) = (7 \text{ g/liter } K_2HPO_4) \times (17 \text{ ml/min})$$
$$= (119 \text{ g/liter } K_2HPO_4) \times (4 \text{ liter water})$$
$$= 476 \text{ g } K_2HPO_4 \text{ added to 4 liters RO water}$$

$$KH_2PO_4 \text{ buffer}: C_1 \times (1 \text{ ml/min}) = (2 \text{ g } KH_2PO_4) \times (17 \text{ ml/min})$$
$$= (34 \text{ g/liter } KH_2PO_4) \times (4 \text{ liter water})$$
$$= 136 \text{ g } KH_2PO_4 \text{ added to 4 liters RO water}$$

1. Add 3 liters RO water and a stir bar to a 4 liter carboy.
2. Add 1 liter RO water to a 2000 ml Erlenmeyer flask and mix the calculated amounts of TSB and buffering agents until dissolved and autoclave for 20 min on liquid setting. Because TSB caramelizes, the concentrated buffers and nutrients are autoclaved separately.
3. Autoclave the 4 liter carboy for 90 min on liquid setting.
4. Aseptically pour contents of the flask into the carboy.

Rinse Water

Rinse water is used to flush any demand on chlorine out of the RDR following the CSTR operation. The rinse water is buffered at the same concentrations as the nutrients (i.e., 7 g/liter K_2HPO_4 and 2 g/liter KH_2PO_4).

---

[18] G. A. McFeters, F. P. Yu, B. H. Pyle, and P. S. Stewart, *J. Indust. Microbiol.* **15**, 333 (1995).

Calculations.

$$7 \text{ g/liter} \times 10 \text{ liters RO water} = 70 \text{ g K}_2\text{HPO}_4$$

$$2 \text{ g/liter} \times 10 \text{ liter RO water} = 20 \text{ g KH}_2\text{PO}_4$$

1. Add 10 liters RO water to a large carboy.
2. Mix the buffering agents until dissolved.
3. Foil and tape tubing attached to the carboy.
4. Autoclave for 90 min on liquid setting.

Validating Chlorine Treatment Concentration

It is necessary to determine the total chlorine concentration of the NaOCl solution before treating the reactor because of the instability of NaOCl. Chlorine demand-free glass and water are used.

1. Using the DPD colorimetric method documented in *Standard Methods for the Examination of Water and Waste Water,* determine the actual concentration of total chlorine.[19]
2. Use the above determined concentration of total chlorine to calculate the volume of NaOCl needed to dose the reactor at 5 mg/liter total chlorine. For example, if the concentration of total chlorine in the NaOCl equals 69,920 mg/liter, use the relationship in Eq. (2),

$$C_1 \times V_1 = C_2 \times V_2 \tag{2}$$

to calculate the volume of NaOCl required to achieve a bulk fluid concentration of 5 mg/liter total chlorine in the reactor.

$$(69{,}920 \text{ mg/liter total chlorine}) \times V_1 = (5 \text{ mg/liter total chlorine})$$
$$\times (500 \text{ ml}) = 0.036 \text{ ml} = 36 \text{ } \mu\text{l NaOCl}.$$

Thus, the reactor is dosed with 5 mg/liter total chlorine by adding 36 μl NaOCl to the bulk fluid.

Preparing Sodium Thiosulfate Neutralization Solution

Note: 1 ml of a 2.1 g/liter sodium thiosulfate solution neutralizes 0.3 mg/liter solution of chlorine.

[19] Anonymous, *in* "AHPA, Standard Methods for the Examination of Water and Wastewater" (A. D. Eaton, L. S. Clesceri, and A. E. Greenberg, eds.), 19th ed., pp. 4–45. AHPA, ANWA, REF, Washington, D.C., 1995.

Calculations

0.3 mg/liter Cl/2.1 mg/liter sodium thiosulfate = 0.14 mg Cl/1.0 mg sodium thiosulfate

For this study, 500 ml of 5 mg/liter total chlorine must be neutralized.

(5 mg Cl/liter) × (1 mg sodium thiosulfate/0.14 mg Cl) = 35.71 mg/liter sodium thiosulfate solution is required to neutralize 5 mg/liter total chlorine.

Test the solution to ensure that it has been properly neutralized.

Conducting the Antibiofilm Test

1. Operate the RDR reactors for 24 hr in batch, followed by 24 hr in CSTR operation (48 residence times).
2. After the 24 hr of CSTR operation, turn off the nutrients and dilution water and connect the rinse water to the dilution water pump. Run the rinse water through the reactors for six residence times (to speed up this process, the flow rate on the pump can be increased to 32 ml/min).
3. Turn the pumps off.
4. Extract 5 ml of the bulk water from each reactor and analyze pH.
5. Clamp off the effluent tubing on each reactor.
6. Add 5 mg/liter total chlorine to the RDR reactor designated for treatment.
7. Immediately collect 5 ml samples from each reactor and measure chlorine concentration.
8. Treat the reactors for 10 min.
9. After 10 min, collect samples from each reactor and measure the chlorine concentration.
10. Neutralize both reactors with 1 ml sodium thiosulfate solution for 2 min.
11. Ensure that the reactors are neutralized by collecting samples from each reactor and analyzing for chlorine.
12. Scrape three randomly chosen coupons from each reactor using the protocol described previously by the authors.[12]
13. Plate samples on R2A agar plates using the drop plate method.
14. Samples incubate for 17–20 hr in 35° incubator.
15. Enumerate the colony forming units (CFU) following incubation, and calculate the number of cfu/cm^2.
16. Calculate the log reduction associated with the treatment (see the next section for details on the calculation).

Repeatability of the Rotating Disk Reactor Tier-One Antibiofilm Test Protocol

Performance of the antibiofilm agent is expressed as a log reduction (LR) value comparing measurements on both antimicrobially treated and untreated biofilm in the same test. After repeated experiments, one can use the variability of the LR values as the basis for determining repeatability; specifically, one can calculate a repeatability standard deviation. An experimental method that produces a small repeatability standard deviation has good repeatability, and a big standard deviation indicates poor repeatability. One difficulty with this approach is that there is no generally accepted cutoff value for the repeatability standard deviation that separates repeatable from nonrepeatable methods.

In this paper, we will follow the terminology of the Association of Official Analytical Chemists—International.[20] The term "repeatability" standard deviation is used for the case where the replicates of the same experiment are all done within one laboratory. The term "reproducibility" standard deviation is used when the replicates are done in different laboratories. The next section shows how to calculate the LR, the standard error of the LR, and the repeatability standard deviation. For illustrative purposes, we apply these techniques to data from the RDR tier-one Antibiofilm Test Protocol using the sodium hypochlorite protocol described in the previous section of this paper.

Calculating the Log Reduction and the Standard Error: Formulas and Examples

Following the RDR antibiofilm test protocol, CFU counts are recorded for control (untreated) coupons and for test (treated with an antimicrobial chemical) coupons. The steps in calculating the log reduction (LR) are as follows:

1. Find the bacterial cell density (CFU/cm^2) for each coupon.
2. Find the mean of the $\log_{10}$ densities for control coupons and the mean of the $\log_{10}$ densities for the treated coupons.
3. Calculate the LR by subtracting the mean log densities for treated coupons from the mean log densities for control coupons. The standard error (SE) formula requires calculation of the control coupon and RDR biofilm coupon variances of log densities, as well as the means.

Calculating the Bacterial Cell Density for Each Coupon. For each dilution in a dilution series, CFU counts are recorded for each "location," where a location is a drop for the drop plate method. The counts need to be scaled up according to the volume represented by the location to provide an estimate of the total viable cells

---

[20] Anonymous, *in* "Official Methods of the Association of Official Analytical Chemists" (K. Helrich, ed.), p. 681. AOAC, Arlington, VA, 1990.

in the beaker, then converted to the number of colony forming units per square cm of the coupon, called the bacterial cell density. The appropriate formula is given by Eq. (3)[21–23]:

$$\text{Density} = \frac{\text{Avg count}}{\text{Volume Plated}} \times \frac{1}{\text{Dilution}} \times \text{Beaker Volume} \times \frac{1}{\text{Surface area}} \quad (3)$$

where

Avg count = average CFU across for a dilution (CFU),

Volume plated = volume plated for each raw data counting location (ml),

Dilution = the 10^{-k} (for 10-fold dilutions), where k is an integer (no units),

Beaker Volume = volume of liquid containing the biofilm removed from the coupon (ml),

Surface area = surface area of the coupon from which the biofilm was removed (cm^2)

Note:

$$\frac{\text{Volume Plated} \times \text{Dilution}}{\text{Beaker Volume}} = \text{fraction of the beaker volume for each raw data count}$$

Example of a Bacterial Cell Density Calculation. An RDR coupon biofilm sample was scraped into a beaker volume of 10 ml; the surface area scraped from the coupon was 1.267 cm^2. The CFUs were counted in each of 10 drop locations using the drop plate method with an individual drop volume of 0.01 ml. At the 10^{-3} dilution, the CFU counts were 27, 32, 22, 22, 25, 21, 14, 24, 17, and 13. The average of these counts is 21.7. Then

$$\text{Density} = \frac{21.7}{0.01} \times \frac{1}{10^{-3}} \times 10 \times \frac{1}{1.267} = 1.71 \times 10^7 \frac{\text{CFU}}{\text{cm}^2}$$

and $\log_{10}$ density = 7.234.

Calculating LR and SE. Let N denote the number of control coupons and M denote the number of treated coupons for which densities have been calculate. The estimated LR for the treatment is to be based on these $N + M$ coupons. Take the $\log_{10}$ transformation of each of the $N + M$ densities. Denote the control coupon $\log_{10}$ densities by $C_1, C_2, \ldots, C_N$, and denote the treated coupon $\log_{10}$ densities

[21] B. Jarvis, "Statistical Aspects of the Microbiological Analysis of Foods." Elsevier, New York, 1989.

[22] F. J. Farmiloe, S. J. Cornford, J. B. M. Coppock, and M. Ingram, *J. Sci. Food Agric.* **5**, 292 (1954).

[23] S. Niemelä, *in* "Statistical Evaluation of Results from Quantitative Microbiological Examinations," 2nd ed. Nordic Committee on Food Analysis, Uppsala, Sweden, 1983.

by $T_1, T_2, \ldots, T_M$. Calculate $\bar{C}$, the mean of the log densities for control coupons, and $\bar{T}$, the mean of the log densities for RDR biofilm coupons, using Eq. (4):

$$\bar{C} = \frac{C_1 + C_2 + \cdots + C_N}{N}; \quad \bar{T} = \frac{T_1 + T_2 + \cdots + T_M}{M} \tag{4}$$

The variances of the log densities are needed to calculate the SE. Calculate S_C^2, the variance of the log densities for control coupons, and S_T^2, the variance of the log densities for treated coupons, using Eq. (5):

$$S_C^2 = \frac{(C_1 - \bar{C})^2 + (C_2 - \bar{C})^2 + \cdots + (C_N - \bar{C})^2}{N - 1};$$
$$S_T^2 = \frac{(T_1 - \bar{T})^2 + (T_2 - \bar{T})^2 + \cdots + (T_M - \bar{T})^2}{M - 1} \tag{5}$$

Calculating the LR and the Associated SE. Calculate the log reduction (LR) using Eq. (6)[24]:

$$LR = \bar{C} - \bar{T} \tag{6}$$

The standard error of LR depends on the variances among the control and treated biofilm coupons. Calculate the SE using Eq. (7)[24]:

$$SE = \sqrt{\frac{S_C^2}{N} + \frac{S_T^2}{M}} \tag{7}$$

This SE formula provides a measure of the inherent, within-test variability, due to the variability among coupons and the variability associated with forming dilutions, plating, and counting CFUs.

Example of LR and SE Calculations. Calculations will be demonstrated using observed $\log_{10}$ densities of 7.79934, 7.23045, and 7.86332 for N = 3 control coupons and 4.41497, 4.85733, and 3.98677 for M = 3 treated coupons. The means of the log densities are $\bar{C} = 7.63104$ and $\bar{T} = 4.41969$. The variances are $S_C^2 = 0.12138$ and $S_T^2 = 0.18749$. Here, as is usual in antibiofilm tests, the variance for treated coupon log densities is larger than the variance for control coupon log densities. The main results are

$$LR = 7.63104 - 4.41970 = 3.21134 \quad \text{and}$$

$$SE = \sqrt{\frac{0.12138}{3} + \frac{0.18749}{3}} = 0.32790$$

It is advisable to carry extra digits during intermediate calculations, then use 1 or 2 decimal places for the final answer; e.g., LR $\pm$ SE = 3.2 $\pm$ 0.3 = the interval (2.9, 3.5).

---

[24] T. A. DeVries and M. A. Hamilton, *Quant. Microbiol.* **1**, 29 (1999).

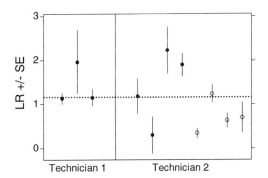

FIG. 3. Log reduction (LR) values for 5 mg/liter chlorine administered for 10 min to biofilms grown on coupons in the rotating disk reactor. Technician 1 conducted 3 experiments and Technician 2 conducted 8 experiments; the vertical line separates the results. Each LR is based on 3 control coupons and 3 treated coupons. The biofilms were grown in either a 500 ml fluid volume (closed circles) or a 200 ml fluid volume (open circles). Over all 11 experiments, the mean LR is 1.16 (horizontal dotted line) and the standard deviation is 0.66. The error bars show the within-experiment standard error (SE).

Repeatability Standard Deviations

The repeatability of an antimicrobial test will depend on the test protocol, the species of organism, and the disinfectant. A literature survey[25] of standard antimicrobial suspension and dried surface tests showed that the repeatability standard deviation of LR values varied between 0.2 and 1.2 with a median repeatability standard deviation of 0.5. It is important to check the repeatability standard deviation of antibiofilm testing protocols. For this reason the LR values for a 5 min application of 5 mg/liter sodium hypochlorite were measured on separate occasions using the RDR protocol (Fig. 3). Figure 3 shows the LR $\pm$ SE for 11 separate repeats of the test. The repeatability standard deviation of the LR values was 0.66. Because two technicians conducted the test, it was possible to conduct a random effects analysis of variance and estimate the three sources of variability that make up the repeatability standard deviation (i) within assay, (ii) among assays within a technician, and (iii) among technicians. On a percentage basis, the variation among repeated RDR tests can be attributed 30% to within-assay variability, 70% to variability among assays within a technician, and 0% to variability among technicians. The lack of variability between the technicians suggests that the RDR protocol can be learned and applied. The relatively large variability among assays within a technician suggests there are factors in the experimental technique that vary when the RDR is set up and run on different days, and that if the method is to be improved, it will be important to find and control those factors. Note, however, that the RDR

[25] N. Tilt and M. A. Hamilton, *J. Assoc. Official Anal. Chem.* **82,** 384 (1999).

protocol is as repeatable as the typical standard suspension or dried surface test, and for that reason, it appears to be sufficiently repeatable for routine use.

The literature survey[25] of standard antimicrobial suspension and dried surface tests also showed that the reproducibility standard deviation of LR values varied between 0.3 and 1.5 with a median reproducibility standard deviation of 0.9. The reproducibility standard deviation is made up of (i) the within-lab variation as measured by the repeatability standard deviation and (ii) lab-to-lab variation. Typically, there was a 50–50% split between these two sources. We have not yet collaborated with other laboratories in round robin testing of disinfectants with the RDR protocol, so we do not know either the lab-to-lab variance or the reproducibility standard deviation for the RDR.

Conclusions

Current tests for evaluating antimicrobials were developed for evaluating the efficacy of antimicrobials on suspended bacteria. Because biofilms have unique characteristics and can be more resistant than suspended bacteria, new and more rigorous test methods must be developed. Methods for evaluating antibiofilm tests must be conducted using relevant biofilm that represents the genetic, transport, and heterogeneity antimicrobial resistance properties of biofilm. Depending on the antibiofilm product claim being tested, the antibiofilm test may need to determine kill, removal, or prevention of regrowth of a biofilm. All of these considerations result in the practitioner having to make more decisions when developing or applying tests for evaluating antibiofilm tests. Additional considerations include whether the test should use a CSTR chemostat approach and how the biofilm organisms are sampled and enumerated.

Because an antibiofilm test is more complicated than traditional antimicrobial tests on suspended bacteria, we expect it will be difficult to develop antibiofilm tests with adequate quality control such as repeatability. The authors have demonstrated the use of a simple rotating disk reactor (RDR) chemostat in evaluating hypochlorite as a model biocide for killing biofilm bacteria. Preliminary work presented by the authors shows that it is possible to achieve an acceptable repeatability of log reduction values for the RDR biofilm test system using NaOCl.

Acknowledgments

The authors acknowledge support from the National Science Foundation Engineering Research Centers Program (Cooperative Agreement #EEC-8907039). Special thanks are also addressed to the following: Montana State University students for their contribution to the data and images appearing in this chapter; Ben Wyffels, Dale Niemeyer, Erin Karakas, and Boloroo Purevdorj.

[25] The MBEC Assay System: Multiple Equivalent Biofilms for Antibiotic and Biocide Susceptibility Testing

By Howard Ceri, Merle Olson, Douglas Morck, Douglas Storey, Ronald Read, Andre Buret, and Barbara Olson

Introduction

Microbial biofilms are now being recognized for their important role in both nature and disease.[1-3] The lack of recognition of biofilms as important components of the microbial world was due in part to the technical difficulties associated with their study. Batch culture techniques provided a ready source of materials for the metabolic characterization of planktonic microorganisms; however, to grow biofilm populations in the laboratory has not been as easy. A number of technologies have been developed to study biofilm growth, as is seen in this and in Volume 310 of *Methods in Enzymology* dealing with biofilms.[4-8] Although these technologies produce reproducible biofilms for the study of biofilm growth, structure, and physiology, they have not been amenable for the routine study of biofilm susceptibility to antibiotics and biocides. For this reason, virtually every antibiotic and biocide available has been selected for activity against planktonic organisms. These drugs often have been found to lack activity against microbial biofilms.[9-12] The MBEC Assay System using the Calgary Biofilm Device[13] provides for the first time an assay easily applicable to screening antibiotics and biocides for activity against microbial biofilms.

---

[1] J. W. Costerton, P. S. Stewart, and E. P. Greenberg, *Science* **284,** 1318 (1999).

[2] B. Dixon, *ASM News* **64,** 484 (1998).

[3] C. Potera, *Science* **284,** 1837 (1999).

[4] A. Kharazmi, B. Giwercman, and N. Hoiby, *Methods Enzymol.* **310,** 207 (1999).

[5] D. G. Alison and P. Gilbert, *Methods Enzymol.* **310,** 232 (1999).

[6] M. Wilson, *Methods Enzymol.* **310,** 264 (1999).

[7] D. J. Bradshaw and P. D. Marsh, *Methods Enzymol.* **310,** 279 (1999).

[8] J. R. Lawrence and T. R. Neu, *Methods Enzymol.* **310,** 131 (1999).

[9] H. Kumon, N. Ono, and J. C. Nickel, *Antimicrob. Agents Chemother.* **39,** 1038 (1995).

[10] T. Larsen and N.-E. Fiehn, *APMIS* **104,** 280 (1996).

[11] D. W. Morck, K. Lam, S. G. McKay, M. E. Olson, B. Prosser, B. D. Ellis, R. Cleeland, and J. W. Costerton, *Int. J. Antimicrobial Agents* **4,** S21 (1994).

[12] P. S. Stewart, *Antimicrob. Agents Chemother.* **40,** 2517 (1996).

[13] H. Ceri, M. E. Olson, C. Stremick, R. R. Read, D. Morck, and A. Buret, *J. Clin. Microbiol.* **37,** 1771 (1999).

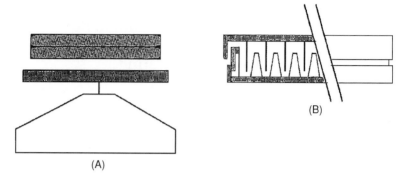

Fɪɢ. 1. MBEC Assay System. (A) MBEC Assay System placed on a rocking table. (B) Cutaway view of the MBEC Assay System showing the pegs of the lid placed in the channels of the trough. (Modified from Ref. 13.)

The MBEC Assay System

The MBEC Assay System is available through MBEC Biofilms Technologies Ltd., Calgary, Alberta [Fax (403) 266-5050 or e-mail mbec@nucleus.com]. This patented assay system involves a reactor for the formation of 96 equivalent biofilms. The MBEC Assay System is ideally suited either for screening new putative antibiotics and/or biocides, or for the determination of both the MIC (minimal inhibitory concentration) and MBEC (minimum biofilm eradication concentration) values in clinical situations for the treatment of chronic, recurrent, or device-related infections. The MBEC Assay System produces 96 equivalent biofilms formed under flow conditions, without the need for pumps. Further, as it is based on the standard 96 well platform it conforms to existing technology available in most laboratories. The MBEC Assay System consists of a two-piece disposable plastic apparatus used for biofilm formation (Fig. 1). The top component consists of a lid that supports 96 pegs placed to sit in the center of each well of a standard 96 well plate. The pegs are formed to allow them to be easily removed, and the lid is sealed such that pegs can be removed without compromising sterility of the system. The bottom piece forms a trough into which the planktonic bacterial suspension is placed. The bottom of the trough is divided into channels in which the pegs of the lid sit. When the apparatus is placed on a tilt table, the media containing planktonic bacteria flows through the channels, causing shear force against the pegs and initiating adhesion of bacteria to form a biofilm on the surface of each peg. The design of the trough is such that an equivalent biofilm forms on each peg as is seen in Fig. 2.

Biofilm Formation

Biofilms can be formed from virtually any organism in monoculture or from specified bacterial combinations in mixed biofilms. They may also be formed from

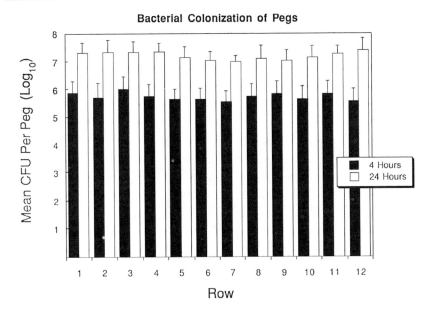

FIG. 2. Reproducibility of the assay: the mean cfu/peg along each row after 4 and 24 hr of growth of *P. aeruginosa* is shown. (Modified from Ref. 13.)

natural sources of bacteria using the medium from which they have been derived as the inoculum. Typically, in order to grow a biofilm, bacteria are grown on agar plates of a suitable growth medium. Colonies picked from the plate are suspended into broth of the same or another suitable growth medium. Inoculum size is standardized to McFarland reference standards and validated by plate counts. Natural populations, such as white water or cooling tower water samples, are added directly to the trough of the MBEC Assay System. The lid is then put in place and the biofilm formed by incubating the apparatus on a rocking table (Red Rocker model; Hoefer Instrument Co.) under appropriate conditions. Incubation is typically carried out at 35° and 95% humidity for patient derived aerobic isolates. Incubation under microaerophilic conditions can be achieved using a CO_2 incubator adjusted to the required oxygen tension, or anaerobic conditions can be achieved by placing the rocker in an anaerobic hood or by placing the MBEC Assay System in an anaerobic jar. The optimal shear force for biofilm formation can be achieved by adjusting the speed of the rocking table. Antibiotic and biocide susceptibility of biofilms of bacteria,[13] fungi,[14] and mycobacteria[15] have been studied using the MBEC Assay System.

[14] H. Ceri, D. Fogg, M. E. Olson, R. R. Read, D. Morck, and A. Buret, *Proc. Interscience Conf. Antimicrob. Agents Chemother.* **38,** 495 (1998).

[15] B. Huddleston, E. Bardouniotis, M. E. Olson, and H. Ceri, *Proc. Biofilms 2000,* Abst. 94, p. 98.

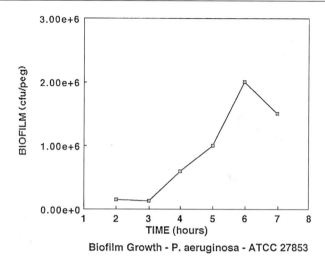

Biofilm Growth - P. aeruginosa - ATCC 27853

FIG. 3. Biofilm growth curve of *P. aeruginosa* in trypticase soy broth on the MBEC Assay System. (Modified from Ref. 13.)

Biofilm Growth Curves

Biofilm growth curves can be derived to study biofilm production or to standardize the inoculum size for susceptibility testing. Once the inoculum is established and growth conditions selected, the growth of the biofilm can be determined by randomly removing pegs from the lid at specific time periods (typically done in duplicate). The pegs are placed in microfuge tubes containing 200 μl of sterile buffer, which is sonicated for 5 min on high using a sonic cleaning system such as the Aquasonic (model 250HT; VWR Scientific). The organisms removed from the biofilm by sonication are then plated on suitable medium to determine viable bacterial number. Reproducible biofilm growth curves as seen in Fig. 3 can be generated in this fashion. Growth curves for the planktonic population can also be determined by plating the liquid growth medium. The reproducibility of the biofilm growth curves allows for the standardization of the assay against fixed bacterial numbers, as in the current MIC procedures.

Susceptibility Testing

Antibiotic or Biocide Preparation

Dilutions of antibiotics or biocides are prepared in suitable media in standard 96 well plates. The choice of medium is dependent on the nature of the organism and/or the solubility of the drug. Typically, for antibiotic testing, stock solutions of 6,200 μg/ml are prepared, in water where possible, and store at −80° until used.

Working solutions of antibiotics are then prepared in cation-adjusted Mueller–Hinton broth (CAMHB; BDH) to a concentration of 1024 μg/ml and serially diluted in CAMHB. This medium was selected in an attempt to standardize the MBEC assay to NCCLS recommended procedures for MIC testing; however, any appropriate medium for the growth of the organism in question can be used. Biocides can be prepared as stock solutions if stable or made fresh when needed, as in the case of chlorine or bromine solutions. Biocide testing can be carried out in solutions of water of defined hardness and pH.

Minimum Biofilm Eradication (MBEC) Assay

Once the test antibiotic or biocide solutions are prepared in the 96 well plates, biofilms formed on the peg-lid of the MBEC Assay System can be tested for susceptibility. The biofilms on the peg lid are first washed in a wash tray with sterile buffer to remove both planktonic cells and any growth medium carried over from the growth phase of the assay. The lid is then placed over the standard 96 well tray containing antibiotics/biocides and allowed to incubate. Incubation time is dependent on the nature of the assay. Antibiotics are typically tested following an overnight incubation period, whereas biocides are exposed to the biofilms for shorter time periods. The MBEC value is defined as the minimum concentration of antimicrobial that eradicates the biofilm. Biofilm eradication can be determined in a number of ways following the incubation in the presence of antibiotic or biocide. The peg lid is removed from the antimicrobial test solution and again washed free of reagents in a 96 well tray containing sterile buffer. The lid is then placed in another 96 well plate containing growth recovery medium, and the apparatus is sonicated as described above. Viable cells in the biofilm can be detected by plating onto appropriate medium, or by measuring increased turbidity over time, using a 96 well plate reader. The 96 well plates are read at 650 nm using the sterile control wells as blanks. The time course of antimicrobial activity can be followed by breaking pins in duplicate at various time periods, washing to remove the active agent, and then sonicating and plating the surviving cells, as in the growth curve described above.

Minimal Inhibitory Concentration Assay

The minimal inhibitory concentration (MIC) of antibiotics can also be determined using the MBEC assay system. Cells shed from the biofilm will grow in the antibiotic test medium, as long as the antibiotic it contains is below the MIC concentration. Therefore, as one conducts biofilm susceptibility testing it is also possible to generate MIC values for the drug. Typically, the test plate to obtain the MBEC value for the antibiotics is incubated overnight. To obtain the MBEC value, the lid is removed from the antibiotic plate and processed as described above. To obtain the MIC values, the antibiotic test plate is read for turbidity at 650 nm as above. MIC values obtained from the MBEC Assay System have been consistent with those values obtained by NCCLS methods[13] in most cases. As the fixed number of bacteria added to the MIC wells is not defined in the MBEC Assay System, difficulties with certain strains and antibiotic combinations can occur, and complete standardization according to the methods of NCCLS is not possible in all situations. The assay can also present difficulties when working with small colony variants where turbidity measurements become an issue. These problems are easily overcome by plating of organisms for viable counts.

MBEC vs MIC Values

The values obtained by MIC and MBEC measurements clearly identify the difference in susceptibility between planktonic and biofilm populations of the same organism (Fig. 4). MBEC values obtained from the MBEC Assay System and via the modified Robbins device can be compared in Table I. The major difference is the time investment in generating the data. The data generated on the MRD were collected over a continuous 8 week period, whereas MBEC Assay System values were generated in less than 1 week.

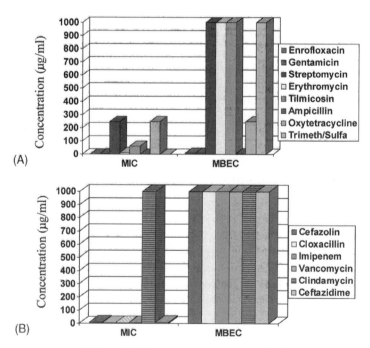

FIG. 4. Comparison of MIC and MBEC values obtained using the MBEC Assay System for (A) *E. coli* isolate from a veterinary biofilm infection; (B) *S. epidermidis* from a human implant infection. In both cases substantial differences can be seen between MIC and MBEC values.

TABLE I
SUSCEPTIBILITY OF A URINARY TRACT ISOLATE OF *E. coli* COMPARING MIC (MINIMAL
INHIBITORY CONCENTRATION) VALUES DETERMINED BY TUBE DILUTION, NCCLS, AND MBEC
ASSAYS WITH MBEC (MINIMAL BIOFILM ERADICATION CONCENTRATION) DERIVED FROM MRD
(MODIFIED ROBBINS DEVICE) AND MBEC ASSAYS

Antibiotic	MIC tube dilution	MIC NCCLS	MIC MBEC derived	MBEC MRD derived	MBEC MBEC derived
Ampicillin	2.0	2.0–4.0	4.0	>16	>1024
Gentamicin sulfate	0.5	0.5–2.0	2.0	>32	16
Trimethoprim sulfamethoxide	1.0	0.1–0.5	0.1–0.5	>256	>256
Fleroxacin	0.16	0.13	0.25	2.0	2.0

Other Functions of MBEC Assay System

1. *Enhancement of biofilm formation:* In many processes using attached bacteria, such as bioremediation, sewage treatment, or fermentation systems, to increase the biofilm mass would enhance the process. The MBEC Assay System affords a simple tool to screen the ability of specific components to increase the production of biofilms (Fig. 5). Growth curves in the presence of additives can be easily carried out to define conditions that optimize biofilm formation. The 96 well plate can also be used as a microfermentor to determine optimal yields in fermentation processes where different combinations of additives are supplied to 96 equivalent biofilms.

2. *Uptake studies:* The uptake of various components including antibiotics into biofilms can be easily studied using the MBEC Assay System. Equivalent biofilms can be exposed to radiolabeled substrates or antibiotics over defined time periods.

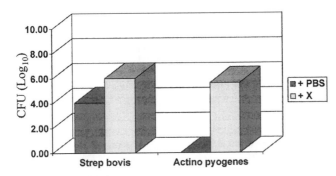

FIG. 5. Enhanced biofilm formation of *S. bovis* and *A. pyogenes* in trypticase soy broth following the addition of specific growth factors, as compared to the same organism grown in the presence of the carrier phosphate buffered saline.

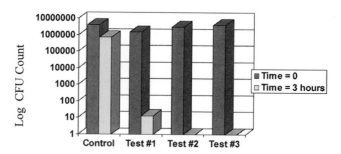

FIG. 6. The effect of a biocide released from a coated material on the reduction of a *P. aeruginosa* biofilm. The T_0 bacterial biofilm density is compared to the biofilm density after 3 hr exposure to noncoated material (control) or to coated materials in three trials.

The pegs can be removed, washed to remove labeled compounds not internalized, and then added directly to extraction compounds to allow for the determination of the kinetics of uptake of the compounds.

3. *Release of antibiotics or biocides bound to surfaces:* The effect of antimicrobials released from surfaces on biofilm formation or viability can be tested in this assay. The bioactive compound can be directly incorporated onto the device, or a coating product may be added to the device for release of the compound. The release of a biocide from a bandage coat is seen in Fig. 6.

4. *Potentiation of antimicrobial activity by additives or combinations of drugs:* The ability to form a large number of equivalent biofilms and to test them simultaneously allows one to screen large numbers of drug combinations for activity against biofilms. This would allow for the custom formulation of reactive agents either by combining a number of antibiotics or biocides to identify those combinations with greatest efficacy, or by looking for additives that may enhance a drug's activity against biofilms. Figure 7 illustrates an example of an additive

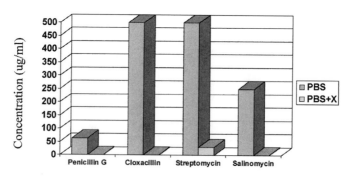

FIG. 7. Activity of penicillin G, cloxacillin, streptomycin, and salinomycin against *S. bovis,* either alone or in combination with an enhancing agent.

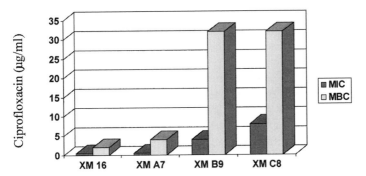

FIG. 8. The development of antibiotic resistance to ciprofloxacin by *S. maltophilia* in biofilms as determined by an increase in MIC and MBC values.

that enhanced antibiotic activity against a bacterial isolate of great significance to veterinary medicine.

5. *Development of antibiotic resistance in biofilms:* Treating biofilms with MIC levels of antibiotics or biocides often results in exposing the biofilm to concentrations of drug that cannot eradicate the biofilm. Exposure of biofilms to MIC or lower levels of drugs can cause changes in the MIC and MBC values. This suggests an increase in antibiotic resistance or a selection for more resistant organisms is taking place within the biofilm. Figure 8 demonstrates the development of ciprofloxacin resistance in *Stenotrophomonas maltophilia.*

[26] Assessment of Antimicrobial Activity against Biofilms

By DAVID DRAKE

Overview

It is well known that bacteria in biofilms are different from their planktonic cell counterparts in many ways, from growth rate to expression of various cell surface proteins. One consequence of this multitude of differences is that bacteria in biofilms have been shown to be substantially more resistant to antibiotics and antimicrobials.[1-3] This can be of paramount importance when one considers that the usual assay methods for determining antimicrobial susceptibility and resistance utilize bacteria in suspension. While standard minimal inhibitory and bactericidal concentration (MIC/MBC) determinations provide a rapid screening method for assessing antimicrobial resistance, the information obtained may not be applicable to the *in vivo* situation where an infection is caused by microorganisms growing in biofilms. To more readily address this disparity in response to antimicrobials, it is important to assess activity of new compounds in a model system that embraces the response of biofilms.[1-3]

We have developed a simple biofilm model system using *Streptococcus mutans*, a primary etiological agent of dental caries. The model system embraces the classical experimental design of McCabe *et al.*[4] Sucrose-driven colonization of a hard surface, in this case glass slides instead of nichrome wires, is the heart of this model, which is described below.

Preparation of *Streptococcus mutans* Biofilms

Primary Cultures

Streptococcus mutans 10449 is initially cultured in invertase-treated trypticase soy broth (Difco) supplemented with 0.5% yeast extract (TSB-YE) in 50 ml volumes. Cultures are incubated in a 5% CO_2 incubator, statically, at 37° for approximately 16 hr. Most of our studies have been with this commonly used strain; however, we have conducted this colonization protocol with other strains and clinical isolates. Profiles of colonization will vary from strain to strain and will have to be completely characterized by the investigator if one wishes to obtain reproducible results. Cells are harvested by centrifugation (5000g), washed twice in TSB-YE, and resuspended in equal volumes of the same medium.

Biofilm Model

The model assembly consists of presterilized glass microscope slides placed into sterile 50 ml centrifuge tubes. The colonization medium is composed of TSB-YE supplemented with 1.0% glucose and 2.0% sucrose. Tubes are inoculated with 0.5 ml of the washed suspensions as described above and incubated statically in 5% CO_2 at 37° for 48 hr. Following these procedures, we are able to achieve biofilms with a viable cell count of approximately 10^7–10^8 cfu/ml from experiment to experiment.

Exposure of Biofilms to Rinses or Dentifrices

Mouth rinses can be used full-strength or at any dilution desired. Dilutions are made in sterile distilled water in twofold increments. Dentifrices are prepared as follows. Slurries are made in sterile distilled water at a ratio of 1.0 g dentifrice to

[1] P. D. Marsh and D. J. Bradshaw, *Adv. Dent. Res.* **11**, 176 (1997).
[2] M. Wilson, H. Patel, and J. H. Noar, *Curr. Microbiol.* **36**, 13 (1998).
[3] J. Pratten and M. Wilson, *Antimicrob. Ag. Chemother.* **43**, 1595 (1999).
[4] R. M. McCabe, P. H. Keyes, and A. Howell, *Arch. Oral Biol.* **12**, 1653 (1967).

1.5 ml water. For most experiments, 10 g dentifrice was mixed with 150 ml distilled water. The resulting slurries are then centrifuged at $7500g$ to obtain supernatants. These fractions are then used in the biofilm exposure assays. Slides with biofilms are removed from their tubes and washed gently in sterile distilled water. Dilutions of mouth rinses or dentifrice supernatants are placed in beakers with stir bars for slow stirring. Slides with biofilms are immersed in the solutions for precisely 30 sec and then washed by repeated dipping (10 times) in two beakers containing 150 ml sterile distilled water. Following washing, biofilms are removed from the slides into sterile TSB-YE by scraping with sterile curettes. Resulting suspensions are vigorously vortexed for 2 min to disrupt cellular aggregates. In some experiments, we have added dextranase (250 μg/ml) to enhance the disruption of microscopic aggregates. These collective procedures, in our hands, have allowed for homogeneous suspensions for determination of numbers of viable cells via standard dilution plating techniques. Suspensions at this point are processed for numbers of viable cells and, in some cases, the ability of the cells to produce acid on a glucose challenge. For the former, suspensions are further diluted in sterile TSB-YE and plated onto mitis-salivarius agar using a spiral-plating system. Viable cell counts are determined following standard spiral-plating methodology. For the latter, a modified procedure of White et al.[5] is used. Suspensions are dispensed into 17×100 plastic test tubes and 250 μl of a 40% stock solution of glucose is added. Tubes are then mixed by vortexing and incubated statically in 5% CO_2 at 37°. At different increments of time, the pH of the suspensions is measured with microelectrodes. An alternative to measuring decreases in pH is an assessment of the amount of lactic acid that is produced. One can remove small aliquots and measure acid produced using a calibrated gas chromatograph. A more simple method is to measure lactate via a standard assay kit (Sigma).[6]

Collectively, these procedures allow one to (1) determine the effects of a brief exposure of S. mutans biofilms to antimicrobials on cell viability, and (2) determine the effects of this brief exposure on the relative physiology of the cells in terms of the ability of the bacteria to produce acid in the presence of glucose. The versatility of this model lies in its implicit simplicity for assaying multiple samples to satisfy appropriate statistical analyses and the ability to perform single or multiple exposure regimens over time. One can expose the biofilms as described above, incubate the biofilms in dilute media or artificial saliva for several hours, and then expose the biofilms to the antimicrobials for another round. In essence, this allows one to simulate the situation whereby plaque communities are exposed to antimicrobials in dentifrices three times over a day. We have done this successfully over a 3-day period with experimental dentifrices and have been able to show cumulative effects upon multiple exposures over time.

[5] D. J. White, E. R. Cox, N. Liang, D. Macksood, and L. Bacca, J. Clin. Dentistry 6, 59 (1995).
[6] M. B. Finnegan, A. Patel, B. M. Ludvigsen, and Z. Zhang, J. Dent. Res. 79, 571 (2000).

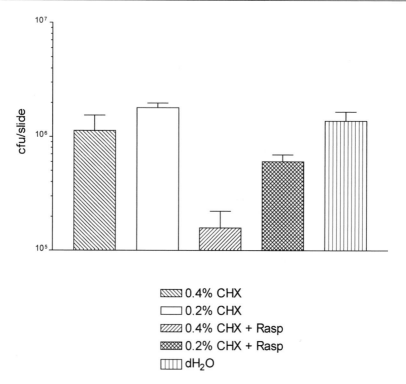

FIG. 1. One 30-second exposure of *Streptococcus mutans* biofilms to 0.4% CHX, 0.2% CHX, 0.4% CHX with Crystal Light Raspberry, and 0.2% CHX with Crystal Light Raspberry.

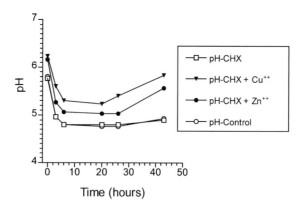

FIG. 2. Glycolysis of 48 hr biofilms of *Streptococcus mutans* after exposures to 0.06% CHX, CHX + Cu^{2+}, and CHX + Zn^{2+}.

An example of the data one can obtain is illustrated in Figs. 1 and 2. Biofilms of *S. mutans* were exposed to experimental chlorhexidine rinses for 30 sec in Fig. 1. One can readily see that the chlorhexidine rinse with raspberry flavoring added exhibited an enhanced level of bactericidal activity against *S. mutans* biofilms. In another experiment, biofilms were exposed to chlorhexidine and Cu^{2+} or Zn^{2+} combinations. The metals were added in the form of cupric acetate and zinc chloride at a concentration of 0.4% wt/vol, respectively. An assessment of the ability of cells to produce acid in the presence of glucose is shown in Fig. 2. It can be seen that the chlorhexidine–Cu^{2+} combination resulted in the lowest decreases in pH following glucose pulsing. Overall, these preliminary data from ongoing studies illustrate the usefulness of this model system for screening relative bactericidal activity and potential effects on the physiology of cells in biofilms of a cariogenic bacterium.

Section VI

Oral Microbial Biofilms

[27] Retrieval of Biofilms from the Oral Cavity

By Robert J. Palmer, Jr., Rosemary Wu, Sharon Gordon,
Cynthia G. Bloomquist, William F. Liljemark, Mogens Kilian, and
Paul E. Kolenbrander

Introduction

The oral cavity is an environment in which many bacteria flourish. Rapid colonization of cleaned tooth surfaces (the normal development of dental plaque) attests to this fact[1]; daily prophylaxis is the primary factor in maintaining oral health. However, the wide variety in physiology displayed by the approximately 500 microbial species known from the oral cavity[2,3] indicates that the conditions within the mouth vary dramatically on a micro scale. Supragingival dental plaque has a different composition than does subgingival plaque,[4] and biofilms on mucosal surfaces are different from those on hard tissues.[5] These different microbial communities result in part from differences in the substratum to which they initially attach, but the greatest source of variation in community composition is likely to result from differences in the environmental conditions experienced by microorganisms as they colonize (some successfully, others unsuccessfully) a particular site. In turn, the organisms themselves are responsible for alteration in the environmental conditions at the site: maturation of dental plaque displays a well-ordered ecological succession of microbial species.[5,6]

Thus, the location within the oral cavity of the sampling site plays a major role in the composition of the microbiota recovered, as does the age of the community when it is sampled. This chapter focuses on retrieval of intact biofilms under *in vivo* conditions that mimic those found in supragingival plaque. Some of the methodology described here is based on an approach developed by Ahrens[7] and Kilian.[8]

[1] B. Nyvad and M. Kilian, *Scand. J. Dent. Res.* **95**, 369 (1987).

[2] W. E. C. Moore and L. V. H. Moore, *Periodontol. 2000* **5**, 66 (1994).

[3] S. S. Socransky, A. D. Haffajee, M. A. Cugini, C. Smith, and R. L. Kent, *J. Clin. Periodontol.* **25**, 134 (1998).

[4] W. E. C. Moore, L. V. Holdeman, E. P. Cato, R. M. Smibert, J. A. Burmeister, and R. R. Ranney, *Infect. Immun.* **42**, 510 (1983).

[5] P. D. Marsh and M. V. Martin, "Oral Microbiology." Wright, Oxford, 1999.

[6] H. Löe, E. Theilade, and S. B. Jensen, *J. Periodontol.* **36**, 177 (1965).

[7] G. Ahrens, *Caries Res.* **10**, 85 (1976).

[8] M. Kilian, M. J. Larsen, O. Fejerskov, and A. Thylstrup, *Caries Res.* **13**, 319 (1979).

Enamel Chips Carried in Intraoral Acrylic Stents

Study Population

The study was reviewed and approved by the Institutional Review Board of the National Institute of Dental and Craniofacial Research, National Institutes of Health (NIDCR clinical protocol 98-D-0116).

Subjects are healthy volunteer outpatients screened for study participation. Experimental procedures are explained verbally and in writing, and informed consent is obtained prior to study enrollment. A complete medical history is elicited, and an oral examination performed using selected sections of the NIDCR NHANES oral examination criteria together with a periodontal assessment of Ramfjord indicator teeth. Subjects are excluded if their medical history indicates factors contributing to altered salivary flow, such as chemotherapy or radiation, autoimmune disorders, diabetes, AIDS, use of certain medications, and self-report of dry mouth. Data are collected on alcohol and tobacco use.

Clinical Procedures

Maxillary and mandibular alginate impressions are made and immediately poured in quickset stone for the formation of opposing articulating casts. The mandibular cast is surveyed and marked for creation of bilateral removable acrylic appliances. The appliance spans the posterior buccal gingival surface and is retained by two ball clasps. The appliances are fabricated by a commercial firm and delivered glutaraldehyde-disinfected. They are adjusted for proper fit in each patient's mouth: the appliance should be comfortable for the volunteer to wear, but it must not loosen or pop up from the jaw during normal mouth movements. Each appliance contains a $3.5 \times 10 \times 1.5$ mm groove to accommodate three $3 \times 3 \times 1$ mm enamel chips. Figure 1 shows two appliances in place within a volunteer's mouth, and a close-up of one appliance showing chip positioning within the appliance. When the appliance is not in the mouth, it is stored in a denture cup (plastic container with tightly fitting lid) containing a few drops of water in the bottom. For long-term storage (between experiments), the appliances in the denture cups are also refrigerated.

Enamel chips are cut from freshly extracted human third molars. The molars are collected in sterile saline, and any adherent tissue is cut away with a scalpel. The teeth are next soaked for 30 min in 2% sodium hypochlorite, rinsed twice with distilled water, then stored in a humid atmosphere and used within 2 weeks. The teeth are brushed using a medium 40-row toothbrush under tap water, then the roots are embedded in Utility Wax (Kerr Manufacturing Co., Romulus, MI. The crown is cut away perpendicular to the sides of the tooth using a Dynamax handpiece (Jelenko, Armonk, NY) equipped with a diamond disk, then a flat region

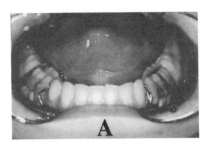 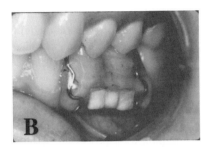

FIG. 1. (A) Intraoral stents in place in a volunteer's mouth, one on each side of the mandible. (B) Close-up of intraoral stent in place showing chip positioning.

on the buccal or lingual surface is identified and 3 × 3 mm regions are delineated with a lead pencil. The chips are freed from the tooth by making incisions with the disk first in the direction from the crown to the root, then by an undercutting incision (parallel to the enamel surface) approximately 1 mm in depth, with the final incision being made at the base of the tooth (parallel to the crown); this order of incisions seems to provide the least propensity for the chips to fly away when released from the tooth. The cut chips are rinsed in sterile water, then ground from the back (dentin) side in a custom-made vise with a redstone polishing burr to a maximum depth of 1 mm. The chips are rinsed again with sterile water, blotted dry, and sterilized by ethylene oxide.

Sterile chips are bonded into the appliance using Sticky Wax (Dentsply International, York, PA). A layer of wax is built up within the appliance channel through repeated application with a flamed cleoid–discoid amalgam carving tool (the wax melts in contact with the probe and hardens at room temperature). Immediately prior to insertion of a chip, the wax in that area of the channel is melted with the tool, then the chip is inserted using flamed forceps. During these procedures, the appliance is handled using sterile gauze. After the chips have been inserted, the appliance is placed in the volunteer's mouth under supervision of a dental hygienist. Subjects undergo prophylaxis immediately prior to appliance insertion. Subjects are asked to avoid eating and drinking for short-term wearing (4 hr). If the volunteer is to wear the appliance longer than 4 hr, eating/drinking are permitted; the volunteer is instructed in removal/insertion, and is requested to remove the appliance prior to eating and to reinsert the appliance only after having rinsed the mouth thoroughly with water. The appliance is stored in the moist denture cup during eating.

After removal of the appliance at the end of the wearing period, it is immediately transported in the denture cup to the microbiology lab (a period of approximately 3 min). The chips are removed using forceps and immediately processed for microscopy.

Samples are processed for scanning electron microscopy (SEM) after attachment (enamel side up) in 24-well plates by attaching the enamel chips to the plate bottom with double-sided tape. The chips are fixed by covering with paraformaldehyde/glutaraldehyde (2% final concentration of each) in 100 mM cacodylate buffer (pH 7.4). Fixative is stored at 4° and discarded after 1 week. Fixed samples are dehydrated through an ethanol series (75%, 90%, 95%, 100%) with final dehydration in hexamethyldisilazane (HMDS). The chips are mounted on aluminum stubs, and coated with gold.

Samples are processed for laser confocal microscopy by various antibody-based techniques or by staining with general bacterial stains such as SYTO stains (Molecular Probes, Eugene, OR) and BacLight LIVE/DEAD (Molecular Probes). All staining steps are performed in micotiter plate wells using approximately 150 μl volumes. The samples are examined using water-immersible optics.

Examples of SEM and confocal micrographs are shown in Fig. 2.

Enamel Pieces Carried on Molars and Premolars

Enamel pieces from extracted bovine teeth are cut into squares approximately $3 \times 3 \times 1$ mm with 45° divergent walls. A notch is cut on the top of the back surface for orientation of the piece. The pieces are measured for surface area using a Schere-Tumica Digital Caliper (Schere-Tumica Inc., St. James, MN) and stored in distilled water at 4° until used.

Custom-made enclosures are fabricated as follows. Alginate impressions are taken of each subject's upper arch. A plaster cast is then constructed and used as a model to form six custom enclosures placed on the buccal surface of upper premolars and first molars (tooth numbers 1.6, 1.5, 1.4, 2.6, 2.5, 2.4). The surfaces of each tooth on the plaster cast are lightly coated with petroleum jelly and a ball-shaped piece of Visio-fil (Espe, Germany) composite resin is applied. Individual enamel pieces are then coated with petroleum jelly and pressed into the ball-shaped composite form until the premeasured outer surface is exposed. An explorer tip is inserted at the top edge of the piece to make a small notch, for subsequent dislodgment of the piece before the enclosure material is light-cured. The Visio-fil enclosure material is then light-cured for 1 min. After curing, the enclosures are separated from the plaster cast and the pieces are removed.

The teeth receiving custom enclosures/pieces are acid etched on the buccal surfaces with Ultra-etch (40% phosphoric acid gel, Ultradent Products Inc., Salt Lake City, UT), rinsed with cold water after 45 sec, then air dried. Concise Orthodontic Enamel Bonding System Resin (3M, St. Paul, MN) is mixed and applied to the etched tooth surface. The custom enclosures are placed on this resin and allowed to harden. The enamel pieces are then inserted into their respective enclosures for each experiment. The test materials fit tightly into their enclosures, preventing the formation of plaque behind or around them (Figs. 3 and 4).

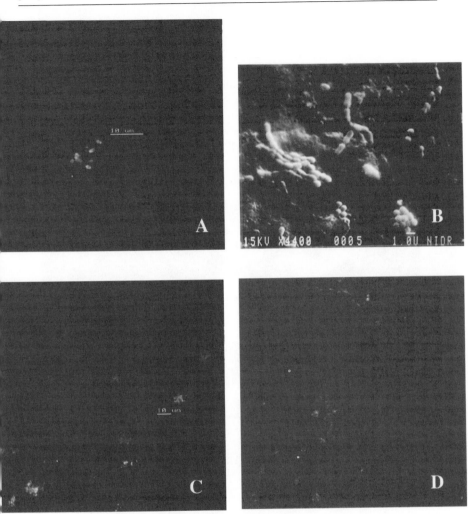

FIG. 2. (A) Confocal laser micrograph of antibody-stained dental plaque on an enamel chip recovered after 4 hr in the oral cavity. IgG raised in rabbits against whole *Streptococcus gordonii* cells was labeled with Alexa Fluor 488 (Molecular Probes) and used to reveal the presence of these cells. The thickness of this microcolony is approximately 3 μm. (B) Dental plaque on an enamel chip recovered after 4 hr in the oral cavity as revealed by SEM. (C),(D). Confocal laser micrograph of BacLight LIVE/DEAD-stained (Molecular Probes) dental plaque on an enamel chip recovered after 4 hr in the oral cavity. Panel C shows green (live cell) fluorescence, whereas panel D shows red (dead cell) fluorescence. The scale bar in panel C is also applicable to panel D. The height of the scanned area is approximately 48 μm.

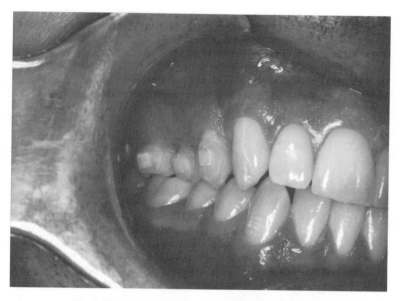

FIG. 3. Enamel pieces mounted in enclosures in the mouth.

FIG. 4. Scanning electron micrograph of the enamel piece in the enclosure.

Removal of Plaque from Enamel Pieces

Experiments of any duration can be done with placements of 4–6 bovine enamel pieces bonded to each subject's teeth. Following the respective time periods the pieces are removed, placed in 1.0 ml prereduced anaerobically sterilized diluent (PRAS)[9] in an anaerobic chamber with an atmosphere of $H_2 : CO_2 : N_2$ (10 : 10 : 80%) (Coy Anaerobic Chamber, Ann Arbor, MI), and sonified for 20 sec (Kontes Cell Disrupter; Kontes, Vineland, NJ). Scanning electron microscopy can be used to monitor complete plaque removal from the enamel pieces. The enamel pieces are dehydrated by freeze-drying, coated with gold/palladium in a evaporator, and observed with a scanning electron microscope. Virtually all plaque bacteria are removed by sonication under these conditions. The plaque samples are serially diluted and plated on elective and/or selective media.

Bacterial Cultivation and Enumeration

Total cultivable anaerobic microbiota is determined using supplemented blood agar medium containing Trypticase soy agar (BBL Microbiology Systems, Cockeysville, MD), 5% defibrinated sheep blood, 0.00005% menadione, 0.0005% hemin, and 0.001% NAD (Sigma Chemical Company, St. Louis, MO) and incubated anaerobically at 37° for 7 days. Dark-pigmented *Bacteroides* spp., which include the species *Porphyromonas gingivalis, Prevotella intermedia,* and *Prevotella melaninogenicus,* are enumerated on the supplemented blood agar. *Fusobacterium* spp. are enumerated as round, entire, smooth iridescent colonies on the supplemented blood agar. *Streptococcus* spp. are enumerated on mitis-salivarius (MS) agar (Difco) with 0.1% potassium tellurite (Difco) added. These plates are incubated anaerobically ($H_2 : CO_2 : N_2,10 : 10 : 80\%$) at 37° for 2 days and then for 24 hr aerobically at room temperature. Adherent hard colonies, which include the species *Streptococcus sanguis, Streptococcus gordonii,* and *Streptococcus oralis* and are referred to as sanguis streptococci, are identified visually by colonial morphology using a stereo microscope and indirect light,[10] but could be identified by a number of other methods.[11] Mutans streptococci are enumerated on MS-bacitracin agar,[12] and *Haemophilus* spp. on modified chocolate (MC) agar.[13] The MC agar is incubated in 10% CO_2 at 37° for 3 days. *Eikenella corrodens* is enumerated on the selective media of Slee and Tanzer[14] after incubation in 10% CO_2 for 4 days. *Neisseria subflava* are identified as cream to yellow round, smooth entire colonies

[9] L. V. Holdemann, E. P. Cato, and W. E. C. Moore, "Anaerobic Laboratory Manual," 4th ed. Virginia Polytechnic Institute and State University, 1977.
[10] J. Carlsson, *Odontol. Rev.* **18,** 55 (1967).
[11] M. Kilian, L. Mikkelsen, and J. Henrichsen, *Int. J. System. Bacteriol.* **39,** 471 (1989).
[12] O. B. Gold, H. V. Jordan, and J. van Houte, *Archs. Oral Biol.* **18,** 1357 (1973).
[13] W. F. Liljemark, C. L. Bandt, C. G. Bloomquist, and L. Uhl, *Infect. Immun.* **46,** 778 (1984).
[14] A. M. Slee and J. M. Tanzer, *J. Clin. Microbiol.* **8,** 459 (1978).

on Trypticase soy agar (BBL) plus vancomycin (Sigma) incubated aerobically for 48 hr as described by Ritz.[15]

Actinomyces spp. were enumerated on CFAT selective agar.[16] Sera obtained from Dr. George Bowden facilitate the species identification.[17]

In the enumeration of bacteria in plaque samples, the ability to detect organisms is dependent on the dilution used in plating. In either cross-sectional or longitudinal studies, the number of organisms enumerated by dark field microscopy dictated the dilution[18]; in developing plaque studies, the subject and time period dictated the dilution used. The following are the minimum levels of detection for organisms in plaque samples: for *Actinomyces, Haemophilus, Veillonella, Neisseria, Eikenella,* and *Streptococcus,* ca. 22 cells per plaque sample (cross-sectional and longitudinal studies) or ca. 11 cells per approximately 8 mm^2 enamel surface (developing plaque studies). Numbers of total cultivable bacteria, and of bacterial species groups, can be expressed per mm^2 of enamel surface or as a proportion of the total cultivable anaerobic microbiota.

Statistical Analyses

Data analyses for plaque studies are done by repeated measures ANOVA between each adjacent time period (e.g., 2 vs 4 hr) for the colony-forming units per mm^2 (CFU/mm^2) of each genus, species, or species group enumerated. When F ratios are significant, the Scheffe test is used to compare individual groups, and statistical significance is set at the 95% confidence level.

Measurement of DNA Synthesis

The incorporation of radiolabeled nucleosides is used as an index of chromosomal replication of bacteria. Incorporation of [methyl-^{3}H]thymidine and [^{14}C]adenine into trichloroacetic acid (TCA) insoluble material is measured using a modification of the method of LaRock *et al.*[19] A volume (0.3 ml) of the sonicated plaque sample from the enamel piece or the entire piece is immediately mixed with 0.1 ml of a mixture of [methyl-^{3}H]thymidine (45,000 mCi mmol^{-1}, Research Products International Corp., Mt. Prospect, IL) and [^{14}C]adenine (270 mCi mmol^{-1}, Amersham Research Products, Arlington Heights, IL) at 150 mCi/ml (100–200 nmol) and 65 mCi/ml (200–250 nmol) or equivalent final concentrations, respectively. The plaque and radiolabel are incubated for 30 min at 37° in the

[15] H. L. Ritz, *Arch. Oral Biol.* **12,** 1561 (1967).
[16] L. J. Zylber and J. V. Jordan, *J. Clin. Microbiol.* **15,** 253 (1982).
[17] G. H. Bowden, J. Ekstrand, B. McNaughton, and S. J. Challacombe, *Oral Microbiol. Immunol.* **5,** 346 (1990).
[18] L. F. Wolff, W. F. Liljemark, C. G. Bloomquist, B. L. Pihlstrom, E. M. Schaffer, and C. L. Bandt, *J. Period. Res.* **20,** 237 (1985).
[19] P. A. LaRock, J. R. Schwartz, and K. G. Hofer, *J. Microbiol. Methods* **8,** 281 (1988).

anaerobic chamber. Radiolabeled plaque samples are removed from the chamber, and 2.0 ml of cold (4°) 10% (wt/vol) TCA added for 30 min at 4°. The TCA-insoluble precipitate is collected onto Gelman GN-6 cellulosic filters (Gelman Sciences, Ann Arbor, MI), rinsed three times with 5.0 ml of cold 10% TCA, and counted in a liquid scintillation counter. Results when plaque is radiolabeled prior to sonication are similar to those when plaque is sonified prior to addition of the radiolabel, the latter method being preferable because plating of radiolabeled bacteria is not required.

Development of Continuous Bacterial Surfaces on Tissue Culture Plates or Bovine Enamel Pieces

Development of a continuous layer of any bacterial species on tissue culture plates or bovine enamel pieces can be accomplished. Parallel studies using scanning electron microscopy (SEM) and light microscopy are conducted to determine the conditions necessary to form a consistent continuous bacterial surface. Preliminary experiments yielded the optimum conditions (as described below) to produce consistent continuous bacterial surfaces.

Tissue culture dishes (35 × 100 mm, Falcon #3001) were coated with 250 μg Cell Tak (BD, Franklin Lakes, NJ) diluted in 5% acetic acid according to the manufacturer's suggestions. All subsequent adsorption steps in the monolayer assay were done at 37°. Two ml of bacterial cells (10^{10}cells/ml) suspended in phosphate buffer were added to each plate, and the plates were centrifuged (IEC PR6000, International Equipment Co., Needham Hts., MA) for 30 min at 1200g. Unadsorbed bacteria were decanted and the plates washed gently twice (pipette). The process was repeated with another 2.0 ml of cell suspension. Unadsorbed bacteria were decanted, the plates were washed gently, and 2.0 ml of buffer was added to the plates, which were then rotated gently at 50 rpm (New Brunswick incubator shaker, model G-25, New Brunswick, NJ) at 37° for 30 min. The buffer was removed, and the plates with bound bacterial cells were rinsed with buffer. Similarly, bovine enamel pieces, measuring approximately 5 × 5 mm, were coated with 25 μg Cell Tak, as described above, and placed in the bottom of a flat-bottomed glass test tube. The pieces were submerged in 1.0 ml of the bacterial suspension and centrifuged (IEC clinical model) at 1500g for 30 min. Unadsorbed bacteria were decanted, the pieces washed twice, a second milliliter of cells added, and the process repeated. The washed pieces with bound bacterial cells were covered with buffer and shaken gently (rotating platform shaker) for 30 min for removal of any loosely adsorbed cells.

Tests for Discontinuous Bacterial Surfaces

To test for possible exposed Cell Tak not covered by the bacterial strains, we used SEM to examine numerous pieces from different experiments. This surface

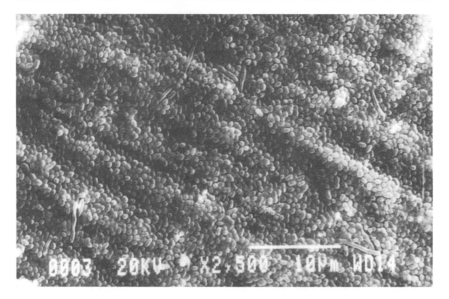

FIG. 5. Scanning electron micrograph of a continuous monolayer of streptococci.

is now ready for a variety of *in vitro* or *in vivo* experiments (see below). Figure 5 shows a streptococcal surface. Several additional genera and species have also been used with good results.

Removable Orthodontic Retainer Method

A second method for mounting enamel pieces or other materials may also be employed in which a removable orthodontic appliance is fitted for each subject. In this case, after formation of a continuous bacterial layer on the material of choice, the pieces are mounted in Registration Material [Vinyl Polysiloxane Express, or Concise (see above) Minnesota Mining and Manufacturing, St. Paul, MN], which is integrated with the appliance as before, leaving only the facial surface of the enamel piece exposed. Four pieces located just facial to the upper first molars and premolars are carried per appliance. These are relatively short-term experiments, because the enamel pieces are on a removable appliance and eating is not possible for the subjects unless the retainer is stored in a holding medium while the subjects eat.

Summary

With the use of the removable stents or bonded enamel piece models with or without a continuous bacterial layer, many *in vitro* or *in vivo* studies can be initiated. For example, studies on salivary pellicle formation, surface characteristics of biomaterials as they affect plaque development, antiplaque agents, the dynamics of adhesion of bacteria, interspecies adhesion of bacteria, the colonization of bacteria, the dynamics of bacterial growth *in vivo,* and the succession of growth in older supragingival plaques[20–23] can be carried out.

Acknowledgment

The NIDCR authors gratefully acknowledge the support of the NIH Clinical Center staff, especially those in the Dental Clinic.

[20] R. J. Skopek, W. F. Liljemark, C. G. Bloomquist, and J. D. Rudney, *Oral Microbiol. Immunol.* **8,** 16 (1993).

[21] R. J. Skopek and W. F. Liljemark, *Oral Microbiol. Immunol* **9,** 19 (1994).

[22] C. G. Bloomquist, B. E. Reilly, and W. F. Liljemark, *J. Bacteriol.* **178,** 1172 (1996).

[23] W. F. Liljemark, C. G. Bloomquist, M. C. Coulter, L. J. Fenner, R. J. Skopek, and C. F. Schachtele, *J. Dent. Res.* **67,** 1455 (1988).

[28] Biofilm Ácid/Base Physiology and Gene Expression in Oral Bacteria

By Robert A. Burne *and* Robert E. Marquis

Introduction

Many bacteria have both a free-living and host-associated existence. Some of these organisms have been shown to have circuits that allow them to alter patterns of gene expression in response to acquisition of a surface or to binding of a target tissue. Examples include the so-called "touch sensors" of *Pseudomonas aeruginosa,*[1] which lead to activation of alginate gene expression after the cells adhere to an inert surface, and the activation of a two-component regulatory system when *Escherichia coli* tip adhesin comes in contact with cells.[2] Induction of the Type III secretion systems and of exported virulence factors by invasive bacteria after contact with host cells has also been well documented.[3] More recently, quorum sensing by adherent bacteria has been shown to be an essential component of biofilm formation by *P. aeruginosa.*[4] For organisms with the capacity to thrive

[1] D. G. Davies, A. M. Chakrabarty, and G. G. Geesey, *Appl. Environ. Microbiol.* **59,** 1181 (1993).

[2] J. P. Zhang and N. S., *Science* **273,** 1234 (1996).

[3] J. H. Brumell, O. Steele-Mortimer, and B. B. Finlay, *Curr. Biol.* **9,** 277 (1999).

independently of a mammalian host, being able to detect and rapidly adapt to ingestion or inhalation by an appropriate host is a necessary step for survival and propagation. However, oral bacteria, in this sense, are distinctly different.

Although well over 300 species of bacteria have been cultivated from the human mouth, the vast majority of oral bacteria, especially those that play meaningful roles in disease or oral biofilm homeostasis, almost exclusively colonize the oral cavity and do not survive well, or in some cases at all, outside of the host. For example, *Streptococcus mutans,* the etiologic agent of dental caries, has no known natural habitat other than the mouth,[5] and this organism does not appear to survive in environmental settings, such as streams or soil, for any extended period. Periodontopathic bacteria, such as *Porphyromonas gingivalis, Prevotella intermedia,* and *Treponema denticola,* also appear to require close association with a mammalian host for persistence. Thus, it has been proposed that the most ecologically important oral bacteria exist in an "obligate biofilm lifestyle." [6]

Because oral bacteria do not undergo as drastic a fluctuation in their environments as some organisms, one might predict that evolutionary pressures have allowed oral bacteria to dispense with many of the elaborate sensing systems needed by other overt or opportunistic pathogens that can occupy multiple, diverse environmental niches. In fact, this appears to be the case. The genomes of most oral bacteria are about half the size of those of many organisms such as *Escherichia coli, Salmonella,* or *Pseudomonas.* Interestingly, as in some overtly pathogenic bacteria, an examination of the *S. mutans* genome sequence, which is about 80% complete at the time of this writing, has revealed that this bacterium has only a few two-component systems for sensing environmental signals. Many oral bacteria also tend to have a narrower temperature range in which they will grow, they do not tolerate high osmolality well, and they cannot grow over as wide a pH range as many other bacteria can, consistent with having evolved in a fairly moderate environment.

Fluctuations in the conditions in the oral cavity are not as extreme in some environs, but the oral cavity is still a very dynamic environment, and biofilm communities are constantly exposed to very rapid and substantial changes in nutrient sources and availability, as well as pH. Sugar concentrations can rapidly change from around 10 μM to greater than 10 mM with an intake of dietary carbohydrate, and this change is accompanied by a rapid plunge in the pH of oral biofilms from values around neutrality to values as low as 4 and below in only a couple of minutes. Also, because of the three-dimensional structure of oral biofilms, steep

[4] D. G. Davies, M. R. Parsek, J. P. Pearson, B. H. Iglewski, J. W. Costerton, and E. P. Greenberg, *Science* **280,** 295 (1998).

[5] S. Hamada and H. D. Slade, *Microbiol. Rev.* **44,** 331 (1980).

[6] R. A. Burne, *in* "Microbial Pathogenesis: Current and Emerging Issues" (D. J. Leblanc *et al.,* eds.), p. 55, Indiana. Univ. Press, Indianapolis, IN (1998).

gradients in nutrients, oxygen tension, and other important factors known to govern gene expression in prokaryotes likely develop. To survive, thrive, and elicit diseases, oral bacteria must respond to these changes rapidly and efficiently. In the case of the oral streptococci, and especially *S. mutans,* tremendous phenotypic plasticity has been retained in order to cope with the changing environment.[7,8] These changes occur at the physiologic and biochemical level, where the organisms utilize allosteric modulation of enzyme activity to control metabolic flux, and they occur at the genetic level, where the expression of a variety of genes is highly responsive to pH, carbohydrate source and availability, oxygen, and other important stimuli.[7-9] Thus, although some of the sensing systems described above may ultimately prove to be less important or superfluous to oral streptococci for survival and virulence, certainly the capacity to modulate their phenotype at the physiologic and genetic level is necessary for these abundant bacteria to compete effectively in oral biofilms and, under the right conditions, to cause disease. Therefore, a thorough understanding of the behavior of oral bacteria when growing in biofilms, and how gene expression and regulation pathways differ between planktonic and sessile populations, is needed to develop novel treatment and prevention strategies for the ubiquitous oral diseases. This chapter will focus on methods to study pH effects on biofilm growth and gene expression, and on techniques to conduct basic studies into the metabolism of simple and complex molecules by adherent populations of oral bacteria.

In Vitro Cultivation of Biofilms of Oral Bacteria

Several systems have been developed for the generation of single- and multi-species biofilms of oral bacteria. These include relatively simple, static, closed systems where organisms are grown on wires, glass slides, or hydroxylapatite disks. Many oral biofilm models, however, use more sophisticated, continuous flow bioreactors, chemostat-based biofilm systems or microflow chambers. In previous communications,[10,11] we discussed the relative merits and shortcomings of a variety of systems that have been used for cultivating oral biofilm bacteria. Basically, the closed, batch cultivation systems have the decided advantage that they are comparatively inexpensive, they are amenable to high throughput, and large numbers of biofilms with very similar characteristics can be generated to obtain statistically meaningful data. Such systems are usually only applicable to simpler mono- or dispecies biofilms, but they are very useful for getting data related to basic

[7] J. Carlsson, *in* "*Cariology*" (B. Guggenheim, ed.), p. 205. Karger, Basel, 1984.

[8] R. A. Burne, *J. Dent. Res.* **77,** 445 (1998).

[9] R. A. Burne, Y. Y. Chen, and J. E. Penders, *Arch. Dent. Res.* **11,** 100 (1997).

[10] R. A. Burne and Y.-Y. M. Chen, *Meth. Cell Sci.* **20,** 181 (1998).

[11] R. A. Burne, R. G. Quivey, Jr., and R. E. Marquis, *Methods Enzymol.* **310,** 441 (1999).

metabolic activities, for assessing the ability of organisms to catabolize or produce various substances, and for the testing of the efficacy of antimicrobial compounds against organisms in a three-dimensional arrangement. However, closed, batch cultivation systems are not at all well suited for studies in which it is desirable to modulate environmental conditions to explore the behavior of the bacteria, nor are they very useful for mixed-population studies of biofilm ecology. In contrast, continuous flow systems are much more powerful for the study of physiology, gene expression, and population dynamics. However, these reactors can be very costly and tedious to work with, additional specialized equipment and expertise may be needed, and the experiments can be time consuming, lasting days or even weeks. A number of methods for forming biofilms, controlling environmental conditions in films, and sampling and analysis of abundant oral gram-positive bacteria are presented below. With some modifications, these approaches should be applicable to multiple microorganisms.

Formation of Oral Biofilms

The general principles and methods for formation of oral biofilms were reviewed in previous communications.[10,11] In this section, the focus will be on manipulation of environmental variables in order to explore how changes in the macro environment affects gene expression and metabolism by adherent populations, with a specific focus on the control of biofilm pH and carbohydrate source.

Control of pH in Biofilms

Because of mass transport limitations, maintenance of a uniform concentration of a given compound throughout an entire biofilm is impractical, unless the compound is completely inert with respect to bacterial metabolism, or to interactions with the cells or extracellular matrix. This is true for the proton, and one of the most difficult challenges is controlling pH in biofilm populations. Controlling pH is further complicated by the fact that bacterial cells and the matrix produced by oral microorganisms have variable buffer capacities and can be effective buffers over different pH ranges. For example, bacterial cell walls buffer efficiently in the pH range 4–6.[12] In the case of the oral streptococci, controlling pH is further complicated by the fact that growth, with only a few exceptions,[13] is obligatorily coupled to glycolysis, and consequently to the production of organic acids.

Basically, there are two approaches to the control of pH. The first is to treat the vessel essentially as one would a chemostat for planktonic populations. In this

[12] L.-T. Ou and R. E. Marquis, *J. Bacteriol.* **101**, 92 (1970).

[13] R. E. Marquis, R. A. Burne, D. T. Parsons, and A. C. Casiano-Colon, *in* "Cariology for the Nineties" (W. H. Bowen and L. A. Tabak, eds.), p. 309. University of Rochester Press, Rochester, NY, 1993.

case, a pH probe is immersed in the vessel and acid or alkali addition is governed by a pump interfaced with a pH controller. This approach has been utilized in most chemostat-based biofilm systems, and the initial results indicate that this is a viable approach,[14] but there may be drawbacks. The first includes fouling of the pH probe. Once this occurs, there is little that can be done other than to stop the run and begin anew. Also, in spite of the fact that the pH of the plank-tonic phase can be controlled effectively, gradients undoubtedly develop within the biofilms,[15] so that the pH of the planktonic phase not only will be different from the average biofilm pH, but may differ substantially from microenvironments in the biofilms.

A second approach to controlling pH is with the use of buffers, which has both positive and negative aspects. Clearly, using buffers, either directly in the medium or pumped in at a rate proportional to the flow of the medium, is very convenient and amenable to use with either batch or continuous flow systems. Once the correct buffer composition has been empirically determined, fairly consistent results can be obtained. However, maintenance of the pH at a given level is sensitive to growth conditions, so, for example, changes in the carbohydrate concentration of the medium will affect the final pH in the vessel, mandating modifications of the buffer composition or concentration. Other shortcomings of the use of buffers include the necessity to alter the ionic strength of the media, increased osmolality, and specific buffer effects.

We have utilized buffered media for the study of the low pH-inducible ure-ase of *Streptococcus salivarius*[16] and more recently for exploring exopolysaccha-ride gene expression in *S. mutans* (Burne, unpublished). In these experiments, it was desirable to have cells maintained at comparatively neutral pH and at a lower pH value, which could be achieved using potassium phosphate buffers. The base medium is a tryptone–yeast extract formulation (TY,[17] 30 g tryptone, 5 g yeast extract) that was diluted twofold and supplemented with 10 mM su-crose. The base medium is prepared and autoclaved, and the carbohydrates and buffers are added after the medium cools as previously detailed.[10] Biofilms are formed in a Rototorque biofilm reactor as previously described.[9] For cultures to be grown at values closer to neutrality, potassium phosphate buffer (KPB) is used at pH 7.8 at a final concentration of 50 mM. For low pH cultures, the medium was supplemented instead with 90 mM KCl to try to keep the potassium concen-trations equivalent. The biofilms are allowed to form and to reach "quasi-steady state," which is defined as being achieved after 10 generations.[10] By adopting equations derived for continuous culture, the generation time (t_g) can be calculated

[14] D. J. Bradshaw, P. D. Marsh, K. M. Schilling, and D. Cummins, *J. Appl. Bacteriol.* **80**, 124 (1996).

[15] D. J. Bradshaw, A. S. McKee, and P. D. Marsh, *J. Dent. Res.* **68**, 1298 (1989).

[16] Y. H. Li, Y. M. Chen, and R. A. Burne, *Environ. Microbiol.* in press.

[17] R. A. Burne, K. Schilling, W. H. Bowen, and R. E. R. E. Yasbin, *J. Bacteriol.* **169**, 4507 (1987).

from the equation

$$t_g = 0.693/D$$

where D is the dilution rate that is calculated by dividing the total volume of medium (and buffer if applicable) that is pumped into the vessel per hour by the total working volume of the vessel.

Determining pH of Biofilms and Planktonic Phase

Others have reported the use of microelectrodes for monitoring various chemical parameters in biofilms, including pH, oxygen, and chloride ion.[18] Coupled with CSLM,[19] this technology can provide very detailed information about very small subcompartments of a heterogeneous biofilm. More practical for gene regulation and physiology studies is to use two methods to get an average measure of biofilm pH. Of course the pH in microcolonies will vary and the pH of water channels will be different from that of the bulk biofilm, but when it is desirable to have an overall picture of the biofilm pH and to contrast that with the planktonic phase, use of a Super-miniature pH electrode (Beetrode) to take multiple samples at different positions in the biofilm, coupled with dispersing the biofilms into deionized H_2O and measuring pH with a standard electrode, offers a rapid and consistent method to explore biofilm pH.[16]

In the past, we have used a Super-miniature, Beetrode pH electrode (Model MEPH3L, World Precision Instruments, New Haven, CT) to obtain pH measurements from multiple sites (>30) in the biofilms.[16] The tip size of the microprobe we have found suitable is 100 μm in diameter. After the cultures reached "quasi-steady state,"[10] slides with biofilms are removed from the vessel and placed on end on a paper towel to allow excess medium to be absorbed. The micro-reference electrode, which can be connected through a Bee-Cal adapter and two cables to the pH probe and a standard pH meter, is positioned in the biofilm to be partially immersed by the biomass. *In situ* measurement of pH is conducted by placing the tip of the pH probe into biofilms. Generally, we have obtained a series of pH readings from about 30 different sites selected at random over the slide. To measure "bulk" biofilm pH, the biofilms can be physically dissociated from the slides into deionized water, the mixture vortexed vigorously, and the pH of the suspension measured immediately using a standard, calibrated\electrode.

Table I shows the results that we have obtained using buffered media to cultivate biofilms of *S. salivarius* and *S. mutans*. Basically, as detailed above, the pH

[18] J. W. Costerton, Z. Lewandowski, D. E. Caldwell, D. R. Korber, and H. M. Lappin-Scott, *Ann. Rev. Microbiol.* **49**, 711 (1995).
[19] J. R. Lawrence, D. R. Korber, B. D. Hoyle, J. W. Costerton, and D. E. Caldwell, *J. Bacteriol.* **173**, 6558 (1991).

TABLE I

pH of Biofilms and Planktonic Phases of Strains of *S. mutans* and *S. salivarius* Cultured to "Quasi-Steady State" in the Rototorque Biofilm Reactor[a]

	With pH control		Without pH control	
Bacterial strain	Planktonic	Biofilm[b] (detached biofilms[c])	Planktonic	Biofilm (detached biofilms)
S. mutans SMS101	6.68 ± 0.015	6.08 ± 0.12 (N.D.)[d]	5.12 ± 0.25	5.32 ± 0.10 (N.D.)
S. salivarius PureIcat	6.70 ± 0.20	6.16 ± 0.18 (6.12 ± 0.15)	5.12 ± 0.25	5.54 ± 0.17 (5.35 ± 0.16)

[a] Biofilms were formed in the Rotorque bioreactor and pH measurements were obtained as detailed in the text. Values shown are averages and standard deviations from at least three independent runs of the Rototorque, and all assays were performed in triplicate.
[b] Values for biofilm cells not in parentheses represent those obtained with the minature electrode.
[c] Values in parentheses represent those obtained using detached biofilms as described in the text.
[d] N.D., Not determined

of the bulk liquid and those values obtained from the biofilms can differ significantly, although one can consistently achieve populations of cells that are exposed to almost a full pH unit difference. Notably, in cultures that are buffered, the biofilm pH is about 0.6 units lower than that measured in the liquid phase. In contrast, the bulk liquid is actually about 0.2–0.3 pH units lower than the biofilm. Since these strains do not produce ammonia or polyamines from amino acids to any appreciable extent, the latter disparity is likely due in part to buffering by bacterial cells, and perhaps by sugar limitation in the deeper reaches of the films. The pH values are the same regardless of whether the microelectrode was used or whether the films were dispersed, but obtaining the measurements with the Beetrode yields a better picture of the heterogeneity of the films in terms of pH.

Use of Reporter Gene Fusions in Biofilms

We have previously reported on the use of two reporter genes that are readily adaptable for use with oral streptococci, β-galactosidase (*lacZ*) and chloramphenicol acetyltransferase (*cat*), and, at least under some circumstances, these genes are compatible with the study of oral biofilms.[10] Since that time, it appears as though other options, such as the fluorescent proteins (e.g., GFP) or β-glucuronidase, may also prove to be useful reporters. Methods for the growth of biofilms, for the recovery of cells, and for the applications of reporter gene technology to biofilm study are detailed elsewhere and in other contributions to this volume.[10,11]

TABLE II
CAT SPECIFIC ACTIVITY OF RECOMBINANT S. *salivarius* STRAINS GROWING IN *in Vitro*
BIOFILMS AT QUASI-STEADY STATE AND FOLLOWING A 25 mM GLUCOSE PULSE

	pH control[a]		No pH control	
Strain/time after glucose pulse	pH after glucose pulse	CAT activity[b] (U/mg protein)	pH after glucose pulse	CAT activity (U/mg protein)
PureICAT				
T_0	6.70 ± 0.20	0.86 ± 0.32	5.12 ± 0.20	9.58 ± 1.40
T_{15}	5.76 ± 0.11	8.04 ± 4.40	4.86 ± 0.12	12.6 ± 1.00
T_{30}	5.21 ± 0.12	17.6 ± 4.28	4.81 ± 0.10	13.3 ± 1.10
T_{60}	4.98 ± 0.13	22.0 ± 1.95	4.72 ± 0.21	16.2 ± 0.60

[a] Biofilms were formed as detailed in the text and in Table I. pH control indicates cells grown in medium supplemented with 50 mM KPO_4, pH 7.8, whereas the no pH control samples were grown in medium supplemented with 90 mM KCl.

[b] CAT activity was measured in biofilm cells as previously described () and was expressed as nmol Cm acetylated min^{-1} mg^{-1} protein. The values shown here are averages and standard deviations from at least three independent runs of the Rototorque, and all assays were performed in triplicate.

Use of pH-Responsive promoter to Follow Changes in Biofilm pH

One of the more challenging aspects of the study of biofilms is the monitoring of pH *in situ*. Microprobes can be placed in a developing biofilm, but the presence of the probe changes biofilm structure and it is not entirely clear that the probes are yielding completely accurate information after mature films form on them. Obviously, probes can be inserted into the films, but this disrupts the architecture. Thus, it would be highly desirable to have a nondisruptive yet reliable method for monitoring changes in biofilm pH. One way to do this would be to fuse a pH-regulated promoter to an easily assayable reporter gene. This conceivably could yield semiquantitative data about the relative pH in different areas of a biofilm. The urease genes of S. *salivarius* are pH-regulated.[20,21] At neutral pH, the genes are tightly repressed, but as the pH is lowered to values below 6.0, urease expression becomes derepressed. Once the genes become derepressed at low pH values, the expression of the urease genes can be further modulated by carbohydrate availability, with the highest levels of expression observed when carbohydrate is present in excess.

Table II shows data obtained in single-species biofilms formed from S. *salivarius* PureIcat, a strain with a chloramphenicol acetyltransferase gene (*cat*) fusion to the pH regulated urease promoter 20 with pH control (50 mM KPB,

[20] Y.-Y. M. Chen, C. A. Weaver, D. R. Mendelsohn, and R. A. Burne, *Infect. Immun.* **180,** 5769 (1998).
[21] Y. M. Chen and R. A. Burne, *FEMS Microbiol. Lett.* **135,** 223 (1996).

pH 7.8) or without pH control (90 mM KCl). What this table best illustrates is that the use of buffers is a reasonable method to achieve differences in pH in fairly thick (100 μm) biofilms in a controlled fashion and that the use of bacterial strains with gene fusions to pH sensitive promoters may be a powerful tool for exploring the development of pH gradients in biofilm model systems. One can also see that fairly rapid changes in the biofilm pH can be elicited, in this case, by pulsing the vessel with a rapidly fermented carbohydrate or, if desired, by addition of organic or inorganic acids.

Sensitivities of Biofilm Cells to Acid Damage

Generally, cells in biofilms are better able to resist inimical environmental influences than are cells in suspensions. Enhanced resistance of cells in biofilms to acid damage might be expected because of the high buffer capacities of films and the slow penetration of solutes into them. In essence, acidification of biofilms is slowed either because incoming protons can be absorbed by biofilm buffers or because the influx of acidogenic substrates such as fermentable carbohydrates can be diffusion limited. Slower acidification gives biofilm cells time to adapt phenotypically to an acid challenge.

Acid damage can be reversible or irreversible. Acid inhibition of glycolysis is a common type of reversible damage, which is part of everyday life for bacteria in dental plaque biofilms. An easy and widely used method to study acid inhibition of glycolysis is by means of so-called acid-drop experiments. In this type of assay, cells are suspended in a nonbuffering medium such as 50 mM KCl plus 1 mM MgCl$_2$ and are supplied with excess fermentable carbohydrate. They then produce acid glycolytically; the pH of the suspension drops rapidly and in time reaches a final value at which the cells can no longer produce acid. For oral streptococci, the acid arrest of the glycolytic system is reversible. If the suspensions are neutralized with alkali, a second round of pH drop ensues. It appears that multiple components of the glycolytic system are sensitive to acidification, including glycolytic enzymes such as enolase,[22] but also the membrane-associated phosphoenolpyruvate: sugar phosphotransferase system for sugar uptake.[23]

When monoorganism biofilms of oral streptococci were compared by means of pH-drop assays with cells of the same organisms in suspensions, pH values for biofilms were somewhat higher than those for cells in suspensions. For example, for S. mutans GS-5, the average final pH value for biofilms was 4.3 ± 0.2 (95% confidence limits, 6 trials) with glucose as substrate, compared with 3.8 ± 0.2 (95% confidence limits, 7 trials) for cells in suspensions. When biofilms were dispersed prior to addition of glucose, the final pH value was nearly the same as that for the

[22] W. A. Belli, D. H. Buckley, and R. E. Marquis, Can. J. Microbiol. **41**, 785 (1995).
[23] G. R. Germaine and L. M. Tellefson, Antimicrobiol. Ag. Chemother. **29**, 58 (1986).

cells from suspension cultures. For these assays, the biofilms were grown in batch cultures on glass slides initially in tryptone–yeast-extract–sucrose medium[24] with daily changes of medium. Then, the day before the experiment, the biofilms were fed with the same medium but with sucrose replaced by glucose. Sugars in the growth medium were in excess, so that growth was limited by acidification rather than by nutrient exhaustion. The desire was to have populations of cells that had undergone acid adaptation prior to the pH-drop assays: in other words, cells with maximal acid tolerance. A variety of other procedures can be used for preparing cells to suit a variety of research objectives. For example, growth can be with continuous flow of medium, at a single, controlled pH value, in catabolite-limited cultures or those limited for other nutrients, or in the presence of various stressing agents.

The higher final pH values of the biofilms in pH-drop assays can be related to lower pH values in the biofilm phase compared with the planktonic phase described in the previous section. However, prolonged incubation of the biofilms did not result in further lowering of pH value measured with an electrode in the suspending medium, and so did not appear to be due purely to diffusion-limited proton movement into and out of the biofilms. Differences in pH between biofilms and planktonic phases may be related to biofilm retention of solutes other than protons and to phenomena of the sort responsible for differences in pH between matrix phases of ion-exchange resins and bulk fluids in which they are suspended. Bacterial cells have a net negative charge, although they are actually amphoteric with both positively and negatively charged groups outside and inside cell membranes.[25] The expectation is that they act predominantly as cationic exchangers, and so protons distribute preferentially into the resin (cell) phase, especially at lower ionic strengths.

A more extreme sort of acid damage is that resulting in cell death. Sensitivities to irreversible acid damage are generally correlated with sensitivities to reversible damage. For example, organisms more resistant to acid inhibition of glycolysis also are more resistant to acid killing. We have found that cells of *S. sanguis* NCTC 10904 and of *Actinomyces naeslundii* ATCC 19246 are less resistant to acid damage than are cells of organisms such as the mutans streptococci. In fact, cells of these acid-sensitive plaque bacteria can be killed in suspensions at pH values somewhat higher than 4, or within the range of plaque pH. However, in biofilms, they proved to be more resistant to acid killing than were cells harvested from the early stationary phase of suspension cultures with excess glucose, i.e., cells that have become acid adapted in batch culture. Exposure of cells in suspensions to pH 3.5 resulted in approximately 5 and 7 log reductions in viable counts of, respectively, *S. sanguis* and *A. naeslundii*. The same organisms in biofilms both

[24] W. A. Belli and R. E. Marquis, *Appl. Environ. Microbiol.* **57**, 1134 (1991).

[25] R. J. Doyle and R. E. Marquis, *Trends Microbiol.* **2**, 57 (1994).

showed only an approximately 1.5 log decrease in viable count in 1 hr after exposure to pH 3.5, followed by tailing and little further reduction in count even over an additional two or three hours. Again, the biofilms were grown in acid-limited batch cultures on glass slides.

A major question regarding enhanced resistance of these biofilm cells to acid damage concerns whether the enhancement is due to the biofilm state or to the physiological characteristics of the biofilm cells. The data presented in Fig. 1 for *S. sanguis* NCTC 10904 and *A. naeslundii* ATCC 19246 show that dispersing the biofilm cells resulted in only a small increase in acid sensitivity, and so enhanced resistance appeared to be due to physiological characteristics.

To disperse the biofilm cells, we first scraped the biofilms from glass slides with a spatula. Then, the biofilms, which remained in large aggregates, were homogenized with a IKA Labortechnik T25 basic homogenizer (Janke & Kunkel GmbH and Co., Staufen, Germany). Next, the homogenized films were sonicated on ice with a Branson Sonifier Cell Disruptor 200 (Branson Sonic Power, Danbury, CT) at full power for 15 sec without glass beads. This treatment is sufficient to obtain suspensions with only single cells visible in the phase-contrast microscope. The cells all were phase-dark, and their appearance indicated that they were not damaged in major ways by the treatment. Disruption of biofilms of gram-negative bacteria requires more trial and error to be sure that the process does not result in damage. We have found that the two-step procedure with initial homogenization and then sonication has an advantage in reducing the sonication time required for dispersal. Dispersal can be carried out with suspensions in protective fluids such as 1% Difco (Detroit, MI) peptone, although the use of peptone results in some foaming during sonication. When dealing with organisms highly sensitive to mechanical damage, such as *Treponema denticola,* we have found[26] that even mild sonication drastically damages the cells. However, treponemes can generally be dispersed simply by gentle shaking. When cells of anaerobes are dispersed, it is wise to avoid exposure to oxygen, so that reduced transport fluids or solutions of thioglycolate or other oxidation–reduction buffers can be used. The suspending medium can be stored under anaerobic conditions or boiled and cooled prior to use to remove traces of oxygen. In the extreme, the apparatus for homogenization and sonication can be placed inside an anaerobic chamber, such as the Coy (Grass Lake, MI) glove-box chamber, to be sure that there is no opportunity for the cells to metabolize oxygen during sonication. Dispersal with high-speed mechanical homogenizers, such as the Wig-L-Bug (VWR Scientific) or the Mini Bead Beater (Cole-Palmer Instruments), can substitute for sonication but do not offer advantages. The objective is dispersal and not cell disruption, so use of glass beads or other adjuncts to enhance shear should be avoided as much as possible. Levels of damage to the cells during dispersal can be assessed enzymatically. For example,

[26] C. E. Caldwell and R. E. Marquis, *Oral Microbiol. Immunol.* **14,** 66 (1999).

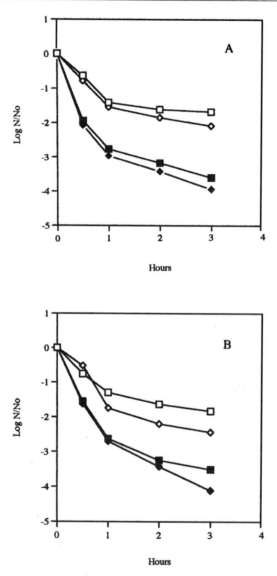

FIG. 1. Acid killing of biofilm cells of *S. sanguis* NCTC 10904 (A) or *A. naeslundii* ATCC 19246 (B). Intact biofilms (□) or dispersed biofilm cells (◇) were suspended in 1% Difco peptone broth at a pH of 3.5 and samples were taken at intervals, diluted in pH-7 peptone broth, and spread plated on Trypticase-soy agar (Difco). The plates were incubated at 37° until colony formation was complete. Addition of millimolar NaF to biofilms (■) or dispersed biofilm cells (◆) enhanced acid killing.

membrane damage that results in increased permeability can be assessed in terms of increased capacities for otherwise intact cells to hydrolyze ATP, which does not normally cross the intact membrane to reach the active sites of ATPases. A commonly used alternative is to assess cell uptake of propidium iodide or other probes (Molecular Probes, Eugene, OR) that form the basis for the commonly used viable staining of bacteria. Disruption of the cell wall can be detected in terms of loss of enzymes such as lactic dehydrogenase. Alternatively, loss of biopolymers such as DNA, which can be assayed chemically, reflects major damage to cells.

The cells in biofilms we used could be sensitized to acid killing by addition of fluoride as shown in Fig. 1. This sensitization has been found previously[22] for cells in suspensions. It is considered that increasing the permeability of the cell membrane to protons by providing fluoride as a weak-acid proton carrier sensitized cells to acid killing. Thus, the biofilm system can be used to assess effects of a variety of membrane-active agents on acid damage to microbial communities. In addition, by disrupting the biofilms and then assessing effects on the dispersed cells, one can distinguish between responses highly dependent on or independent of the biofilm state.

Summary

Environmental pH is one the major factors affecting the composition, biological activities, and pathogenic potential of the biofilms colonizing supragingival surfaces. In periodontal diseases, small changes in pH from the metabolism of amino acids and urea may influence the activity of proteolytic enzymes of host and bacterial origin.[8] Still, there is a significant void in the understanding of pH-dependent gene expression in bacteria, in general, and this is of course a more acute problem when one considers there is virtually no information about gene expression in response to pH in biofilms. The development of new methods and applications of some of the techniques detailed above should help to ameliorate this situation and to generate much-needed data about the role of pH in biofilm composition, stability, and activity.

Acknowledgments

This work was supported by Grants DE13239, DE10362, DE12236, and DE06127 from the U.S. National Institute for Dental and Craniofacial Research.

[29] *In Vitro* Modeling of Biofouling of Dental Composite Materials

By DAVID J. BRADSHAW, JAMES T. WALKER, BERND BURGER,
BERND GANGNUS, and PHIL D. MARSH

Introduction

Dental plaque is the name given to the microbial communities found on the surfaces of the teeth, embedded in polymers of bacterial and host origin. Dental plaque is found in the mouth in both health and disease. However, unlike most bacterially mediated diseases, the majority of dental diseases (caries and periodontal diseases) result not from simple monoculture infection, but rather from imbalances in the commensal oral microbiota, which allow disease-producing species to predominate.

Dental plaque may become calcified, forming hard deposits called calculus on tooth surfaces. When calculus forms on dental composite materials, it can cause staining that may ultimately result in the need to remove and replace the material.

We have developed an *in vitro* model system to include relevant, removable, and replaceable surfaces on which biofilms may develop.[1,2] The model was subsequently modified further to a two-stage system, in which biofilms were developed in an aerated second-stage vessel.[3] The objective of this study was to evaluate the utility of this model system to compare the degree of oral microbial biofilm development on six commercial, light-curing dental composite materials supplied "blind" by ESPE Dental AG. Biofilm formation was assessed by several criteria, including qualitative and quantitative viable counts of biofilm bacteria, percentage coverage analysis, and an assessment of the strength of microbial adhesion. The results obtained in the *in vitro* model system were compared in a preliminary evaluation carried out *in vivo* at ESPE.

Preparation of Test Coupons

Square specimens of the indirect composite resin listed in Table I were prepared according to the manufacturers' instructions for use with the aid of a Delrin mold (10 × 10 × 2 mm).

Two-Stage Chemostat System

A two-stage chemostat system was constructed (Fig. 1) using twin microprocessor (Brighton Systems, Hove) controllers, with a conventional first-stage culture vessel (500 ml working volume) connected via an overflow weir to the

TABLE I
INDIRECT COMPOSITE RESINS INVESTIGATED

Material	Batch	Manufacturer	Abbreviation
Sinfony E3	001	ESPE Dental AG	SI (7)
Visio Gem E71	043F	ESPE Dental AG	VG (11)
Zeta LC EN 11	4129	Vita Zahnfabrik GmbH	VZ (10)
Solidex Incisal 59	BN079685	Shofu Inc.	SO (8)
Targis Dentin 210	821446	Ivoclar AG	TA (9)
Artglass EM	024	Heraeus-Kulzer GmbH	AG (12)

second-stage "biofilm" vessel. The second-stage vessel (1450 ml working volume) had a modified lid with multiple ports to allow the insertion and removal of the test materials (Table I) for biofilm formation. This type of system has proved to be of great value in allowing biofilm formation in the second stage of the system, whereas the first-stage provides a constant source of organisms.[3,4] The defined conditions of a chemostat have been shown to provide a high degree of reproducibility and control in biofilm development.[2]

Growth Conditions and Medium

The growth medium in both stages of the chemostat was BM/5 plus 2.5 g liter^{-1} mucin.[5] Mucin is the major source of carbon and nitrogen for oral bacteria in the mouth, and its inclusion in the medium thus reflects the *in vivo* nutritional environment, while also providing a relevant conditioning film to aid bacterial attachment. The culture was maintained at 37°, and the pH was maintained at 7.0 ± 0.1 by the automatic addition of 1 M NaOH. The gas phase in the first-stage vessel was 5% (v/v) CO_2 in nitrogen.

Once the first stage culture had reached steady state (4–5 days, at dilution rate, $D = 0.1$ hr^{-1}), the overflow weir of the first stage was connected to the second-stage "biofilm" chemostat. The gas phase in the second chemostat was 5% CO_2 in air, sparged at 200 ml min^{-1}, to simulate the conditions in the mouth. Sucrose (0.5% w/v, final concentration) was added three times a day (by manual pulsing) to the second stage vessel at 9.30 A.M., 1.45 P.M., and 4.15 P.M. on each working day to encourage the growth of *S. mutans*. The pH of the culture in the second stage was allowed to fall for a period of 30 min following each sucrose pulse and the culture

[1] C. W. Keevil, D. J. Bradshaw, T. W. Feary, and A. B. Dowsett, *J. Appl. Bacteriol.* **62**, 129 (1987).
[2] D. J. Bradshaw, P. D. Marsh, K. M. Schilling, and D. Cummins, *J. Appl. Bacteriol.* **80**, 124 (1996).
[3] D. J. Bradshaw, P. D. Marsh, C. Allison, and K. M. Schilling, *Microbiology* **142**, 623 (1996).
[4] J. T. Walker, A. B. Dowsett, P. J. L. Dennis, and C. W. Keevil, *Int. Biodeterioration* **27**, 121 (1991).
[5] D. J. Bradshaw and P. D. Marsh, *Methods Enzymol.* **310**, 279 (1999).

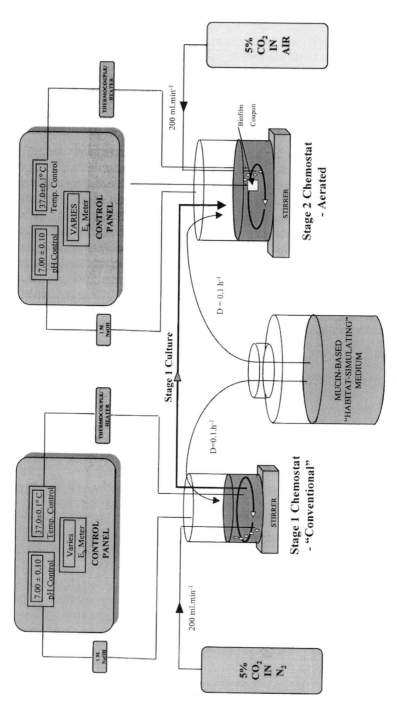

FIG. 1. Schematic of dental biofilm continuous culture chemostat.

returned to pH 7.0 after the 30 min period, by the addition of 1 M NaOH. This pulsing regime simulates the environmental carbohydrate and pH changes that occur *in vivo*. The dilution rate in the second stage was also $0.1 \, hr^{-1}$, corresponding to a mean generation time of 6.9 hr. Dissolved oxygen tension (dO_2) and redox potential (E_h) were monitored in the second-stage vessel.

Bacterial Strains

Nine bacterial species were inoculated into the first stage from frozen inocula.[6] The bacterial strains inoculated were *Streptococcus mutans* R9, *S. oralis* EF 186, *S. sanguis* SK 150, *Lactobacillus rhamnosus* AC413, *Actinomyces naeslundii* WVU627, *Neisseria subflava* A1078, *Veillonella dispar* ATCC 17745, *Prevotella nigrescens* T588, and *Fusobacterium nucleatum* ATCC 10953. The inoculum thus included representative facultative anaerobes (*N. subflava, S. mutans, S. oralis, S. sanguis, A. naeslundii, L. rhamnosus*) and obligate anaerobes (*V. dispar, P. nigrescens, F. nucleatum*) that are commonly found in dental plaque.

Biofilm Formation and Viable Counts

After steady-state was reached, five sterilized coupons of each of the composite resins were then inserted aseptically into the steady-state cultures, 1–1.5 cm below the surface of the liquid phase. Biofilms were allowed to develop for 14 days. After this time, three coupons of each material were removed, excess fluid drained off, and the biofilm on the surface scraped into a 4.5 ml volume of diluent,[7] using a stainless steel dental probe. The resulting suspension was vortexed for 30 sec, and then serially diluted in decimal steps. Samples (100 µl) of each dilution were then spread on a range of selective and nonselective agar media, as described previously,[6] to enable the detection and enumeration of each member of the consortium. The fourth coupon of each material was used for image analysis, and the fifth for the assessment of strength of adhesion (see sections below).

Microscopy and Image Analysis

Coupons were stained for 1 min with 50 µl pre-filtered (Sartorius, UK, 0.2 µm) propidium iodide (Sigma, UK; 1 mg ml^{-1} stock in sterile distilled water) before being gently rinsed in nonflowing sterile distilled water (twice) to remove excess propidium iodide. Coupons were visualized using a Nikon Labophot 2 microscope with episcopic fluorescence.[8] Eight representative images were captured

[6] D. J. Bradshaw, A. S. McKee, and P. D. Marsh, *J. Dent. Res.* **68**, 1298 (1989).

[7] A. S. McKee, A. S. McDermid, D. C. Ellwood, and P. D. Marsh, *J. Appl. Bacteriol.* **59**, 263 (1985).

[8] C. W. Keevil and J. T. Walker, *Binary* **4**, 92 (1992).

as computer files (*.tif) for image analysis of percentage coverage using Optimas Software (Optimas, Datacell, UK).

Assessment of Strength of Adhesion

Eginton et al.[9] described a method to determine the easily removable fraction of organisms rather than merely the total number attached to surfaces. The coupons were rinsed in three successive volumes of sterile diluent (20 ml) before being placed on the surface of predried (1 hr at 55°) Columbia blood agar plates. After 1 min, the tiles were removed to fresh plates and a spreader used to distribute the biofilm across the plate. This process was repeated through a succession of 15 agar plates. Strength of adhesion is defined as being in proportion to the number of repeated transfers after which colonies can still be detected.

Statistical Analysis

Results are presented as $\log_{10}$ colony forming units (cfu) per cm^2 of surface area. Comparisons between log-transformed counts on all test materials were carried out using the analysis of variance. Significant differences between pairs of materials were investigated using the Scheffé test. Comparisons of percentage coverage of test materials were made using the nonparametric Kruskal–Wallis test. Statistical analyses were carried out using Statgraphics software (STSC Inc., Rockville, MD).

Planktonic Population of Test Vessel

The bacterial strains inoculated into the seed vessel became established in both the seed and test vessel with a total viable count in the planktonic phase of greater than $\log_{10} 8.0$ cfu ml^{-1}.

Biofouling of Dental Composite Materials

Viable Counts

More than $\log_{10} 6.0$ cfu cm^{-2} were recovered from each test material (Fig. 2). A significantly greater number of bacteria were recovered from Targis ($p < 0.05$, Scheffé test, following one-way analysis of variance). The differences between the other materials were $< 0.5 \log_{10}$ cfu cm^{-2}, and these were not significant at the $p = 0.05$ level. A diverse community developed on each of the materials. *F. nucleatum*, *Streptococcus mutans*, and *Lactobacillus rhamnosus* were the most numerous organisms recovered from the materials, with slightly lower numbers of *P. nigrescens*. The materials did not show any marked strain selectivity.

[9] P. J. Eginton, H. Gibson, J. Holah, P. S. Handley, and P. Gilbert, *J. Ind. Microbiol.* **15**, 305 (1995).

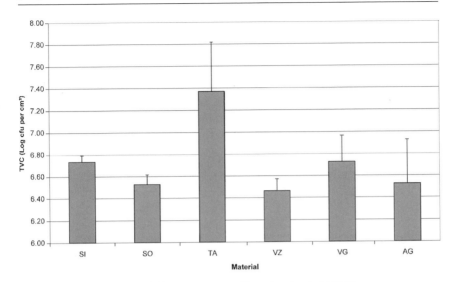

FIG. 2. Total viable counts recovered from dental material biofilms.

Percentage Analysis (Microscopy) of Biofouling Coverage

The image analysis data results indicated that Visio Gem was fouled to a greater extent than the others (Fig. 3). Sinfony, Targis, and Zeta LC were the least fouled. The degree of fouling could be ranked as Visio Gem > Solidex > Artglass >

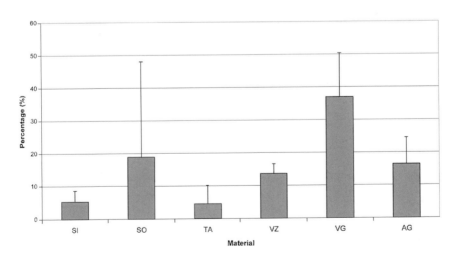

FIG. 3. Percentage coverage (biofouling) of test materials.

Sinfony > Zeta LC > Targis. Sinfony and Targis were found to have only small microcolonies on their surfaces, whereas only single cells were observed on Zeta LC. On Artglass, a similar single cell pattern was observed, but the cells were more densely packed. Microcolonies and a layer of single cells were observed on Solidex, and these were denser on the Visio Gem, which was the most fouled. Statistical analyses were carried out on the percentage coverage results using the nonparametric Kruskal–Wallis test (Statgraphics software, STSC Inc). Sinfony, Targis, Zeta LC, and Artglass were found to be statistically significantly less colonized than Solidex and Visio Gem at the $p < 0.05$ confidence levels. Targis[9] was significantly less fouled than Artglass, but not statistically significantly different from Sinfony and Zeta LC ($p > 0.05$).

Strength of Adhesion Assay

Using the adhesion strength assay,[9] Sinfony and Artglass displayed the weakest adhesion between the bacteria and the material (greatest number of bacteria removed onto the agar plates). Zeta LC showed the strongest adhesion between the bacteria and the material followed by Visio Gem, Solidex, and Targis (Fig. 4).

Advantages and Limitations of Experimental Systems

One of the main uses of indirect dental composite materials is their use as esthetic tooth colored veneer on metal substructures of (partially) removable dentures and dental implant work. One of their main disadvantages compared with ceramic materials is a greater susceptibility to accumulate dental plaque, which gets mineralized and can then cause external staining of the restoration by absorbing nutrient colorants, tobacco smoke, etc.[10,11] Many studies have led to the conclusion that materials with lower surface energy (hydrophobic) accumulate less plaque than materials with high surface energy (hydrophilic).[12–15] However, all materials within the oral cavity eventually approach a surface energy comparable to that of enamel.[16] This effect is caused by deposition of a biopolymer layer on the surface (pellicle formation) to which the bacteria attach.[17]

[10] H. Weber, L. Netuschil, and Z. Zahnaerztl, *Dtsch. Zahnaerztl. Z.* **4,** 278 (1992).

[11] K.-P. Wefers, *Zahnarzt Magazin* **89,** 2732 (1999).

[12] M. Quirynen, M. Marechal, H. Busscher, A. Weerkamp, J. Arends, P. Darius, and D. van Steenberghe, *J. Clin. Periodontol.* **17,** 138 (1989).

[13] M. Quirynen, M. Marechal, H. Busscher, A. Weerkamp, J. Arends, P. Darius, and D. van Steenberghe, *J. Dent. Res.* **68,** 796 (1990).

[14] J. Van Dijk, F. Herkstroeter, H. Busscher, A. Weerkamp, H. Jansen, and J. Arends, *J. Clin. Periodontol.* **14,** 300 (1987).

[15] E. Weiss, M. Rosenberg, H. Judes, and E. Rosenberg, *Curr. Microbiol.* **7,** 125 (1982).

[16] H. De Jong, P. De Boer, H. Busscher, A. van Pelt, and J. Arends, *Caries Res.* **18,** 408 (1984).

[17] I. Al-Hashimi and M. Levine, *Archs. Oral Biol.* **34,** 289 (1989).

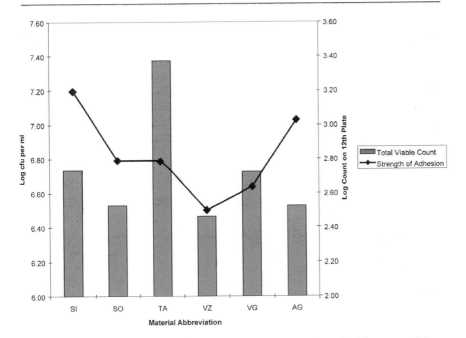

FIG. 4. Strength of microbial adhesion, compared with viable count for each of the test materials.

On hydrophilic surfaces, such as natural teeth, ceramics, or inorganic fillers, the protein adsorption is reversible, whereas on hydrophobic surfaces it is irreversible. It is assumed that because of the irreversible adsorption of proteins on such surfaces the resulting plaque layers can better withstand hydrodynamic shear forces in the oral cavity and get preferentially mineralized and then stained.

Since commercial composite materials contain a variety of different monomers and a selection of different fillers (e.g., quartz, dental glass, pyrogenic silica, or organic fillers), the resulting surfaces are not homogeneous on a microscopic scale. Macroscopic methods for surface characterization such as contact angle or ζ-potential measurements give only an averaged result over the surface, whereas proteins and bacteria will see the microscopic picture and may predominantly absorb on either hydrophobic resin or hydrophilic filler areas. The absolute values resulting from these macroscopic methods for surface characterization may therefore not be correlated with assays of bacterial adhesion as far as composite materials are concerned.

The presence of salivary proteins is also important in studies concerned with dental plaque accumulation because of the proteins[1] role in providing receptors for bacterial attachment. However, predictions on the amount and strength of bacterial adhesion that will result *in vivo* are not yet possible. The strength of adhesion may be particularly important, since it will be likely to have significant effects on ease

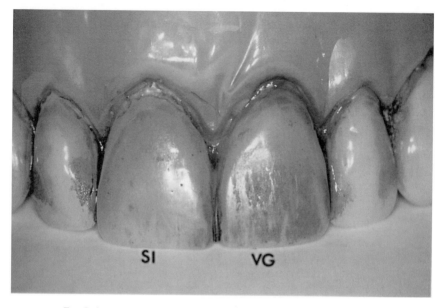

FIG. 5. Comparison of *in vivo* staining of Sinfony (SI) and Visio Gem (VG).

of cleaning of surfaces in the mouth. The development of an *in vitro* screening system in preliminary evaluation of materials is therefore invaluable.

This study has shown that it is not feasible to use a single method to assess the degree of biofouling of a surface. When total cfu per cm^2 was used as the criterion, Targis had significantly more biofilm than other materials (Fig. 2), and yet this resin had the lowest percentage coverage as assessed by microscopy (Fig. 3). In contrast, the strength of attachment of the organisms within the biofilm was much greater on other materials such as Sinfony or Artglass (Fig. 4). These findings, therefore, imply that different materials can significantly alter the nature and properties of the biofilm that develops upon them, in terms of numbers of bacteria, percentage surface coverage, and strength of attachment. Such differences may have important clinical implications.

Nevertheless, the results of this work do correlate well with the outcome of a preliminary split-mouth *in vivo* experiment, where the two incisor teeth of complete dentures were cut back and were veneered with different indirect resins. It could be shown that after a period of 6 weeks without special oral hygiene instructions, Sinfony displayed a 67% less plaque-covered area compared to Visio Gem.[18] Figure 5 shows an example of such a complete denture after staining (Plaque-Check DS2, Hu-Friedy, Germany). The analysis of biofouling in the *in vitro*

[18] B. Burger, B. Gangnus, and R. Guggenberger, *German Offenlegungsschrift DE* **197**, 48 (1997).

system described here also showed Sinfony to be less fouled than Visio Gem (Fig. 3).

The system described here allowed a threefold approach to determining the biofouling potential of each material, i.e., in terms of (i) qualitative and quantitative assessment of the bacterial species attached to the surfaces (cfu per unit area), (ii) a visual (microscopic) analysis of the degree of biofouling, and (iii) quantification of the relative strength of adhesion of bacteria to the different test materials. Previous studies have shown that this comprehensive approach is the only suitable way to evaluate accurately the degree of biofouling of test materials.[19]

---

[19] R. Leech, P. D. Marsh, and P. Rutter, *Arch. Oral Biol.* **24**, 379 (1979).

---

[30] An *in Vitro* Model for Studying the Contributions of the *Streptococcus mutans* Glucan-Binding Protein A to Biofilm Structure

By JEFFREY A. BANAS, KARSTEN R. O. HAZLETT, and JOSEPH E. MAZURKIEWICZ

Introduction

There are several models for the generation of bacterial biofilms, and several of them have been used to investigate the oral bacteria that make up dental plaque.[1] The choice of a model depends on the objective and focus of an investigation, and the advantages and disadvantages inherent in the individual models.[2] No model is fully capable of replicating the *in vivo* environment, though several sophisticated designs have been developed to simulate, as closely as possible, the natural conditions for biofilm development. The biofilm model described here was chosen for an examination of a *Streptococcus mutans* knockout strain that no longer secreted the glucan-binding protein A (GbpA). Previous experiments had indicated that the knockout strain differed from the wild-type in both virulence (gnotobiotic rat model) and *in vitro* adhesion properties.[3] In an effort to explain these results a method was developed for examining whether the loss of GbpA resulted in a change in the overall plaque biofilm architecture.[4] This method is

---

[1] C. H. Sissons, *Adv. Dent. Res.* **11**, 110 (1997).
[2] J. W. T. Wimpenny, *Adv. Dent. Res.* **11**, 150 (1997).
[3] K. R. O. Hazlett, S. M. Michalek, and J. A. Banas, *Infect. Immun.* **66**, 2180 (1998).
[4] K. R. O. Hazlett, J. E. Mazurkiewicz, and J. A. Banas, *Infect. Immun.* **67**, 3909 (1999).

a simple culturing design that incorporates elements of continuous culture and nutrient flow. Although it had its limitations, the method nonetheless provided results consistent with what had been previously observed regarding the properties of the GbpA knockout. Thus, the method represents a simple, inexpensive means for investigating the contributions of extracellular proteins to homogeneous plaque structure, and may also be useful as a basis for choosing strains to examine in more sophisticated biofilm generators or in animal models.

Generation of the Biofilm

Growth Conditions

The biofilms were generated in tissue culture-treated microtiter dishes using bacteria grown in chemically defined media (CDM, JRH Biosciences, Lenexa, KS) at 37° in a 5% CO_2 incubator. The CDM was supplemented with sodium bicarbonate and sucrose. Several variables of growth were tested, and these will be discussed individually below.

Microtiter Dishes

Four different systems were used; these included tissue culture-treated polystyrene 96-well (product no. 25860, Corning) or 24-well (product no. 3524, Costar) microtiter dishes, or Nunc Lab-Tek 16-well or 4-well detachable chamber slides. The chamber walls could be removed from the chamber slides after growth of the biofilm and were used when the biofilms were to be analyzed by confocal laser scanning microscopy. The 16-well apparatus had a glass slide, whereas the 4-well apparatus contained a Permanox plastic slide. Satisfactory biofilms were generated in all systems, though the differences between the mutant and GbpA knockout were more subtle on the Permanox than on the polystyrene or glass. When the wells were coated with hydroxylapatite prior to inoculation with bacteria, the 24-well polystyrene dishes worked best. The hydroxylapatite coating seemed to be more brittle on the chamber slides or in the smaller wells of the 96-well dishes.

Hydroxylapatite Coating

The initial application of this model[4] employed hydroxylapatite-coated dishes. The method of Schilling et al.[5] was used to coat the microtiter wells with hydroxylapatite. The coated plates were then sterilized in a UV cross-linker ($\lambda = 254$) for

[5] K. M. Schilling, R. G. Carson, C. A. Bosko, G. D. Golikeri, A. Bruinooge, K. Hoyberg, A. M. Waller, and N. P. Hughes, Colloids Surf. 3, 31 (1994).

17 min. Subsequent experiments compared biofilm formation on hydroxylapatite-coated and uncoated dishes. Figure 1A shows the *S. mutans* biofilm formed after 2 days in uncoated polystyrene wells and wells coated with hydroxylapatite. As observed previously[4] the wild-type biofilm had larger aggregates, whereas the biofilm formed by the GbpA knockout produced a more even coating of the surface of the well. No appreciable difference could be discerned between hydroxylapatite-coated or uncoated wells. The coating of hydroxylapatite may be more important when using oral bacterial species that preferentially adhere to the acquired enamel pellicle.

Rotation

In a previous investigation of the GbpA knockout and wild-type *S. mutans*[3], it was noted that quantitative differences in adhesion only occurred when cultures of the organisms were grown with a slow, constant rotation. This motion, which might simulate salivary flow and other physical motion in the oral cavity, appeared necessary for the contribution of GbpA to be distinguished. To incorporate the motion into the biofilm model, the microtiter dishes were placed on a rotator at an angle of 60° less than horizontal with a rotation of approximately 10 rpm. Figure 1B shows a comparison between the biofilms formed in dishes that were on the rotator and those formed in dishes that were incubated in the same incubator but without rotation. Although differences between *S. mutans* wild-type and GbpA knockout strains can be noticed under both conditions, the biofilms grown with the constant rotation displayed more profound differences between the two strains and the differences were readily apparent even without magnification. The biofilms shown in Fig. 1B were grown in wells without a hydroxylapatite coating. However, when tested, the biofilms formed in hydroxylapatite-coated wells without rotation often became partially dissociated from the flat bottom of the well such that the biofilm and hydroxylapatite formed a severely undulating surface that complicated the interpretation of the biofilm architecture.

Length of Incubation

Biofilms were allowed to develop for 1, 2, or 4 days. The medium was changed every 24 hr by gentle aspiration of the spent medium followed by gentle introduction of fresh, prewarmed CDM. No additional inoculum was introduced during the changes in medium. The biofilms increased in mass each additional day of incubation (Fig. 1C). However, by 2 days the differences between the wild-type and knockout *S. mutans* strains were readily apparent without magnification. Four-day biofilms provided the starkest contrast between the two strains, but the risk of contamination also increased with the manipulations that were necessary to propagate the biofilms for longer periods.

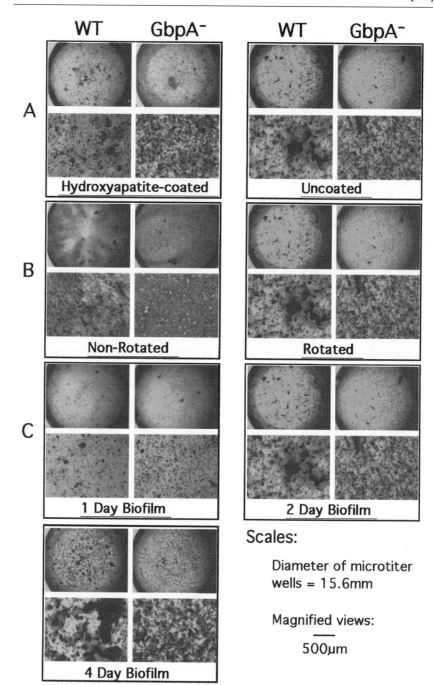

Scales:

Diameter of microtiter wells = 15.6mm

Magnified views:

500μm

Sucrose Concentration

Sucrose concentrations of 5% or 10% were tested to determine if they had an effect on either the overall biofilm structure or on the length of time necessary to attain a particular degree of biofilm formation. No obvious differences were detected in biofilms generated with either sucrose concentration, though the aggregate sizes may have been slightly greater in biofilms grown in CDM with 10% sucrose. In both instances 2 days of growth still appeared to be necessary to reveal clear differences between the strains tested.

Inoculum Size

All inocula into the microtiter dishes were equalized by optical density at 540 nm using freshly prepared overnight cultures grown in CDM. Twofold differences in inoculum size did not seem to alter the biofilm structure or the length of time necessary for the biofilm to develop.

Analysis of the Biofilm

Staining of the Biofilm

To stain the biofilm for easier visualization under low magnification, a 1% solution of crystal violet was used as described previously.[6,7] Following generation of the biofilms the media were aspirated and the wells were washed with 1 ml of phosphate-buffered saline (PBS). Twenty-five microliters of the 1% crystal violet solution was then added to 1 ml of PBS within the wells and allowed to stain for a minimum of 15 min. Alternatively, 25 microliters of Schiff's reagent (Sigma) was added to 1 ml of PBS and allowed to stain for a minimum of 30 min. The staining with crystal violet was satisfactory on all biofilms. The Schiff's staining was difficult to see on 1-day biofilms, but was satisfactory on the older biofilms. The Schiff's stain also displayed some variations in color intensity within a biofilm, which aided the analysis of biofilm differences between strains.

[6] G. A. O'Toole and R. Kolter, *Mol. Microbiol.* **28**, 449 (1998).
[7] C. Y. Loo, D. A. Corliss, and N. Ganeshkumar, *J. Bacteriol.* **182**, 1374 (2000).

FIG. 1. Comparison of biofilm structure formed by wild-type and GbpA knockout strains grown under varying conditions. In each panel the wild-type biofilm is on the left and the GbpA knockout biofilm is on the right. The top views in each panel show the appearance of the entire microtiter well; the lower views have been magnified 40×. All biofilms had in common that they were grown in 24-well microtiter dishes at 37° in a CO_2 incubator. (A) The effect of coating the microtiter dishes with hydroxylapatite prior to inoculation with *S. mutans* was compared with biofilms generated in uncoated microtiter dishes. (B) The effect of constant slow rotation on biofilm deposition was compared with biofilms grown in the absence of rotation. (C) The effect of age on the generation of the biofilms was examined after 1, 2, or 4 days.

A

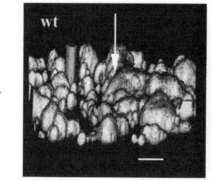
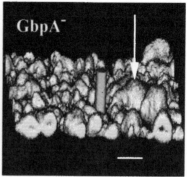

3D-Solid rendering of biofilms. Arrow = position of Y-Z slice in figure below. Solid vertical rod = height of 100 μm; horizontal bar = 50 μm.

B

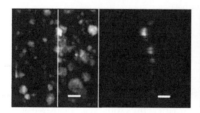
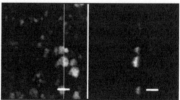

The left panel in each pair is a maximum intensity projection image of the entire stack. The right panel is a Y-Z re-slice through that stack along the line in the left panel. Bar = 50 μm.

C

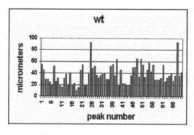
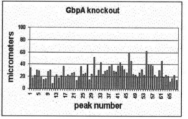

wt Biofilm	GbpA⁻ Biofilm
Avg. Ht. 37.42 μm*	Avg. Ht. 29.84 μm *
Max. Ht. 93.25 μm	Max. Ht. 79.74 μm
Min. Ht. 10.81 μm	Min. Ht. 8.12 μm
# 199	# 157

* significant difference p < 0.001

Confocal Microscopy

Confocal laser scanning microscopy was used to analyze the biofilms produced by the wild-type and GbpA knockout strains. The objective was to corroborate our visual analysis of crystal violet–stained biofilms with respect to biofilm topology, and to add a level of quantitation. For a comprehensive description of confocal microscopy applications in the areas of microbial ecology and biofilm analysis, the reader is referred to any of several review articles.[8–10]

Sample Preparation. Biofilms were generated in microtiter wells that could be detached from a chamber slide. Prior to detachment, the bacteria were stained with the LIVE *Bac*Light Bacterial Gram Stain (Molecular Probes, Eugene, OR) fluorescent dye mixture (5 μM SYTO 9, 7 μM hexidium iodide, and 0.3% dimethyl sulfoxide in 10 mM Tris buffer, pH 7.0) in a total volume of 0.5 ml. After staining for 15 min in the dark, the stain was aspirated off and the biofilm washed with 0.5 ml Tris buffer. The microtiter wells were detached from the chamber slide, a shallow well was created with a thin ring of silicone adhesive sealant (Permatex Clear RTV) by outlining the perimeter of the biofilm, 100 μl of Tris buffer was added to the biofilm, and it was covered with a glass coverslip. The biofilms were then examined with a Noran OZ confocal laser imaging system (Noran Instruments, Madison, WI) on a Nikon Diaphot 200 inverted microscope equipped with a 20 × 0.75 N.A. objective lens and a Kr/Ar laser. Attempts to fluorescently label the biofilm matrix by growing the organisms in the presence of FITC-labeled dextran (10,000 molecular weight) resulted in an altered biofilm topology in which the peak heights of aggregates increased substantially. Therefore, this option was not pursued further.

---

[8] D. E. Caldwell, D. R. Korber, and J. R. Lawrence, *Adv. Microb. Ecol.* **12**, 1 (1992).

[9] S. Singleton, R. Treloar, P. Warren, G. K. Watson, R. Hodgson, and C. Allison, *Adv. Dent. Res.* **11**, 133 (1997).

[10] J. R. Lawrence and T. R. Neu, *Methods Enzymol.* **310**, 131 (1999).

FIG. 2. Analysis of the wild-type and GbpA knockout biofilms by confocal laser scanning microscopy. All biofilms had in common that they were generated on chamber slides coated with hydroxylapatite, and incubated for 2 days with rotation at 37° in a CO_2 incubator. (A) Three-dimensional solid renderings of the biofilms that include a 100 μm tall solid vertical rod placed in the depiction to orient the viewer to the floor of the biofilm. The arrows indicate the sites of Y–Z slices shown in part B of the figure. (B) Each panel shows a maximum intensity projection image on the left and includes a white line indicating the plane of the Y–Z reslice that appears on the right. Several individual peak heights through regular interval Y–Z reslices were measured and presented in part C of the figure. (C) The upper panels represent the peak height distributions of a subset of the 199 (wild-type) or 157 (GbpA knockout) peaks measured. The maximum, minimum, and average peak heights for all the data (wild-type and GpbA knockout) are presented in the lower panels. The average peak heights were compared by a two-tailed Student's *t* test and determined to be statistically different ($p < 0.001$).

Image Collection Paradigm. Optical sections of fluorescently-tagged bacteria in mature biofilms were taken at 1 μm steps through the full thickness of each biofilm, some of which exceeded 110 μm. To optimize the "confocality" of the image a 10 μm slit was used. The signal-to-noise ratio was optimized by using a pixel dwell time of 800 nsec and a jump average of 16 images for each image in the stack. Laser intensity was set as low as possible to minimize photobleaching and saturation of the image. Using the Noran INTERVISION 3D module, three-dimensional volumes were rendered as solids and as maximum intensity projection (MIP) images.

Topological Analysis

The solid 3D-volume rendering technique projects a simulated ray through the stack of images starting from the viewer's perspective, until it strikes the surface of the volume, at which point that voxel is shaded to render the solid. Solid rendering permits the viewer to study the biofilm from any direction, thus providing the viewer with an appreciation of the complexity of the biofilm's surface. In Figure 2A the solid images were tilted and rotated to the right to provide a perspective from above and slightly from the left. A solid rod of known height was projected within each solid to be used as a visual gauge of the relative heights of the respective biofilms.

The surface of the wild-type biofilm was very heterogeneous in height and in microcolony morphology in comparison to the biofilm formed by the GbpA knock-out strain. These biofilms were generally of lower height with microcolonies of somewhat uniform morphology. When viewed from the top, the wild-type microcolonies appeared to cover less of the surface area than did the GbpA knockout microcolonies.

Quantitation of Biofilm Height

Maximum intensity projection (MIP) images generated from the confocal image stack provided a top-down view of the microcolonies in a biofilm and resembled the view seen through the inverted microscope using bright-field or phase-contrast optics (see panels in Fig. 1). For our purposes, however, the MIP images provided quantitative information directly related to biofilm height. The 3D MIP volumes were subjected to reslicing along the Y–Z axis to create two-dimensional views along planes that were normal to the MIP images. TIFF (tagged image format) images of 10 Y–Z slices were taken at regular intervals (30–33 μm) from left to right in the MIP image. Biofilm heights were measured on these images using Media Cybernetics' Image Pro Plus software. Examples of a MIP image and one Y–Z slice for each biofilm type are presented in Fig. 2B.

The mean aggregate heights of the wild-type biofilms were significantly greater than those of the GbpA knockout biofilms. In Fig. 2C the distribution of peak heights is presented graphically and in tabular form. A total of 199 wild-type peaks and 157 GbpA knockout peaks were measured from regular-interval Y–Z slices. Peak height averages from different Y–Z slices within the same biofilm did not differ statistically. The wild-type films showed a greater range and heterogeneity in peak height than did the knockout films, which tended to be of a more uniform lower height.

Antimicrobial Susceptibility

Beyond the visualization and quantitation of biofilm structural differences, it is also possible to use this model to document these differences by comparing the susceptibilities of the biofilm bacteria to antimicrobial agents. We previously reported[4] that fewer organisms in a wild-type biofilm were killed by a 2-hr exposure to penicillin than were killed in a GbpA knockout biofilm.

Summary

The method described here for analyzing biofilms was sensitive enough to allow the detection of differences formed by pure cultures of *S. mutans* or a GbpA knockout strain. Other strains have also been tested, and the differences in biofilm structure were sometimes even more extensive (data not shown). The advantages of this method are that it is quick, inexpensive, and adaptable to almost any laboratory setting. The constant rotation of the cultures, which was employed to simulate salivary flow, appears to be a critical element for establishing biofilm differences. An analysis of protein profiles confirmed that the biofilm bacteria were metabolically distinct from the planktonic phase bacteria.

For the strains tested, the variations in biofilm architecture could be visualized with or without magnification. Staining of the bacteria was not required, though we typically stained the biofilms with either crystal violet or Schiff's reagent. Altogether, this *in vitro* method for generating biofilms allowed the evaluation of visual, quantitative (confocal microscopy), and functional (antimicrobial susceptibility) differences. We have employed these methods in a reductionist approach to understanding the contribution of individual proteins to dental plaque development. These methods may also be useful in the screening of mutants that would be of greatest interest for testing in multispecies biofilms, animal models, or more complex biofilm models.

Acknowledgments

This work was supported by grant number DE10058 from the National Institute of Dental and Craniofacial Research and RR12894 from the National Center for Research Resources. We gratefully acknowledge the technical support of Eva Skarshinski, Meghan Kelly, Justin Miller, and David Lynch.

[31] Detection of Streptococcal Glucan-Binding Proteins in Biofilms

By SOMKIAT LUENGPAILIN, JIRAPON LUENGPAILIN, and RON J. DOYLE

Introduction

In the presence of sucrose, some streptococci produce tenacious biofilms on various surfaces.[1] Extracellular and cell-bound glucosyltransferases (GTFs) produce α-1,6 and/or α-1,3 glucans from the sucrose. The glucans seem to be required for biofilm development because α-glucanases are able to suspend the adherent cells. Glucan-binding proteins[2-5] (Chapter 30 of this volume) contribute to the integrity of the biofilm matrix. The glucan-binding proteins (GBPs), including a glucan-binding lectin (GBL),[2] appear to be both cell-associated and secreted. Although there are reports on GBPs and GBLs of planktonic streptococci, there is no information on biofilm glucan-specific proteins. An enhanced chemiluminescence (ECL) method for GBPs developed by Galperin et al.[6] is now adapted to study sucrose-dependent biofilms of Streptococcus sobrinus.

Preparation of Growth Medium and Inoculum

S. sobrinus 6715 (serotype g) is grown in a 250-ml Erlenmeyer flask containing 50 ml of trypticase soy broth (TSB; BBL Microbiology Systems, Cockeysville, MD) at 37° in a 5% CO_2 incubator for 16 hr. To eliminate traces of dextran and sucrose before use, TSB is incubated at 37° for 2 hr with 20 units of dextranase (EC 3.2.1.11, Cat. No. D-5884, Sigma Chemical Co., St. Louis, MO) per g of dry medium and then incubated at 55° for 2 hr with 2500 units of invertase (EC 3.2.1.26, Cat. No. I-4504, Sigma) per g of dry medium. The enzymatic treatments are necessary to prevent glucan-induced autoagglutination during biofilm formation. Phenylmethylsulfonyl fluoride (PMSF, 0.5 mM) is added into the growth medium after autoclaving in order to reduce proteolytic activity. The PMSF has no effect on bacterial growth.

---

[1] I. Ofek and R. J. Doyle, in "Bacterial Adhesion to Cells and Tissues." Chapman and Hall, New York, 1994.

[2] Y. Ma, M. O. Lassiter, J. A. Banas, M. Y. Galperin, K. G. Taylor, and R. J. Doyle, J. Bacteriol. **178,** 1572 (1995).

[3] W. Haas and J. A. Banas, Adv. Exp. Med. Biol. **418,** 707 (1997).

[4] D. Drake, K. G. Taylor, A. S. Bleiweis, and R. J. Doyle, Infect. Immun. **56,** 1864 (1989).

[5] P. D. Bauer, C. Trapp, D. Drake, K. G. Taylor, and R. J. Doyle, J. Bacteriol. **175,** 819 (1993).

[6] M. Y. Galperin, M. O. Lassiter, Y. Ma, K. G. Taylor, and R. J. Doyle, Anal. Biochem. **225,** 185 (1995).

Cultures of Biofilm and Planktonic Cells and Recovery of Bacterial Cells

Two liters of the treated TSB are prepared as described above and divided into eight 2.8-liter low-form Erlenmeyer flasks, each containing 250 ml medium. The culture flasks are precleaned with nitric acid prior to use. To induce biofilm formation, sucrose (2%) is supplemented into four flasks. All flasks are inoculated with 1% of the above overnight culture and incubated at 37° in a 5% CO_2 atmosphere.

After incubation for 16 hr, thick biofilms have formed on the flask bottoms of the sucrose-supplemented cultures, but not on those of the sucrose-free cultures. All of the cell cultures are placed on ice for 15 min and shaken at 100 rpm for 2 min in order to remove nonadherent cells. Cell suspensions are transferred to centrifuge bottles and spun at 10,000 g, 4°, for 5 min. Extracellular proteins are isolated from culture supernatants as detailed below.

The biofilms are washed two times to remove loosely bound cells with 100 ml of 0.1 M phosphate-buffered saline (PBS, 0.15 M NaCl, pH 6.0) on an orbital shaker at 100 rpm for 10 min. Only limited amounts of bacteria are suspended in the washing buffer, the vast majority of biofilm remaining intact. To recover adherent cells, the biofilm is incubated with 100 ml of the same buffer containing 250 units of dextranase and 50 μl of protease inhibitor cocktail (Cat. No. P-8340, Sigma). Dextranase catalyzes the endohydrolysis of 1,6-α-glucosidic linkages in dextran synthesized from the added sucrose. One unit of activity causes the release of 1 μmol isomaltose from dextran per min. Almost all of the bacteria become free of the biofilm matrix after a 2-hr incubation period in a gyratory shaker at 75 rpm, 37°.

The planktonic cells are treated in parallel. Briefly, planktonic cells are washed two times and suspended in 100 ml of PBS containing dextranase and protease inhibitors. The mixture is then transferred back to the 2.8-liter Erlenmeyer flasks and shaken at 75 rpm, 37°, for 2 hr.

Isolation of Cell Surface Proteins

The detached cells from biofilms and the planktonic cells of 250-ml cultures are centrifuged at 10,000 g, 4°, for 5 min, washed two times, and suspended in 0.02 M PBS (pH 7.2) to yield an OD_{540} of 1.0 (Spectronic 21 spectrophotometer; Milton Roy Co., Rochester, NY). Each cell suspension is divided into 120 ml portions in 250-ml centrifuge bottles and 12 ml portions in 50-ml centrifuge tubes, separately spun, and transferred to microtubes. The bigger pellet is suspended in 1 ml of 2% sodium dodecyl sulfate (SDS), whereas the smaller one is suspended in 100 μl of Laemmli sample buffer (pH 6.8) (Bio-Rad Laboratories, Hercules, CA) containing 2% SDS, 25% glycerol, 0.01% bromphenol blue, and 0.063 M Tris-HCl. The mixtures are incubated at 37° for 2 hr and spun at 13,000 rpm for

30 sec. Extract of the bigger pellet is dialyzed (MW cutoff 6000) at 4° for 6 hr against 0.02 M PBS (pH 7.2) with buffer changes every 2 hr. Protein concentrations are determined by Coomassie Plus-200 protein assay reagent (Pierce, Rockford, IL).

Isolation of Extracellular Proteins from Culture Supernatants

Culture supernatants of biofilm and planktonic cells are separately pooled, approximately 1 liter each, into two 2-liter beakers. Proteins in the supernatants are salted out by slow addition of ammonium sulfate to 65% saturation with gentle stirring at 4° overnight. The precipitates are collected by centrifugation at 25,000 g, 4°, for 30 min, dialyzed, and assayed for protein as described above.

Electrophoresis

SDS–PAGE is carried out as described by Laemmli[7] using a Mini-PROTEAN II Dual Slab Cell (Bio-Rad Laboratories, Richmond, CA). The dialyzed samples are mixed with two volumes of Laemmli-sample buffer and incubated at 37° for 2 hr. Samples (10 μg protein of cells and 3 μg protein of supernatant) are loaded on a 4–15% linear gradient Ready Gel (Bio-Rad). The gel is electrophoresed at constant 150 volts for 45 min in electrode buffer, pH 8.3, containing 0.1% SDS, 0.19 M glycine, and 0.025 M Tris.

Western Blotting

Western blotting is performed according to Towbin et al.[8] using a GENIE blotter (Idea Scientific Co., Minneapolis, MN). The electrophoresed gel and filter papers (Scheicher & Schuell, Keene, NH) are soaked for 15 min in Towbin buffer, pH 8.3, containing 20% methanol, 0.19 M glycine, and 0.025 M Tris. An Immun-Blot PVDF (polyvinylidene difluoride) membrane (Bio-Rad) is submerged in absolute methanol for 15 sec, in distilled water for 3 min, and subsequently in Towbin buffer for 10 min. The separated proteins on the gel are electrophoretically transferred onto the membrane, at constant 24 volts, for 30 min. The blot is then submerged in absolute methanol for 10 sec and air-dried for 15 min on a filter paper.

Enhanced Chemiluminescence

Detection of GBPs by the enhanced chemiluminescence (ECL) technique is performed as previously described by Galperin et al.[6] The blot is incubated

---

[7] U. K. Laemmli, *Nature* **227,** 680 (1970).
[8] H. Towbin, T. Staehelin, and J. Gordon, *Proc. Natl. Acad. Sci. U.S.A.* **76,** 4350 (1979).

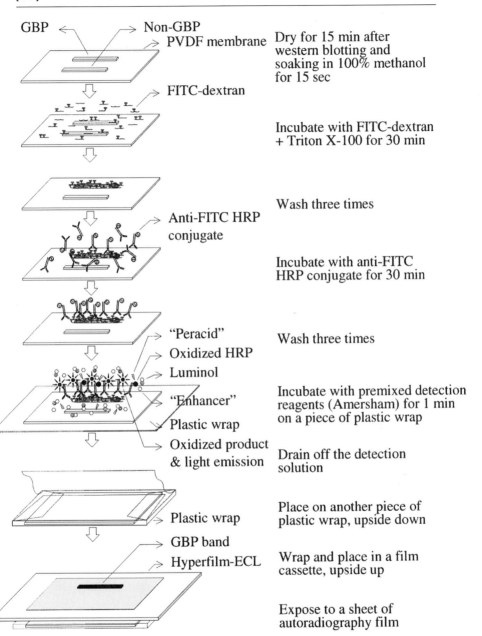

GBP ← → Non-GBP
→ PVDF membrane   Dry for 15 min after western blotting and soaking in 100% methanol for 15 sec

→ FITC-dextran

Incubate with FITC-dextran + Triton X-100 for 30 min

Wash three times

→ Anti-FITC HRP conjugate

Incubate with anti-FITC HRP conjugate for 30 min

→ "Peracid"   Wash three times
→ Oxidized HRP
→ Luminol
→ "Enhancer"   Incubate with premixed detection reagents (Amersham) for 1 min on a piece of plastic wrap
→ Plastic wrap
→ Oxidized product & light emission   Drain off the detection solution

→ Plastic wrap   Place on another piece of plastic wrap, upside down

→ GBP band
→ Hyperfilm-ECL   Wrap and place in a film cassette, upside up

Expose to a sheet of autoradiography film

FIG. 1. GBP detection by the ECL method.

for 30 min with 15 ml of 0.02 M PBS (pH 7.2) containing 5 μM fluorescein isothiocyanate-labeled dextran (FITC-dextran, MW 19,600, Cat. No. FD-20S, Sigma) and 0.5% Triton X-100. Unbound FITC-dextran is removed by washing for 10 min two times with 150 ml of the PBS and for 10 min one time with 150 ml of 0.1 M Tris-HCl, pH 7.5, 0.15 M NaCl. The membrane is then incubated for 30 min with 15 ml of the Tris-HCl buffer containing a 1000-fold dilution of anti-fluorescein horseradish peroxidase conjugate (Cat. No. RPN 3022, Amersham, Arlington Heights, IL) and 0.5% bovine serum albumin. Unbound conjugate is removed by washing for 10 min three times with 150 ml of the Tris-HCl buffer containing 0.1% Tween 20. Excess buffer is drained off from the washed membrane by holding the membrane vertically and touching the edge of the membrane against tissue paper. The membrane is placed onto a piece of plastic wrap, protein side up. Equal volumes (3 ml) of detection reagent 1 are mixed with detection reagent 2 (Cat. No. RPN 2109, Amersham). The detection mixture is added to the protein side of the membrane and incubated for 1 min at room temperature without agitation. Excess detection solution is drained off. The membrane is placed onto another piece of plastic wrap, protein side down without air bubbles. After wrapping, the membrane is taped in the film cassette, protein side up. Under red safelights in a dark room, a sheet of autoradiography film (Hyperfilm-ECL, Amersham) is placed on top of the membrane. The cassette is then closed for 30 sec. The film is removed and immediately developed using a Kodak X-OMAT 1000A Processor (Eastman Kodak Company, Rochester, NY). Second exposures may vary from 1 to 10 min. The developed film is scanned by a Fluor-S MultiImager system (Bio-Rad).

Results from Planktonic Cultures and Biofilms

Figure 1 provides a schematic for the ECL method. Proteins were recovered from cells and supernatants of planktonic cultures and biofilms, subjected to SDS–gel electrophoresis, and transferred onto PVDF for ECL analysis. Biofilm cells contain numerous GBPs ranging in molecular weight from 40,000 to >150,000 (Fig. 2). Supernatants of biofilm cultures had only one major protein, suggesting that most of the GBPs are involved in biofilm formation. The ECL technique allows for the detection of multiple GBPs and permits a rapid comparison between planktonic and biofilm conditions.

Reliabilities and Limitations of the ECL Method

The ECL method can be used for any polysaccharide-binding protein, providing the protein can be renatured following SDS–PAGE. In addition, it is required that the polysaccharide be successfully derivatized with FITC. Controls include use of oligomers of the glucans to prevent GBP–glucan interaction and the use of

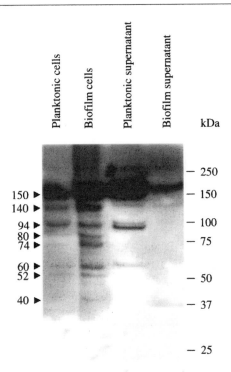

FIG. 2. Cell-associated and secreted GBPs isolated from of planktonic and biofilm cultures of *S. sobrinus*. After SDS–PAGE (4–15% linear gradient gel) and Western blotting, renatured GBPs were detected by incubating with FITC–dextran, anti-FITC HRP conjugate, and ECL detection solution, respectively. Numbers indicate apparent molecular weights of GBPs (left panel) and prestained markers (right panel). Amounts of proteins loaded on the gel were 10 μg for cell extracts and 3 μg for supernatants.

tryptophan in the reaction mixes to minimize any hydrophobic effects involving the FITC probe. Another way to ensure specificity is to purify GBP fractions from Sephadex affinity chromatography before SDS–PAGE.

Acknowledgment

Support from the NIH, the Jewish Hospital Foundation, and the Royal Thai Government is appreciated.

Author Index

Subject Index

A

Adhesion assays, bacteria, *see also* Atomic force microscopy
events in adhesion, 276–277
limitations of counting assays, 277–278
Staphylococcus epidermidis attachment studies
counting of attached bacteria, 48, 50
flow system
continuous culture chemostat, 45–49, 62
modified Robbins device, 47, 49–50
mathematical modeling, 52–53
rate factors, 43–44
simulations, 53–54, 62
statistical analysis, 51–52
test materials and adhesion conditions, 50–51
time-lapse digital imaging, 315–317
AFM, *see* Atomic force microscopy
Assimilable organic carbon, *see* Biodegradable organic matter
Atomic force microscopy
adhesion assay for bacteria
advantages, 284
controls, 282
data analysis and interpretation, 281, 283–284
force measurements, 281
immobilization of bacteria, 280
instrumentation, 278–279
limitations, 284–285
potable water biofilms, 246, 251–253, 255
softness measurement of *Serratia marcescens*
fibrillated versus nonfibrillated strain biofilms, 271
unsaturated biofilms, 140–141

B

Biodegradable organic matter
assays

overview of steps, 145–146
selection of technique, 170
assimilable organic carbon assays
ATP assay, 149–151
comparison with biodegradable dissolved organic carbon assays, 162–164
overview, 146–148
biodegradable dissolved organic carbon assays
bacteria attached to sand assay
apparatus, 153
calculations, 154
inoculum preparation and storage, 153–154
principle, 152
bioreactor assay
apparatus, 154
colonization, 154–155
maintenance of bioreactor, 155
principle, 154
sampling, 156
comparison with assimilable organic carbon assays, 162–164
overview, 146–147, 149
parameters affecting results, 158–160
suspended bacteria determination
glassware preparation, 151
principle, 151
sample inoculation and incubation, 152
water collection and preparation, 151–152
variation between assays, 160–162
biofilm nutrients, 144–145
nutrient levels in waters, 156–158
removal during water treatment for bacteria control
biofilm density studies, 169–170
coagulation, 165–166
filtration, 166, 168–169
membrane technologies, 166–167
ozone treatment, 167, 169
target levels, 165

ISBN 0-12-182238-9

9 780121 822385

90051